Urban Forest and Urban Microclimate

Urban Forest and Urban Microclimate

Editors

Thomas Rötzer
Stephan Pauleit
Mohammad A Rahman
Astrid Reischl

MDPI • Basel • Beijing • Wuhan • Barcelona • Belgrade • Novi Sad • Cluj • Manchester

Editors

Thomas Rötzer
TU München
Freising
Germany

Stephan Pauleit
TU München
Freising
Germany

Mohammad A Rahman
TU München
Freising
Germany

Astrid Reischl
TU München
Freising
Germany

Editorial Office
MDPI AG
Grosspeteranlage 5
4052 Basel, Switzerland

This is a reprint of articles from the Special Issue published online in the open access journal *Forests* (ISSN 1999-4907) (available at: https://www.mdpi.com/journal/forests/special_issues/urban_tree_design_and_urban_microclimate).

For citation purposes, cite each article independently as indicated on the article page online and as indicated below:

Lastname, A.A.; Lastname, B.B. Article Title. *Journal Name* **Year**, *Volume Number*, Page Range.

ISBN 978-3-7258-1511-1 (Hbk)
ISBN 978-3-7258-1512-8 (PDF)
doi.org/10.3390/books978-3-7258-1512-8

Contents

forests

MDPI

Editorial

Urban Forest and Urban Microclimate

Thomas Rötzer [1,*], Astrid Moser-Reischl [1,2], Mohammad A. Rahman [2] and Stephan Pauleit [2]

1 Chair for Forest Growth and Yield Science, Technical University of Munich, Hans-Carl-von-Carlowitz-Platz 2, 85354 Freising, Germany; astrid.reischl@tum.de

2 Chair for Strategic Landscape Planning and Management, Technical University of Munich, Emil-Ramann-Str. 6, 85354 Freising, Germany; ma.rahman@tum.de (M.A.R.); pauleit@tum.de (S.P.)

* Correspondence: thomas.roetzer@tum.de; Tel.: +49-8161-714667

Citation: Rötzer, T.; Moser-Reischl, A.; Rahman, M.A.; Pauleit, S. Urban Forest and Urban Microclimate. *Forests* **2023**, *14*, 2391. https://doi.org/10.3390/f14122391

Received: 24 April 2023
Accepted: 1 December 2023
Published: 7 December 2023

1. Introduction

Urban environments are challenging places for urban greenspaces, especially for trees, which have the greatest impact on ecosystem service provisions. High temperatures and high levels of radiation, reduced water availability, limited above- and below-ground growing space, and high levels of pollutants are just some of the challenges urban greenspaces are facing and are expected to face more frequently in the near future [1–5]. In addition to these site conditions, cities themselves are often warmer than their rural surroundings due to the urban heat island effect [6–8]. Future climate change and increasing urbanization are expected to significantly exacerbate the UHI effect and its associated environmental problems such as changes in local precipitation, the spread of disease from warmer climates, and air pollution [9]. Despite these conditions, urban greening is still one of the most feasible strategies to address major urban challenges: urban greening can be considered a nature-based solution to improve quality of life, conserve biodiversity, and enhance climate resilience [10,11]. In particular, urban trees and forests are highly capable of mitigating the urban microclimate and ameliorating the effects of urban heat islands and ongoing climate change [12–14]. They provide cooling through evapotranspiration and shading, store carbon, reduce runoff, and improve air quality.

The provision of ecosystem services is highly dependent on many factors, including climate; the type of urban greening (trees, vertical greening, roof greening, shrubs, etc.); tree species, age, and vitality; as well as the drought tolerance of a species (e.g., [12,15–17]). In addition, the location within a city and surrounding urban structures should be considered when quantifying the provision of ecosystem services by urban greenery. However, knowledge regarding the ecosystem service provision of different types of urban greening and especially urban tree growth in relation to these factors and conditions is still insufficient. There is a great need for such knowledge for the sustainable planning and management of urban greenspaces. Therefore, detailed knowledge of dimensional changes, growth rates, and ecosystem services in urban greenspaces in general as well as urban trees in particular as a function of their age and environmental conditions is needed.

2. Outline of Urban Forests and the Urban Microclimate

This Special Issue, "Urban Forest and Urban Microclimate", addresses the above-mentioned topics that influence the growth and ecosystem services of urban greenspaces and urban trees, the species characteristics that affect growth patterns and ecosystem service provision, potential sustainable designs of urban trees, and the functions of urban trees in improving the microclimate. Ten research papers form this Special Issue, and they can be divided into the following topics:

(1) Tree growth and vitality assessments across multiple urban space designs using allometric studies;
(2) Benefits of the cooling effects from urban greenspaces at different spatial and temporal scales for indoor and outdoor thermal comfort;

(3) Assessment of hydrology in urban areas and soil properties;
(4) Understanding and mapping urban greenspaces across scales to promote multi-functional landscapes and resilient cities with focus on climate vulnerability and drought tolerance.

The articles by Amer et al. (2023) [18], Franceschi et al. (20122) [19], and Wang et al. (2023) [20] can be classified within the first topic of tree growth assessment, while the articles by Li and Zheng (2022) [21], Bao et al. (2022) [22], and Zhang et al. (2022) [23] deal with ecosystem service provision and thermal comfort. The articles by Schütt et al. (2022) [24] and Liang et al. (2022) [25] can be assigned to the third theme on soil property studies, and the articles by Shu et al. (2022) [26] and Dervishi et al. (2022) [27] address the fourth topic of mapping urban greenspaces as well as drought tolerance analysis. In the following, each study is briefly summarized.

The work of Amer et al. (2023) [18] addresses the allometric relationships between the dimensions of urban trees, their aboveground biomass as carbon stores, and their shading potential. This study is of great interest because it focuses on three typical species growing in the arid city of Jericho, Palestine. The study presents a novel quantitative approach to estimating the ecosystem services of urban trees in arid cities that will be most affected by climate change. Such studies are very important for extending our knowledge base on sustainable urban development. Similarly, the study by Franceschi et al. (2022) [19] also focuses on tree allometry, but in a contrasting temperate climate. The main aspect of this study is the canopy shape of common European urban tree species and its influence on the calculation of canopy volume and the provision of ecosystem services. Franceschi et al. (2022) [19] demonstrated that urban tree crowns are mostly ovoid, but that aspects of the tree's environment, such as buildings in close proximity, can also influence their crown shape. In addition, the work by Wang et al. (2023) [20] addressed the effects of canopy size and shape and tree cover on cooling performance. Using a combination of field experiments and modeling approaches, it was shown that small canopies have better cooling performance than large canopies for the same amount of cover, but these relationships can change with different levels of cover.

In the study by Li and Zheng (2022) [21], the effects of vertical greening on indoor thermal comfort were analyzed using different modeling approaches. Through a more realistic approach, it was shown that vertical greening often does not have as positive an impact on indoor comfort as expected. Therefore, the use of vertical greening should be carefully considered. Bao et al. (2022) [22] conducted another study published in this SI focusing on human thermal comfort. Here, the effects of plant community characteristics on temperature and humidity in urban areas were investigated. The results showed that thermal comfort (i.e., cooling and humidity) is affected by tree canopy density and greenery, as well as by different types of plant communities. Tree–grassland and tree–shrub–grass species had the most significant effects on thermal comfort, so plant communities should be considered when designing parks or urban greenspaces. On the other hand, Zhang et al. (2022) [23] presented a LiDAR approach to study thermal comfort under street trees, which is influenced by the morphological structure of trees and microclimatic factors in the lower canopy. The results show a strong negative correlation between tree structures such as canopy volume, area, and diameter and air temperature, humidity, and brightness, which is also dependent on tree species and canopy shape. Therefore, Zhang et al. (2022) [23] recommended planting tall oval- and peaked-crown tree species with a dense and wide canopy and dense foliage to maximize the effects on the understory microclimate.

In the work of Schütt et al. (2022) [24], the hydrological properties of artificial urban planting soils were investigated. Using an experimental field, different structural soils were tested for their effects on tree growth as well as the influence of tree morphological characteristics on ecosystem service provisions. Urban soils reduced tree growth and thus the provision of ecosystem services. The study suggested that trees with finer rooting systems may be planted in sandy soils, while others may respond to drought stress by reducing their water potential. Liang et al. (2022) [25] also studied the diurnal and seasonal variations in soil heat flux in urban riparian areas. In particular, the relationships between soil heat flux and net radiation were analyzed in relation to different time points, soil moisture levels, and vegetation conditions. The results show a large influence of leaf area index, soil water content, and net radiation on soil heat flux in relation to canopy cover and forest type.

In their paper, Shu et al. (2022) [26] described how to unify urban tree assessment and improve information transfer/communication between different professionals. Using a TIM modeling approach, they showed how tree structure data such as the topological geometry of trunk and branches can be included and how the information can be evaluated and used by other users. Dervishi et al. (2022) [27] conducted a dendrochronological analysis of common urban tree species in Central Europe and showed the influence of climate on tree growth. Overall, trees had 8.3% lower DBH at 100 years of age in dry climates than in wet climates. In general, drought-tolerant tree species showed less or no influence of soil aridity. In contrast, drought-sensitive tree species were negatively affected by a dry climate.

3. Concluding Remarks

The papers published in this Special Issue, "Urban Forest and Urban Microclimate", cover a wide range of topics on growth patterns and drought tolerance, microclimatic effects and thermal comfort, and green management of urban greenery, including urban trees. Geographically, different climatic zones and continents are covered: one study was conducted in the arid climate of the Middle East, three studies in subtropical China, two studies in continental regions of Germany and China, one study with a temperate climate in Germany, and one study covering different climatic zones worldwide as well as a modeling study. Overall, the articles provide information about important aspects of urban green infrastructure and are essential for sustainable green management focusing on thermal comfort and climate adaptation.

Conflicts of Interest: The authors declare no conflict of interest.

References

1. Bernhofer, C.; Matschullat, J.; Bobeth, A. *Das Klima in der Regklam-Modelregion Dresden*; Regklam Publikationsreihe Heft 1; Rhombos: Berlin, Germany, 2009.
2. Böll, S.; Schönfeld, P.; Körber, K.; Herrmann, J.V. Stadtbäume unter Stress—Projekt »Stadtgrün 2021« Untersucht Stadtbäume im Zeichen des Klimawandels. *LWF Aktuell* **2014**, *98*, 4–8.
3. IPCC. *Climate Change 2021: The Physical Science Basis. Contribution of Working Group I to the Sixth Assessment Report of the Intergovernmental Panel on Climate Change*; Masson-Delmotte, V., Zhai, P., Pirani, A., Connors, S.L., Péan, C., Berger, S., Caud, N., Chen, Y., Goldfarb, L., Gomis, M.I., et al., Eds.; Cambridge University Press: Cambridge, UK; New York, NY, USA, 2021.
4. Morgenroth, J.; Buchan, G.D. Soil Moisture and Aeration Beneath Pervious and Impervious Pavements. *Arboric. Urban For.* **2009**, *35*, 135–141. [CrossRef]
5. Rötzer, T.; Moser-Reischl, A.; Rahman, M.; Hartmann, C.; Paeth, H.; Pauleit, S.; Pretzsch, H. Urban tree growth and ecosystem services under extreme drought. *Agric. For. Meteorol.* **2021**, *308–309*, 108532. [CrossRef]
6. Oke, T.R. The energetic basis of the urban heat island. *Q. J. R. Meteorol. Soc.* **1982**, *108*, 455. [CrossRef]
7. Akbari, H.; Pomerantz, M.; Taha, H. Cool surfaces and shade trees to reduce energy use and improve air quality in urban areas. *Sol. Energy* **2001**, *70*, 295–310. [CrossRef]
8. Day, S.; Wiseman, P.E.; Dickinson, S.; Harris, J.R. Contemporary Concepts of Root System Architecture of Urban Trees. *Arboric. Urban For.* **2010**, *36*, 149–159. [CrossRef]
9. Oliveira, S.; Andrade, H.; Vaz, T. The cooling effect of green spaces as a contribution to the mitigation of urban heat: A case study in Lisbon. *Build. Environ.* **2011**, *46*, 2186–2194. [CrossRef]
10. Millennium Ecosystem Assessment. *Ecosystems and Human Wellbeing—Health Synthesis*; World Health Organization: Washington, DC, USA, 2005.
11. Rahman, M.A.; Pawijit, Y.; Xu, C.; Moser-Reischl, A.; Pretzsch, H.; Rötzer, T.; Pauleit, S. A comparative analysis of urban forests for storm-water management. *Sci. Rep.* **2023**, *13*, 1451. [CrossRef]
12. Rahman, M.A.; Hartmann, C.; Moser-Reischl, A.; von Strachwitz, M.F.; Paeth, H.; Pretzsch, H.; Pauleit, S.; Rötzer, T. Tree cooling effects and human thermal comfort under contrasting species and sites. *Agric. For. Meteorol.* **2020**, *287*, 107947. [CrossRef]
13. Rötzer, T.; Rahman, M.; Moser-Reischl, A.; Pauleit, S.; Pretzsch, H. Process based simulation of tree growth and ecosystem services of urban trees under present and future climate conditions. *Sci. Total Environ.* **2019**, *676*, 651–664. [CrossRef]

14. Gómez-Baggethun, E.; Barton, D.N. Classifying and valuing ecosystem services for urban planning. *Ecol. Econ.* **2013**, *86*, 235–245. [CrossRef]
15. Moser-Reischl, A.; Rahman, M.A.; Pauleit, S.; Pretzsch, H.; Rötzer, T. Growth patterns and effects of urban micro-climate on two physiologically contrasting urban tree species. *Landsc. Urban Plan.* **2018**, *183*, 88–99. [CrossRef]
16. Shashua-Bar, L.; Pearlmutter, D.; Erell, E. The influence of trees and grass on outdoor thermal comfort in a hot-arid environment. *Int. J. Clim.* **2011**, *31*, 1498–1506. [CrossRef]
17. Rahman, M.A.; Moser, A.; Gold, A.; Rötzer, T.; Pauleit, S. Vertical air temperature gradients under the shade of two contrasting urban tree species during different types of summer days. *Sci. Total Environ.* **2018**, *633*, 100–111. [CrossRef] [PubMed]
18. Amer, A.; Franceschi, E.; Hjazin, A.; Shoqeir, J.H.; Moser-Reischl, A.; Rahman, M.A.; Tadros, M.; Pauleit, S.; Pretzsch, H.; Rötzer, T. Structure and Ecosystem Services of Three Common Urban Tree Species in an Arid Climate City. *Forests* **2023**, *14*, 671. [CrossRef]
19. Franceschi, E.; Moser-Reischl, A.; Mohammad, A.; Rahman, S.P.; Pretzsch, H.; Rötzer, T. Crown Shapes of Urban Trees-Their Dependences on Tree Species, Tree Age and Local Environment, and Effects on Ecosystem Services. *Forests* **2022**, *13*, 748. [CrossRef]
20. Wang, H.; Cai, Y.; Deng, W.; Li, C.; Dong, Y.; Zhou, L.; Sun, J.; Li, C.; Song, B.; Zhang, F.; et al. The Effects of Tree Canopy Structure and Tree Coverage Ratios on Urban Air Temperature Based on ENVI-Met. *Forests* **2023**, *14*, 80. [CrossRef]
21. Li, J.; Zheng, B. Does Vertical Greening Really Play Such a Big Role in an Indoor Thermal Environment? *Forests* **2022**, *13*, 358. [CrossRef]
22. Bao, Y.; Gao, M.; Luo, D.; Zhou, X. The Influence of Plant Community Characteristics in Urban Parks on the Microclimate. *Forests* **2022**, *13*, 1342. [CrossRef]
23. Zhang, X.; Lei, Y.; Li, R.; Ackerman, A.; Guo, N.; Li, Y.; Yang, Q.; Liu, Y. Research on Thermal Comfort of Underside of Street Tree Based on LiDAR Point Cloud Model. *Forests* **2022**, *13*, 1086. [CrossRef]
24. Schütt, A.; Becker, J.N.; Reisdorff, C.; Eschenbach, A. Growth Response of Nine Tree Species to Water Supply in Planting Soils Representative for Urban Street Tree Sites. *Forests* **2022**, *13*, 936. [CrossRef]
25. Liang, A.; Xie, C.; Wang, J.; Che, S. Daily Dynamics of Soil Heat Flux and Its Relationship with Net Radiation in Different Urban Riparian Woodlands. *Forests* **2022**, *13*, 2062. [CrossRef]
26. Shu, Q.; Rötzer, T.; Detter, A.; Ludwig, F. Tree Information Modeling: A Data Exchange Platform for Tree Design and Management. *Forests* **2022**, *13*, 1955. [CrossRef]
27. Dervishi, V.; Poschenrieder, W.; Rötzer, T.; Moser-Reischl, A.; Pretzsch, H. Effects of Climate and Drought on Stem Diameter Growth of Urban Tree Species. *Forests* **2022**, *13*, 641. [CrossRef]

forests

MDPI

Article

Structure and Ecosystem Services of Three Common Urban Tree Species in an Arid Climate City

Alaa Amer [1,2,3,*], Eleonora Franceschi [1], Amgad Hjazin [4,5], Jawad H. Shoqeir [3], Astrid Moser-Reischl [1], Mohammad A. Rahman [5], Maher Tadros [6], Stephan Pauleit [5], Hans Pretzsch [1] and Thomas Rötzer [1]

1 Forest Growth and Yield Science, TUM School of Life Sciences, Technical University of Munich, Hans-Carl-von-Carlowitz-Platz 2, D-85354 Freising, Germany
2 AL-Quds Public Health Society (AQPHS), Ibn Batouta Street, Jerusalem P.O. Box 20760, Palestine
3 Soil & Hydrology Research Lab (SHR), Earth and Environmental Sciences Department, Al-Quds University, Jerusalem P.O. Box 20760, Palestine
4 Environment and Climate Change Research Directorate, National Agricultural Research Center (NARC), Al-Balqah 19381, Jordan
5 Strategic Landscape Planning and Management, Technical University of Munich, Emil-Ramann-Straße 6, D-85354 Freising, Germany
6 Department of Natural Resources and Environment, Jordan University of Science and Technology, P.O. Box 3030, Irbid 22110, Jordan
* Correspondence: alaa.amer@tum.de

Abstract: Urban forests play a critical role in improving the quality of life in cities, but in arid environments, little is known about the potential benefits and growth conditions of different tree species. Our study aimed to fill this gap by investigating the relationships between tree dimensions, above-ground biomass carbon storage, and shading potential in three common urban trees in the arid city of Jericho, Palestine, (i.e., *Ficus nitida*, *Delonix regia*, and *Phoenix dactylifera*). The trees were chosen according to their distribution in urban locations and tree vitality, with ages ranging from 20 to 90 years. Based on the results from tree structure measurements, the carbon storage and shading potential were calculated using the City Tree model. The results indicate a moderate to strong relationship between tree height, crown diameter, and crown volume for *F. nitida* and *D. regia* (R^2 = 0.28–0.66), but no relationship for *P. dactylifera* (R^2 = 0.03–0.06). The findings suggest that the analyzed tree species can considerably contribute to the potential benefits of trees in improving the climate of an arid city: *D. regia* shows a higher median of above-ground biomass carbon storage of 155 kg C tree^{-1}, while *P. dactylifera* 91 kg C and *F. nitida* 76 Kg C. *D. regia* and *F. nitida* have a higher median of shading potential, (31 m^2–41 m^2), respectively. Information on the ecosystem services from urban trees and their relationships in terms of species, age, and tree planting urban location are very important for city planners, in relation to sustainable urban green spaces in arid cities.

Keywords: crown dimension; arid city; *Delonix regia*; *Ficus nitida*; *Phoenix dactylifera*; urban trees; carbon storage; shade potential; tree pit surface area; leaf area index

Citation: Amer, A.; Franceschi, E.; Hjazin, A.; Shoqeir, J.H.; Moser-Reischl, A.; Rahman, M.A.; Tadros, M.; Pauleit, S.; Pretzsch, H.; Rötzer, T. Structure and Ecosystem Services of Three Common Urban Tree Species in an Arid Climate City. *Forests* **2023**, *14*, 671. https://doi.org/10.3390/f14040671

Academic Editor: Chi Yung Jim

Received: 10 February 2023
Revised: 12 March 2023
Accepted: 15 March 2023
Published: 24 March 2023

1. Introduction

Urban trees are an essential component of urban green spaces, playing a crucial role in enhancing the well-being of city inhabitants. Urban trees offer myriad benefits, including reducing the urban heat island effect (UHI), mitigating the effects of climate change by removing atmospheric CO_2 [1,2], moderating microclimates [3], and providing shade by reducing the temperatures on surfaces under tree canopies, particularly in the summer months in arid cities [4]. Additionally, these urban green spaces covered by trees also offer a variety of social and cultural benefits, including recreational opportunities, aesthetic value, and potential inspiration for the arts and other creative endeavors [5]. Furthermore, urban trees ameliorate the thermal environment of surroundings, and provide cooling effects through evapotranspiration and shading, thereby regulating local and regional climates [6–9].

Urban streets, particularly in semi-arid regions, can experience a significant increase in temperature, ranging from 3 to 6 °C compared to the surrounding rural environment [10]. Semi-arid regions account for 42% of the total global land area and support approximately 38% of the global population, and are often located in developing countries [11]. The provision of tree benefits largely depends on tree growth, which can vary with a range of microenvironmental and other site-specific factors [12], for instance, anthropogenic disturbances such as mechanical injury [13,14], low soil quality [15], sealed surfaces reducing water availability for tree roots [16], and limited rooting space [17], soil compaction [18], and reduced nutrient resources and soil aeration [19,20].

These disturbances are often location-dependent, and the risks they pose to tree vitality can vary substantially over small areas—depending, for instance, on planting locations in parking lots, gardens, squares, or streets. Rötzer et al. [21] have found that streets, paved squares, rooftops, and car parks limit the growth of trees, while larger gardens and public green spaces, such as parks and cemeteries, can provide ideal habitats for trees. Sanders et al. [22] demonstrate that planting space has a significant impact on tree growth, with trees planted in reduced space exhibiting reduced maximum size.

In semi-arid regions specifically, irregular rainfall, poor tree management practices, and drought stress can also negatively impact urban tree growth [23–25], and could influence the benefits trees are able to provide. Because the effective management of urban trees depends on a detailed understanding of the effects of growing environment, a substantial and growing literature seeks to evaluate the effects of climate change on urban tree growth rates in various climate zones [23,24,26]. Several factors can reduce tree growth in arid and warm areas where water resources are limited [27,28]. In contrast, a few studies observed that some factors may increase the urban tree growth rate compared to rural trees, e.g., [29–32], including, for instance lower ozone concentration, larger annual atmospheric N deposition, and higher CO_2 concentration [31,33].

Considering the various factors influencing urban tree growth and their ecosystem services, recent research on tree growth and structure in urban green spaces has focused on monitoring and understanding these changes. By studying the relationships between structural variables such as leaf area index, crown dimension, tree height, and stem diameter, it is possible to model growth patterns and predict ecosystem services provisioning. This information can aid in the improvement of planning and management practices for urban landscapes [34]. However, urban tree growth in arid cities is poorly understood, which impedes modelling and limits the available evidence base for planners and managers.

City planners, for instance, must take into account the ability of urban trees to acclimate to their surroundings and the structural variables that affect their future growth in order to optimize their benefits and ensure their long-term survival in an urban environment [3]. As such, the structural development of urban trees, including size and shape, is closely linked to the benefits they provide [35]. For instance, the area and density of shading from solar radiation is largely a function of the shape and volume of tree crowns [36], while carbon sequestration and storage are driven by biomass and growth increment [37,38].

Moser et al. [39] developed a regression equation to predict future structural dimensions through direct field measurements based on tree diameter and age. Issa et al. [40] used crown dimensions to create an allometric equation to calculate total biomass, serving as a basis for remote sensing prediction and biomass assessment.

Understanding the relationship between structural variables of trees such as tree height, diameter at breast height, crown dimensions, and crown volume is essential to predict growth and ecosystem services [41,42]. Typically, diameter at breast height (dbh) is used to estimate tree growth based on the pipe model theory and functional carbon balance theory [43–45]. These theories allow for the derivation of tree structure and biomass from basic tree measurements.

Pretzsch et al. [46], and Watt et al. [47] use dbh as an explanatory variable to predict crown dimensions. Although allometric equations for urban tree species have been developed for tropical and temperate regions [48,49], studies about the structural dimensions and ecosystem services of urban trees in arid cities are scarce. Despite limited research on the tree growth patterns of urban trees in arid cities [50], there is a growing need to understand the factors that influence their growth and survival in these challenging environments. This research can provide a basic understanding of the structural dimensions and ecosystem services of urban trees in arid cities. We therefore analyzed the structural variables of urban trees in an arid city and estimated their carbon storage and shading potential (shaded area and shade density) as ecosystem services. We also aimed to examine the influence of site conditions, such as tree planting urban location and total unsealed area (tree pit surface area), on the tree structural variables, to understand the relationship between commonly planted urban tree structural variables and their effect on selected ecosystem services. The following hypotheses were tested:

Hypotheses 1 (H1). For each of the tree species, *Delonix regia*, *Ficus nitida* and *Phoenix dactylifera*, significant different relationships exist in terms of

(a) Tree height and crown dimensions with diameter at breast height (dbh, independent parameter).
(b) dbh, tree height, and crown dimensions with tree age (independent parameter)
(c) dbh, tree height, and crown dimensions with leaf area index (LAI, independent parameter)
(d) dbh, tree height, and crown dimensions with tree pit surface area (independent parameter).

Hypotheses 2 (H2). Tree planting urban location has a significant influence on tree structural variables (tree height, dbh, crown dimension), and ecosystem services.

Hypotheses 3 (H3). The ecosystem services of carbon storage and shading potential of the three tree species differ significantly from each other.

2. Materials and Methods Tab

2.1. Study Site

Tree structural data was collected in the city of Jericho, located in the eastern part of the West Bank, Palestine (coordinates: 31.8611° N, 35.4618° E). Jericho is one of the oldest cities in the world, dating back to 7000 BC [51], with an elevation of 252 m below sea level. The climate is hot semi-arid with an average annual precipitation of 145 mm, and a mean annual temperature of 22.5 °C for the period 1991–2020 [52]. Trees were sampled by following urban transects (starting from the city center to the edge of the city boundary in all four cardinal directions (north, south, east, and west) (see Figure 1).

2.2. Categorization of Trees Based on Sites

The selected trees were classified based on their urban planting location and divided into four categories (due to their uneven distributions): (a) street trees, located on both sides of roads; (b) public place trees, planted in gardens with semi-vegetation-covering and semi-surrounded by buildings; (c) trees standing in parking lots, located in car parking areas; and (d) square trees, located downtown, where most social activity occurs. Young to old trees, and only healthy and vital trees were selected, as determined through visual inspection, and rated using a scale according to Roloff [53]. Trees that were heavily pruned or damaged, as well as those with low-forking branches, were excluded, followed [39,54]. It is worth noting that the Jericho City Garden Department prunes the trees annually to prevent negative effects on pedestrians. The tree data collection was conducted from June to November 2020, resulting in a total of 212 commonly available trees being measured, of which 69 were *D. regia*, 73 were *F. nitida*, and 70 were *P. dactylifera* (see Table 1).

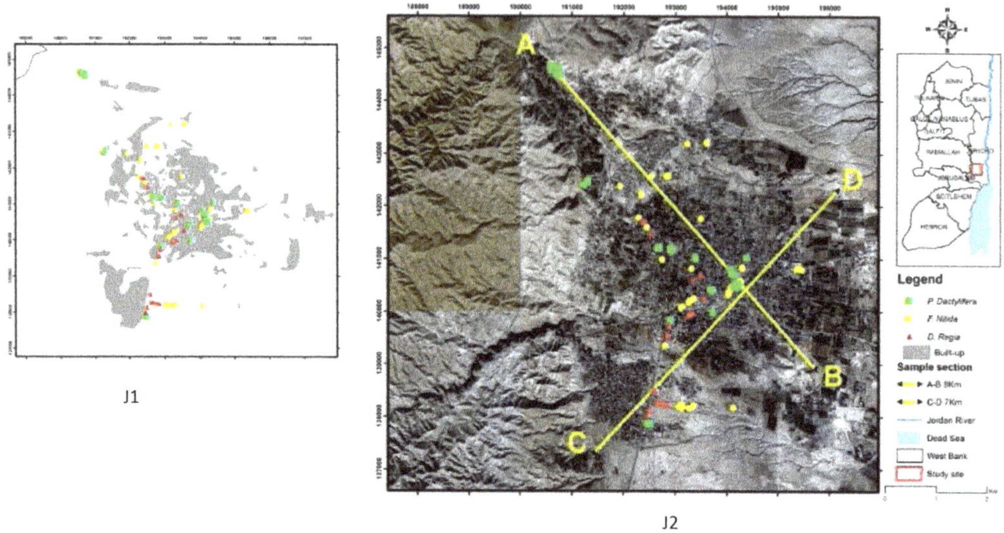

Figure 1. Spatial distribution of the measuring site following a north–south and east–west transect within Jericho. (**J1**) shows the distribution of the selected urban trees along the transects encompassing the urban area. (**J2**) shows an aerial image of the boundaries of Jericho, depicting the northwest–southeast transect (A,B) with a length of 7 km and the southwest–northeast transect (C,D) with a width of 5 km.

Table 1. The number of measured tree species for four different planting categories: public space, parking lot, street, and square.

Tree Species	Public Space	Parking Lot	Street	Square	Sum
Delonix regia	0	7	62	0	69
Ficus nitida	15	19	39	0	73
Phoenix dactylifera	50	0	15	5	70

2.3. Plant Species Description

Three common urban trees were selected in the arid city of Jericho, Palestine: the common fig tree (*Ficus nitida*), royal poinciana (*Delonix regia*) and date palm (*Phoenix dactylifera*). According to the Jericho municipality, by 2020, the city area had planted approximately 1000 *F. nitida* trees, 3000 *D. regia* trees and an unknown number of *P. dactylifera* trees. *F. nitida* is a common ornamental [55], large evergreen fig tree species [56], native to vast areas worldwide, particularly in warm tropical and subtropical regions [57]. These trees can reach a height of up to 10 m [58] and present a gray and smooth bark [59], are moderately drought tolerant, tolerant to different soil formations, rapid growth and salinity tolerant [60], and need full sunlight to partial shade [61]. *Delonix regia* (*D. regia*) is a common species, has been historically grown as an ornamental tree [50,62], and is commonly grown in the tropics and subtropics [63]. The trees are umbrella-shaped [64], with a maximum height of 10–15 m, a girth of up to 2 m, and have large trunks [50,62]. They are grown in public gardens, along roadsides, in parks, between buildings and in residential areas [65]. It is a light-demanding species, develops sluggishly and unevenly in the shadows [64], and is intolerant to heatwaves and high solar radiation. Nevertheless, it can tolerate many types of soil formation, although sandy soils are more functional for growth [66]. *Phoenix dactylifera* (date palm) is a diploid and monocotyledonous plant [67]. It is one of the oldest fruit crops [68]. It can be described as a tall plant with an average height range of 15–20 m [69] and lives on average for over 100 years [70]. The palm tree's trunk can reach up to 30 m in length and is enclosed in fiber for protection (e.g., to protect the trunk from herbivorous insects and animals) and reducing water loss [71]. *P. dactylifera* species tolerate harsh growth conditions, high temperatures, droughts and high levels of salinity [72].

2.4. Measured Tree Variables

A global positioning system (GPS) (eTrex Vista ® CX Garmin) was used to record the tree positions (longitude, latitude, and elevation). Diameter at breast height (dbh) was measured for all species using a measuring tape. For *F. nitida* trees, where the trunk height was lower than 130 cm, the diameter was measured at 70 cm instead of 130 cm. A Leica Disto D510 Laser Distance Measurer was used to measure the crown radii and the tree pit. The crown radii were measured from the center of the tree trunk to the end of the longest branch, whereas the tree pit surface area was measured starting from the center of the tree trunk and up to the end of the unsealed area. The total unsealed (tree pit surface area) area was calculated based on the City Tree model [12]. Crown radii and tree pit surface area were measured in eight intercardinal directions (N, NE, . . . , NW) following Moser et al. [39]. True-Pulse 200 Rangefinder laser technology was used to measure tree height (h) and height-to-crown base (hcb) (e.g., the distance between the lowest branch and the ground). Crown length (cl) was derived by measuring the distance between the lowest branch and the top of the tree. Crown diameter (cd), crown projection area (cpa), and crown volume (cv) were calculated using equations used from the literature [54]. A crown reduction shape factor Fc = 0.5 was applied for parabola-shaped crowns of *F. nitida* and *D. regia* to calculate the crown volume [21]. *P. dactylifera* crown volume was calculated based on a spherical crown shape. All tree ages were used based on the agricultural tree records retrieved from Jericho City.

2.4.1. Leaf Area Index (LAI) and Ecosystem Services

The LAI of the trees was derived from hemispheric photographs taken between August and October using a Nikon D7500 camera SIGMA Circular Fisheye EX DC HSM 4.5 mm 1:2.8 fisheye lens. WinSCANOPY (Regent Instruments, INC) was used to analyze the resulting hemispherical photos, i.e., to derive the LAI for *D. regia*, *F. nitida* and *P. dactylifera*, following Moser et al., [39]. Some trees were excluded from the leaf area index (LAI) analysis, including one *F. nitida* tree and 22 *P. dactylifera* trees. These exclusions were due to factors such as foliage loss during a long drought period in 2020, which was exacerbated by an inconsistent irrigation system and pruning.

2.4.2. Ecosystem Service Calculation

We estimated the ecosystem services (i.e., above-ground biomass carbon storage (*Csa*) (Kg C) and shading potential (SP) (shaded area and shade density) for *D. regia* and *F. nitida* according to the City Tree model [12]:

The above-ground biomass carbon storage is calculated by

$$Csa = Csfol + Csbt + Csstem \quad (1)$$

where *Csfol* = foliage biomass carbon, *Csbt* = branches and twigs biomass carbon, *Csstem* = stem biomass carbon. They can be calculated with the following equations:

$$Csfol = (LAI \times cpa/sla) \times 0.5 \quad (2)$$

$$Csbt = (\, exp\ (a + b \times 0.95 \times Ln\ (dbh))) \times 0.5 \quad (3)$$

where $a = -3.7299$, $b = 2.33$, which is obtained from [12].

$$Csstem = (volume \times specific\ wood\ density) \times 0.5 \quad (4)$$

According to El-Khatib et al. [73] and Agrawal et al. [74], the specific leaf area (*sla*) for *F. nitida* is 9.433 m^2/kg, and for *D. regia* it is 8.1 m^2/kg. The specific wood density for *F. nitida* (690 kg dw/m^3) [75], and for *D. regia* (510 kg dw/m^3) was obtained from Orwa et al., [64]. Stem volume was calculated from dbh, height and crown length according to [12] by assuming a cylindrical stem form.

To obtain the above-ground biomass carbon storage for *P. dactylifera*, we followed Issa et al., [40] using an allometric equation and considering that the maturity stages of our samples age exceeded 10 years. The above-ground biomass carbon storage of *P. dactylifera* can be estimated by:

The above-ground biomass carbon storage (*Csa*) = trunk biomass carbon storage (Cst) + crown biomass carbon storage (Csc)

$$Csa = Cst + Csc \tag{5}$$

Trunk biomass carbon storage (Cst) = fresh trunk biomass $(ft\ bm) \times 0.37 \times 0.9331 \times 0.58$

$$Cst = ft\ bm \times 0.37 \times 0.9331 \times 0.58 \tag{6}$$

$$ft\ bm = 40.725 \times Ht\text{\^{}}0.9719 \tag{7}$$

H_t: trunk height; 0.37 conversion factor from fresh crown biomass to dry weight (kg. dw); 0.9331 conversion factor to organic matter; and 0.58 as a conversion factor to carbon storage (kg C).

Carbon storage crown (*Csc*) = fresh crown biomass $(fc\ bm) \times 0.41 \times 0.9243 \times 0.58$

$$Csc = fc\ bm \times 0.41 \times 0.9243 \times 0.58 \tag{8}$$

$$fc\ bm = 14.034 \times e\ (0.0554\ x\ CA) \tag{9}$$

where *CA* is a crown area [m^2] calculated by the following equation

$$CA = \pi cd2/4 \tag{10}$$

Conversion factor from fresh crown biomass to dry weight (kg. dw): 0.41, conversion factor to organic matter: 0.9243, and conversion factor to carbon storage (kg C): 0.58.

The shade area and shade density for *D. regia*, *F. nitida*, and *P. dactylifera* were calculated according to the City Tree model [21].

The City Tree model, which took into consideration the crown shape, was followed to calculate a tree's shade area, shade density, and shade index. To determine the shade area, the average shade area between 8 a.m. and 6 p.m. on the 21st of June, the longest day in the northern hemisphere, was calculated. The shade area was calculated using the crown shade projection area formulas (cspa), with the crown diameter and shade length (instead of crown length) applied. To calculate the shade length, crown length, and cotangent for the hour, the location of the sun's height was considered.

$$aveA\ shade = \left(\sum_{i=8}^{18} shade\ area\ i\right)/11 \tag{11}$$

(i): representing the hour of the day, and 11: representing the total number of hours that are taken into consideration.

The shade density (dshade) was calculated following [21], for each tree by:

$$dshade = LAI \times cpa/cv \tag{12}$$

2.5. Statistical Analysis

The crown dimension variables were calculated in Microsoft Office Excel 365. All statistical analyses and figures were generated using R software, version 3.6.3 [76]. To test the normality of the data, we used the Shapiro–Wilk test (Shapiro and Wilk, 1965) [77], and log-transformed data were used when necessary. To test H1 (a), tree height and crown dimension are significantly dependent on dbh and H1 (b, c, d); dbh, tree height, and crown dimension are significantly dependent on leaf area index and tree age. Correlation-regression analyses with ordinary least squares (OLS) were performed by using log-transformed data following Pretzsch et al., Stoffberg et al., and Peper et al. [46,78,79]. Equation (13) for H1(a), and Equation (14) for H1(b, c, and d).

$$ln\ (y) = a + b \times ln\ (x) \tag{13}$$

$$(y) = a + b \times ln\ (x) \tag{14}$$

Through OLS regression, the response (y) is calculated from the predictor (x). When applying the models, we selected OLS instead of reduced major axis or moving average regression [80]. The second hypothesis (H2), the influences of different tree planting urban locations on tree structure and selected ecosystem services, was tested using a one-way ANOVA followed by the post hoc Tukey HSD test. In addition, it was used to test the third hypothesis (H3). The ecosystem services related to carbon storage and shade potential varied considerably among the three species. To visualize the structural variables, the impact on ecosystem services was considered. A linear mixed model (LMM) with random effect was used by using the "lme4" package in the R software, i.e., above-ground biomass carbon storage and shade area was used as the outcome variable, and the tree structure was used as the fixed effect, while tree pits and tree planting sites were considered random effects.

3. Results

3.1. Dependency of Tree Structure on dbh and Tree Age

All measured and calculated tree structural mean values and related standard deviation are given in ascending age classes for *F. nitida* and *P. dactylifera*, but for *D. Regia*, the ages of all samples ranged between 20 and 25 years. Table S1 provides valuable information on the characteristics of three tree species, including their age, dbh and crown dimension. The data highlights significant variations in these characteristics, both between species and within age categories, providing useful insights for researchers and practitioners in forestry and related fields. The limited age of trees in the city can be attributed to their recent planting and the fact that they constitute a significant proportion of the urban forest in the city.

The results show tree height and crown dimension are strongly correlated with the diameter at breast height (dbh) for *D. regia* and *F. nitida*, (see Table 2 and Figure 2). However, for *P. dactylifera*, the relationship between h and dbh is not significant, and the correlation between the crown volume and crown diameter and the dbh is weak. The strongest dependency was found between dbh and tree height, crown volume, and crown diameter for *F. nitida*, and between dbh and crown diameter for *D. regia*. However, there is no relationship between dbh with tree height or crown dimension for *P. dactylifera*.

Table 2. Results of linear regression analyses using dbh as a predictor variable and h, cd, and cv as response variables. The equation used was ln(y) = a + b ln(x). The abbreviations used were (dbh) diameter at breast height; (h) tree height; (cd) crown diameter; (cv) crown volume; a and b for regression coefficients; T for *T*-test value; P for *p*-value (with levels of significance indicated by symbols such as *** and *); R^2 for coefficient of determination; F for F-test value; and df for degree of freedom and standard error (SE).

Species	Parameter	*n*	a	b	T	P	SE	R^2	F	df
D. regia	ln(dbh) vs. ln(h)	69	0.17	0.38	5.15	<0.001 ***	0.07	0.28	26.54	67
	ln(dbh) vs. ln(cd)	69	0.10	0.52	8.20	<0.001 ***	0.06	0.50	67.18	67
	ln(dbh) vs. ln(cv)	69	−0.06	1.31	7.94	<0.001 ***	0.04	0.48	63.08	67

Table 2. *Cont.*

Species	Parameter	*n*	a	b	T	P	SE	R^2	F	df
F. nitida	ln(dbh) vs. ln(h)	73	0.13	0.46	11.63	<0.001 ***	0.04	0.66	135	71
	ln(dbh) vs. ln(cd)	73	0.01	0.57	10.15	<0.001 ***	0.06	0.59	103	71
	ln(dbh) vs. ln(cv)	73	−0.58	1.71	11.10	<0.001 ***	0.15	0.63	123	71
P. dactylifera	ln(dbh) vs. ln(h)	70	0.66	0.28	1.45	0.15	0.19	0.03	2.10	68
	ln(dbh) vs. ln(cd)	70	1.46	−0.47	−2.09	0.04 *	0.22	0.06	4.37	68
	ln(dbh) vs. ln(cv)	70	4.11	−1.41	−2.09	0.04 *	0.67	0.06	4.37	68

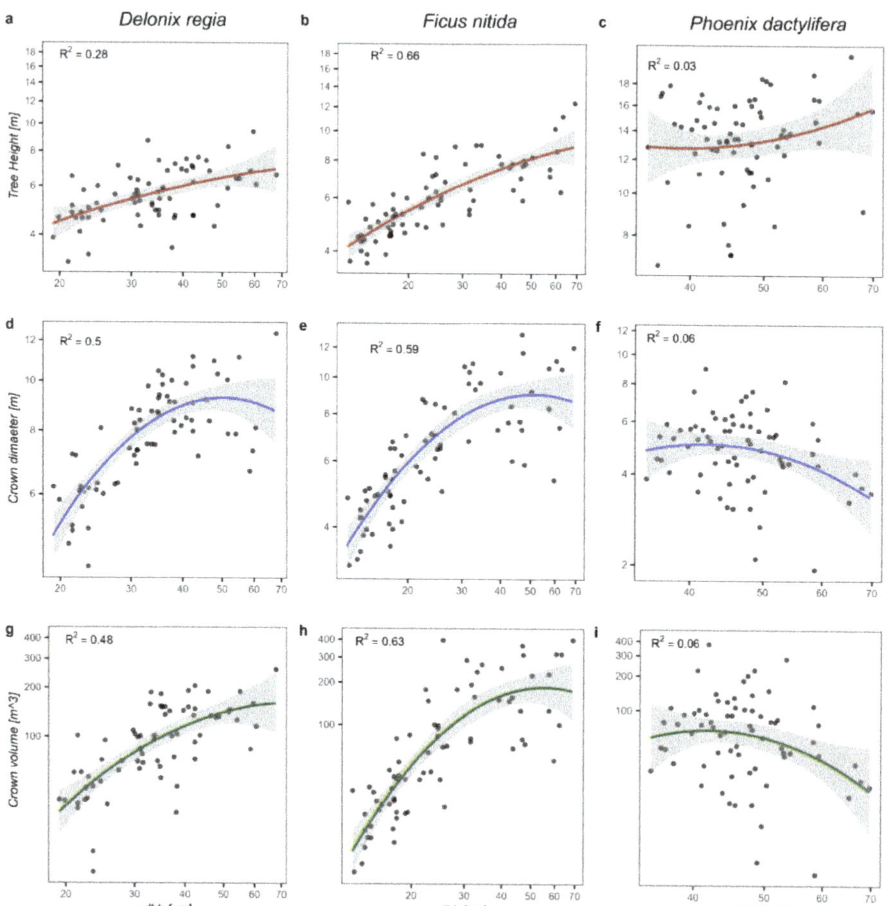

Figure 2. The relationships between dbh and tree height (**a**–**c** in the top row), crown diameter (**d**–**f** in the middle row), and crown volume (**g**–**i** in the bottom row) for the investigated species.

The relationship between dbh, tree height, and crown dimension with the age of the three tree species were studied by the outcomes of linear regression analysis and shown in Table 3. The results show a significant relationship of dbh with age for *D. regia*, but all other variables are not significant. *F. nitida*, shows a strong to a moderate relationship with age, particularly to dbh ($R^2 = 0.61$). Finally, the *P. dactylifera* results revealed a non-significant variance for all tree variables ($R^2 \leq 0.05$).

Table 3. Allometric linear relationships between age and tree height, crown diameter, and crown volume as a response, and the regression equation for *F. nitida*, *D. regia*, and *P. dactylifera*. Abbreviations: (dbh) diameter at breast height; (h) tree height; (cd) crown diameter; (cv), crown volume; regression coefficients (a, b); coefficients of determination (R^2); standard errors (SE); and F-values, as well as P for *p*-value (with levels of significance indicated by symbols such as *** and *).

	Parameters	*n*	a	b	T	P	SE	df	R^2	F
D. regia	Age vs. ln(dbh)	69	2.36	0.05	2.09	0.04 *	0.02	67	0.06	4.38
	Age vs. ln(h)	69	1.17	0.03	1.38	0.17	0.01	67	0.03	1.89
	Age vs. ln (cv)	69	4.32	0.01	0.14	0.89	0.05	67	<0.01	0.02
	Age vs. ln (cd)	69	3.1	−0.34	−0.8	0.43	0.45	67	<0.01	0.6
F. nitida	Age vs. ln(dbh)	73	2.68	0.03	10.61	<0.001 ***	<0.01	71	0.61	112.6
	Age vs. ln(h)	73	1.51	0.014	7.86	<0.001 ***	<0.01	71	0.47	61.81
	Age vs. ln (cv)	73	3.3	0.05	6.16	<0.001 ***	<0.01	71	0.36	37.92
	Age vs. ln (cd)	73	1.57	0.01	5.41	<0.001 ***	<0.01	71	0.3	29.28
P. dactylifera	Age vs. ln(dbh)	70	3.73	0.01	1.9	0.06	<0.01	68	0.05	3.64
	Age vs. ln(h)	70	2.6	0.00	−0.59	0.56	0.01	68	0.01	0.34
	Age vs. ln (cv)	70	4.6	−0.01	−1.58	0.12	<0.01	68	0.04	2.5
	Age vs. ln (cd)	70	1.7	0.00	−1.577	0.12	<0.01	68	0.04	2.48

3.2. Dependency of LAI on Tree Species and Tree Structure

Linear regression analysis was used to investigate the relationship between leaf area index (LAI) and the variables dbh, h, cv, and cd for three tree species (*F. nitida*, *D. regia*, and *P. dactylifera*). However, we found no significant relationships between LAI and any of the variables (see Supplementary Table S2). The analysis showed that *F. nitida* and *D. regia* had significantly higher LAI values than *P. dactylifera* ($p < 001$ ***), with mean LAI values of 5.3 ± 0.22 and 5.8 ± 0.20, respectively, compared to the *P. dactylifera* mean LAI value of 2.9 ± 0.15. The standard errors for the mean LAI values for *F. nitida*, *D. regia*, and *P. dactylifera* were 0.22, 0.20, and 0.15, respectively. LAI may be an important factor to consider when comparing these three species. The sample sizes were 72, 69, and 48 for *F. nitida*, *D. regia*, and *P. dactylifera*, respectively.

3.3. Impact of Tree Urban Location and Tree Pit Surface Area on a Tree Structure

3.3.1. Tree Planting Urban Location

The results revealed that the dbh of *D. regia* and *F. nitida* exhibit significant variations across different site categories. Furthermore, the crown volume of *F. nitida* and *P. dactylifera* also showed significant variations as detailed in Table 4. The results also indicate that the tree height and age in *D. regia* differ across different sites (this might be due to different planting times), while the crown projection area and crown diameter of *F. nitida* is also significantly affected by the site. However, all other tree structural variables for the three tree species were found to not be significantly impacted by the site. We calculated the mean tree pit surface area of three species (*F. nitida*, *D. regia*, and *P. dactylifera*) in three different sites (a street, a parking lot, and a public place) along with the standard error. The statistical analyses show that the mean values of *F. nitida* and *D. regia* species are significantly different across different sites, as indicated by the *p*-values, $p \leq 0.001$ and 0.009, respectively. On the other hand, the mean values of *P. dactylifera* species do not show significant differences across the sites, as indicated by the *p*-value of 0.36, (See Table S3 in the supplementary section).

3.3.2. Tree Pit Surface Area

Weak and significant differences were found in the variables dbh, h, cd, and cv of *F. nitida* in relation to the tree pit surface area, as well as in the variables dbh, cd, and cv *of P. dactylifera* (Refer to Table 5). However, none of the previously mentioned *D. regia* variables were found to have significant differences in the tree pit surface area.

Table 4. Mean of the trees' structural data: age, dbh, h, hcb, cl, and related SD in response to the growth site for *D. regia*, *F. nitida*, and *P. dactylifera*, as well as the *p*-value (with levels of significance indicated by symbols such as ***, ** and *) for each ANOVA. The mean in the same column differs significantly when followed by different letters. Abbreviations: (dbh) diameter at breast height; (h) tree height; (cl) crown length; (cd) crown diameter; (cpa) crown projection area; (cv) crown volume; SD, standard deviation.

Site	*n*	Age	dbh [cm]	h [m]	hcb [m]	cl [m]	cd [m]	cpa [m²]	cv [m³]
		Mean ± SD	Mean ± SD	Mean ± SD	Mean ± SD	Mean ± SD	Mean ± SD	Mean ± SD	Mean ± SD
D. regia		$p \leq 0.01$ ***	$p < 0.01$ **	$p = 0.02$ *	$p = 0.2$	$p = 0.12$	$p = 0.53$	$p = 0.69$	$p = 0.26$
Parking lot	-	-	-	-	-	-	-	-	-
Public place	7	24.1 ± 1.5 a	46.8 ± 7.6 a	6.7 ± 1.8 a	2.2 ± 0.3 a	4.5 ± 1.9 a	8.4 ± 0.9 a	55.6 ± 12.5 a	122 ± 42.7 a
Street	62	21.5 ± 1.1 b	33.98 ± 10.7 b	5.6 ± 1.1 b	1.8 ± 0.7 a	3.7 ± 1.2 a	7.9 ± 1.8 a	52.02 ± 22.98 a	98.7 ± 52.4 a
F. nitida		$p = 0.20$	$p = 0.012$ *	$p = 0.098$	$p = 0.14$	$p = 0.09$	$p = 0.02$ *	$p = 0.02$ *	$p = 0.01$ *
Parking lot	19	19.1 ± 21.5 a	25.5 ± 15.6 ab	5.60 ± 1.5 a	1.93 ± 0.88 a	3.67 ± 1.0 a	5.84 ± 1.2 b	27.94 ± 11.8 b	53.7 ± 31.6 b
Public place	15	14 ± 4.9 a	20.1 ± 5.4 b	5.89 ± 1.4 a	1.76 ± 0.59 a	4.13 ± 1.4 a	5.92 ± 2.1 ab	30.73 ± 77.2 ab	73.9 ± 77.2 ab
Street	39	21.6 ± 13.5 a	32.7 ± 15.6 a	6.58 ± 2.0 a	1.99 ± 0.76 a	4.59 ± 1.7 a	7.6 ± 3.1 a	52.98 ± 45.8 a	154.4 ± 187.4 a
P. dactylifera	.	$p = 0.11$	$p = 0.13$	$p = 0.8$	$p = 0.67$	$p = 0.83$	$p = 0.06$	$p = 0.05$	$p = 0.02$ *
Parking lot	0	-	-	-	-	-	-	-	-
Public place	50	67.1 ± 21.4 a	46.3 ± 6.2 a	13.46 ± 3.0 a	10.43 ± 3.1 a	3.03 ± 1.7 a	4.92 ± 1.4 a	20.78 ± 11.6 ab	78.7 ± 69.7 ab
Street	15	67.6 ± 22.5 a	51.6 ± 10.8 a	14.10 ± 3.2 a	10.79 ± 2.7 a	3.31 ± 1.7 a	4.74 ± 1.0 a	16.87 ± 35.9 b	56.3 ± 35.90 b
square	5	73.3 ± 18.2 a	46.1 ± 9.0 a	13.25 ± 3.6 a	9.9 ± 3.0 a	2.74 ± 1.7 a	4.88 ± 1.7 a	21.60 ± 13.9 a	87.2 ± 85.5 a

Table 5. Results of the summary of the regression analysis of tree pit surface area, the predictor variables, and diameter at breast height (dbh), tree height (h), crown diameter (cd), and crown volume (cv), as a response, and the regression equation $(y) = a + b \times \ln(x)$. The table below lists the determination of R^2, residual standard error, and *p*-values. The R^2 value and the *p*-value (with levels of significance indicated by symbols such as ***, ** and *)for each ANOVA show the relationship between the tree structural variables and the tree pit surface area for the species.

Species	Parameter	*n*	A	b	*t*-Value	*p*-Value	RSE	df	R^2	F-Value
D. regia										
	Tree pit surface area vs. ln(dbh)	69	3.56	−0.01	−0.95	0.35	0.31	67	0.01	0.9
	Tree pit surface area vs. ln(h)	69	1.72	0	−4.08	0.97	0.22	67	<0.01	0
	Tree pit surface area vs. ln(cd)	69	2.03	0	0.65	0.52	0.22	67	0.01	0.42
	Tree pit surface area vs. ln(cv)	69	4.4	0.02	0.75	0.46	0.58	67	0.01	0.56
F. nitida										
	Tree pit surface area vs. ln(dbh)	73	3.43	−0.12	−6.21	<0.01 ***	0.37	71	0.35	38.54
	Tree pit surface area vs. ln(h)	73	1.87	−0.05	−4.08	<0.01 ***	0.25	71	0.19	16.64
	Tree pit surface area vs. ln(cd)	73	1.97	−0.07	−4.66	<0.01 ***	0.32	71	0.23	21.72
	Tree pit surface area vs. ln(cv)	73	4.45	−0.17	−3.56	<0.01 ***	0.95	71	0.15	12.68
P. dactylifera										
	Tree pit surface area vs. ln(dbh)	70	3.88	0	−1.1	0.04 *	0.15	68	0.05	0.04
	Tree pit surface area vs. ln(h)	70	2.58	0	−0.04	0.70	0.25	68	<0.01	0.15
	Tree pit surface area vs. ln(cd)	70	1.44	0	3.24	<0.001 **	0.27	68	0.12	10.47
	Tree pit surface area vs. ln(cv)	70	3.46	0.02	5.741	<0.001 ***	0.72	68	0.32	32.66

3.4. Ecosystem Services of F. nitida, D. regia, and P. dactylifera

The relationship between above-ground biomass carbon storage (Csa), and shaded area with tree structure was analyzed using LMM, with Csa and shaded area as outcome variables, and dbh, crown diameter, and tree height as fixed factors for *D. regia* and *F. nitida* (Figure 3a,b), and h, cd, and age of *P. dactylifera* (Figure 3c). The results indicated that dbh, h, and cd were significant predictors of Csa ($p < 0.001$) with a positive effect on the Csa of *D. regia* and *F. nitida*. The model showed high goodness-of-fit with a marginal R^2 of 99% and a conditional R^2 of 99% (Supplementary Table S4). The random effects of the tree pit surface area and tree planting urban locations were found to have zero additional variation in Csa, suggesting that the variation in above-ground biomass carbon storage can be fully explained by the fixed factors. The *D. regia* and *F. nitida* models (15 and 16) fit the data well.

$$\text{Ln (Csa)} = 2.35 - \ln(\text{dbh}) \times 4.44 + \ln(\text{h}) \times 0.48 + \ln(\text{cd}) \times 0.23 \quad (15)$$

$$\text{Ln (Csa)} = 2.81 - \ln(\text{dbh}) \times 4.52 + \ln(\text{h}) \times 0.89 + \ln(\text{cd}) \times 0.27 \quad (16)$$

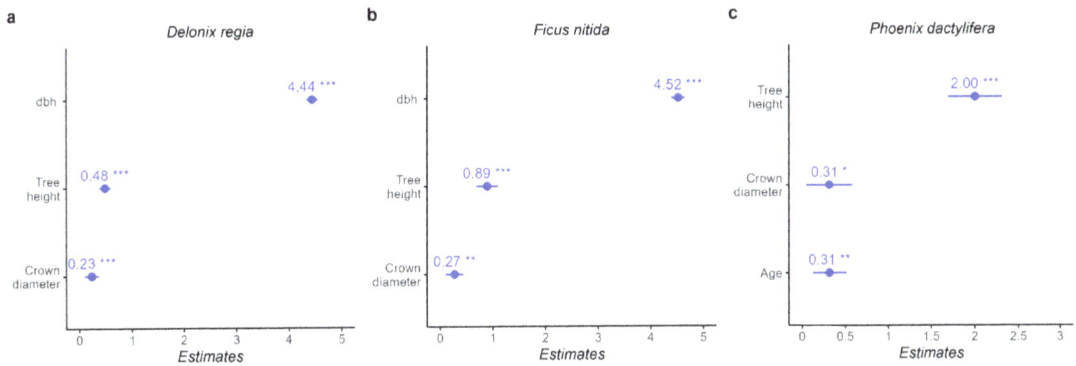

Figure 3. Impact of the tree structural variables on above-ground biomass carbon storage of three tree species in three plots using a linear mixed model. the *p*-value (with levels of significance indicated by symbols such as ***, ** and *) for each ANOVA shows the significant impact of the tree structure variables and above-ground biomass carbon storage for the species.

Model (17) quantifies the relationship between tree height, crown diameter, and age for above-ground biomass carbon storage. We used random effects. Above biomass carbon storage for *P. dactylifera* can be applied based on the following model:

$$\text{Ln (Csa)} = 1.48 + \ln(\text{h}) \times 1.99 + \ln(\text{cd}) \times 0.31 + \ln(\text{Age}) \times 0.31 + \varepsilon \quad (17)$$

The results of the LMM analysis indicated that h and cd, the fixed factors, positively impacted the shaded area in both *D. regia* and *F. nitida* (Figures 4a and 4b, respectively). Conversely, for *P. dactylifera*, the effect of the fixed factors was statistically insignificant and negative (Table S5) in the Supplementary Materials section. Tree height and crown diameter are statistically significant as predictors of the shaded area. The results are depicted in Figure 4, which displays the fixed effect of the shaded area, with point estimates and 95% confidence intervals, and the significance of each predictor variable (*p*-value). The results suggest that increasing h and cd values lead to an increase in shaded areas in both *D. regia* and *F. nitida*. The LMM regression analysis for *D. regia* explained 18.9% of the response variable variation. The conditional R^2 accounted for 64.4% of the variation in the response variable due to random effects. The regression results of *F. nitida* showed a high goodness of fit for both marginal R^2 and conditional R^2, with the model explaining 94.3% and 95.1% of the response variable variation, respectively. The variation in the shaded area can therefore be fully explained by the fixed factors and other random effects, as reflected in the *D. regia* model (18) and the *F. nitida* model (19), but not in the case of *P. dactylifera*.

$$\text{Ln}\ ({}_{\text{ave}}\text{A}_{\text{shade}}) = 1.29 + \ln(\text{h}) \times 1.19 + \ln(\text{cd}) \times 1.63 + \varepsilon \quad (18)$$

$$\text{Ln}\ ({}_{\text{ave}}\text{A}_{\text{shade}}) = 0.52 - (\text{h}) \times 2.95 + \ln(\text{cd}) \times 2.25 + \varepsilon \quad (19)$$

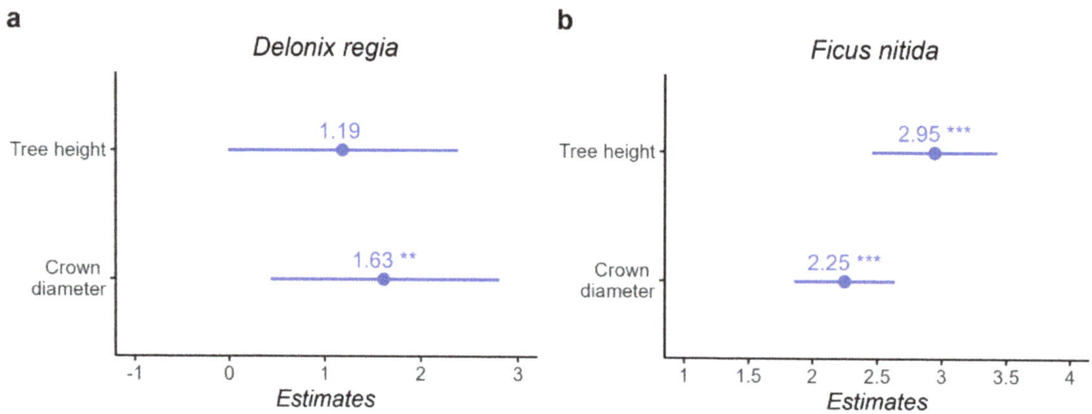

Figure 4. Impact of the tree structure variables on shaded area of three tree species in three plots using a linear mixed model. *D. regia* (**a**), *F. nitida* (**b**), with fixed effects of tree species, and random effects accounting for tree pit surface area and tree planting site. the *p*-value (with levels of significance indicated by symbols such as *** and **) for each ANOVA shows the significant impact of the tree structure variables and shaded area for the species.

The results of the study on above-ground biomass carbon storage and shading in three species (*D. regia*, *F. nitida*, and *P. dactylifera*) are shown in Figure 4. The results reveal a significant difference in Csa among the species, with *F. nitida* and *P. dactylifera* being significantly different but not from *D. regia* (Figure 5L). In terms of shading, a significant difference was also found among the species, with *D. regia* and *F. nitida* being similar but different from *P. dactylifera* (Figure 5R).

The average above-ground biomass carbon storage Csa, for *D. regia* trees was 179 kg C with an average shaded area of 42 m^2. (See Table 6.) Significant differences were found for the Csa of *F. nitida* amongst the age categories ($p < 0.001$), with an average ranging from 35 to 420 kg C. The shaded area of *F. nitida* increased from 20 m^2 for young trees (<15 years) to 69 m^2 for old trees (>15 years). The difference in shade density *for F. nitida* was not significant between age categories ($p = 0.29$). The above-ground biomass carbon storage of *P. dactylifera* did not show significant differences between age categories ($p = 0.11$), with an average above-ground biomass carbon storage ranging between 77.7 and 93 kg C. The average shade area for *P. dactylifera* showed a significant difference between age categories ($p < 0.001$), but shaded density was not significant ($p = 0.33$).

The main effects of plant growth site for *D. regia* on above-ground biomass carbon storage were significant ($p = 0.01$) but were not significant for shaded area and shade density ($p = 0.28$ and $p = 0.16$), respectively (see Table 7). Similarly, the effects of plant growth site on the ecosystem services of *P. dactylifera* were not significant for above-ground biomass carbon ($p = 0.88$), shaded area ($p = 0.84$), and shade density ($p = 0.37$), respectively, across different plant sites such as street trees, parking lot trees, and public place trees. Nevertheless, the effects of the plant growth site on above-ground biomass carbon storage and shaded area were significant ($p = 0.03$) for *F. nitida*, but not on shade density ($p = 0.76$).

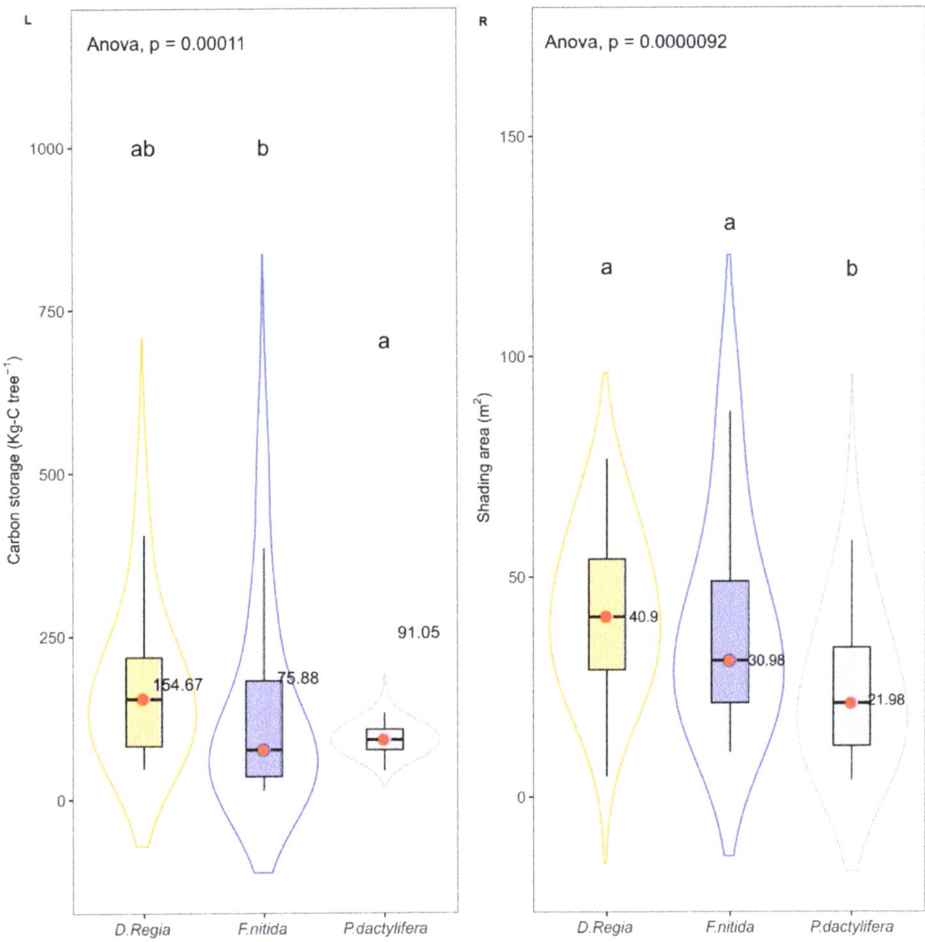

Figure 5. Ecosystem services (above ground biomass carbon storage figure (**L**) and shading area (**R**)) of *D. regia*, *F. nitida*, and *P. dactylifera* in the arid city of Jericho. Letters indicate the results of post hoc Tukey test. Different letters denote significant differences.

Table 6. Mean, minimum, maximum, and related standard deviation as well P for *p*-value (with levels of significance indicated by symbols such as ***) of Csa above-ground biomass carbon storage, shaded area, and shade density for *D. regia*, *F. nitida*, and *P. dactylifera* for different age classes. Means within the columns differ significantly when separated by different letters.

Age	*n*	Csa kg [C]			Shaded Area [m^2]			Shade Density [m^2/m^3]		
		Mean	Max	Min	Mean	Max	Min	Mean	Max	Min
D. regia										
20–25	69	178.6 ± 118.8	591.6	46.8	41.8 + 17.4	76.8	4.6	3.58 ± 2.6	19.0	0.9
F. nitida					$p \leq 0.001$ ***					
<15	30	35 ± 12.5 a	75.6	17.4	20.40 ± 6.9	43.0	10.2	3.88 ± 1.3	6.3	1.4
16–24	34	163.23 ± 129.3 b	500	13.2	58.41 ± 35.5	211.1	22.5	3.34 ± 1	4.7	0.8
>25	9	420.3 + 178.7 c	711.2	180.12	69.23 ± 28.1	107.1	37.2	1.67 ± 0.6	2.70	0.6

Table 6. *Cont.*

Age	*n*	Csa kg [C]			Shaded Area [m²]			Shade Density [m²/m³]		
P. dactylifera					$p = 0.33$					
<29	10	77.7 ± 5 a	87.4	68.0	38.25 + 8.5	50.4	26.4	0.74 + 0.20	1.1	0.4
50–70	22	98.95 ± 23.7 a	166.5	60.1	29.24 + 21.6	75.4	3.8	0.87 + 0.4	1.7	0.4
>80	38	92.64 ± 23.5 a	132.8	43.6	17.21 + 11.7	49.1	5.02	1.00 + 0. 6	2.7	0.5

Table 7. Means and SD of the ecosystem services above-ground biomass carbon storage Csa, shaded area, and shade density for *D. regia*, *F. nitida*, and *P. dactylifera* in response to growth site, as well as the *p*-value (with levels of significance indicated by symbols such as ** and *)for each ANOVA. The mean values in the same column differ significantly when followed by different letters.

Site	*n*	Csa [kg C]	Shaded Area [m²]	Shade Density [m²/m³]
		Mean	Mean	Mean
D. regia		$p = 0.01$ **	$p = 0.28$	$p = 0.16$
Public place	7	134.01 ± 35.5 a	35.5 ± 10.2 a	4.9 ± 2.1 a
Street	62	182.0 ± 125.5 b	42.8 ± 17.4 a	3.5 ± 2.6 a
F. nitida		$p = 0.03$ *	$p = 0.03$ *	$p = 0.76$
Parking lot	19	121.8 ± 170.1 ab	30.7 ± 12.5 b	2.9 ± 1.2 a
Public place	15	61.3 ± 44.5 b	36.8 ± 24.8 ab	3.2 ± 1.2 a
Street	39	183.4.6 ± 173.2 a	53.5 ± 39.8 a	2.8 ± 1.5 a
P. dactylifera		$p = 0.88$	$p = 0.84$	$p = 0.37$
Public place	45	92.5 ± 24 a	24.2 ± 17.4 a	0.89 ± 0.5 a
Street	14	97.31 ± 20.8 a	24.3 ± 15.7 a	1.11 ± 0.5 a
Square	5	90.8 ± 19.9 a	29.3 ± 11.8 a	0.70 ± 0.1 a

4. Discussion

A quantitative understanding of the structure and dimensions of urban trees is critical to better predict tree ecosystem services. However, the relationships between tree structure and ecosystem services in arid regions are poorly understood. Therefore, we applied several possible numerical approaches to calculate the structure and ecosystem services of trees. We analyzed the dependency of tree structure on dbh and age and the dependency of LAI and tree structure on three common urban trees in the arid city of Jericho. We also studied the effect of the different urban planting locations and tree pits on urban trees' dimensions and on their ecosystem services.

The study outcomes provide a basic understanding for further research on the relationship between urban trees structure and ecosystem services in arid regions. It offers valuable insights into the growth patterns of arid urban trees, (e.g., dbh, crown dimension, and age) and their ability to acclimate (by showing growth efficiency that is not native to this region, for example, *F. nitida* and *D. regia*). Additionally, an allometric model was built to visualize the impacts of the tree structural variables on the ecosystem services, such as above-ground biomass carbon storage and the shade potential of urban trees based on the relationship between tree structure and ecosystem services. The study highlights the important role of urban trees in providing ecosystem services in arid regions and offers valuable insights for city planners and urban managers in their efforts to improve urban tree selection and create sustainable and resilient urban ecosystems in arid cities.

4.1. Relationship between Structural Tree Parameters (dbh, Age, Tree Pit Surface Area, and Tree Urban Location)

The results indicated a moderate to strong relationship between age and tree structure for *F. nitida* (R^2 = 0.3–0.61), which is slightly weaker than the relationships obtained by Moser et al. [39] for three different urban tree species in central Europe. Our results for *P. dactylifera* and *D. regia* show a weak and nonsignificant proportion of variance between age and tree structural variables in both species ($R^2 \leq 0.06$).

The availability of resources limited annual precipitation, competition for above- and below-ground space, and poor soil quality, influence the relationship between age and dbh [41,81]. The stem diameter at breast height with tree height and crown dimension shows strong to moderate relationships for *F. nitida*, but the relationship was slightly weaker in *D. regia*, as a light-demanding and shade-intolerant tree, whereas *F. nitida* is light-demanding but partially shade-tolerant [61]. Light-demanding tree species have weaker stem diameter and crown volume relationships [3]. The growth allocation of trees can greatly change in response to light availability [82], which also supports our results. The results indicate that street trees, which are often planted in close proximity to one another, experience increased competition for sunlight, particularly when their crowns come into contact with each other. Light availability is a critical factor that can influence the growth and development of trees. Light-demanding tree species, such as those that typically grow in open habitats, require high levels of sunlight to thrive. Specifically, these trees may allocate more resources to the production of leaves and branches, which can increase their ability to capture sunlight and produce energy. This may result in weaker stem diameter and crown volume relationships [3].

The tree structural relationships of *D. regia* illustrate a moderate trend that is slightly weaker than those of the studies conducted by Arzai et al. [50], who investigated the connections among canopy width, tree height, and dbh of various urban tree species, finding a strong correlation between tree height and crown diameter with dbh, as an adaptive tree species. This difference is possibly based on the natural climate of the study area, which is tropical [83].

Many other factors, such as annual pruning to shape the tree, especially at an early stage [84], and the removal of damaged, dead, dried, and crossing branches [85], can also affect crown dimension–dbh relationships. Pruning mature trees may be for reasons of shape, tree health, aesthetics, safety, or clearance from infrastructure [86]. The correlation between stem diameter and the crown dimension of *P. dactylifera* was nonsignificant. As a monocotyledonous plant, *P. dactylifera* lacks the ability to form a vascular cambium, a meristem tissue that allows for secondary growth in dicotyledonous plants. The vascular cambium is responsible for the formation of new layers of xylem and phloem, which contribute to the increase in diameter of the plant's stem or trunk over time. Without the formation of a vascular cambium, the date palm does not undergo regular secondary growth and does not exhibit the characteristic increase in diameter [64]. This is in line with the results of Issa et al., [40], whose regression coefficient shows weak but significant relationships between dbh and crown area for *P. dactylifera*.

Generally, tree samples were selected from different urban locations, that typically suffer from a scarcity of water due to the lack of a regular irrigation system. Our results show a significant difference in dbh in the tree planting site for *F. nitida*, and a significant difference in dbh and age in the tree planting site for *D. regia*.

In Jericho City, many irrigation patterns exist (water transportation tanks, manual plastic tubes, normal irrigation systems, and normal water buckets). Additionally, some street trees are situated close to agricultural farms that provide them with resources (water and nutrition).

However, the research of Coombes et al. [87] found that the site factors had very little effect on the allometric relationship between dbh and crown diameter. However, the results presented showed that the difference in irrigation patterns and the distribution of nutrient resources for trees in Jericho may lead to different growth patterns in urban areas; therefore, this may be the reason for the different ratio of tree structural relationships. In addition, the results showed differences in *F. nitida,* in canopy diameter, and volume between parking lot and street trees due to tree size variations. For *D. regia* trees in public places, the trees vary in size as well. The trees in the public place (e.g., garden) are older than the trees in the street, but there were no significant differences in *P. dactylifera* at all, and the reasons behind the fact that the overall mean of *P. dactylifera* tree ages in different urban locations of the city are not significantly different. Furthermore, the findings revealed that the relation between the tree pit surface area and tree structure for *F. nitida* and *P. dactylifera* are weak but statistically significant, but is not significant in *D. regia*. The possible reason behind that uneven distribution of tree samples selected, e.g., 62 of *D. regia*, is that most of the street trees had a very small tree pit surface area. Even if they were irrigated by the above irrigation patterns, the amount of water to reach the plant would be very small, especially in summertime with high evaporation rates.

4.2. Leaf Area Index of the Three Urban Tree Species

The results show a nonsignificant and weak proportion of variance between LAI and structural parameters. The R^2 values were close to zero for all variables. Özbayram et al. [88], in their research, studied the correlation between LAI values and tree variables in Turkey, and a negative correlation in black pine stands was found (i.e., stand age, mean diameter) and a positive correlation in red pine (i.e., stand age, mean diameter, top height, green tree height, and basal area). Özbayram et al. concluded that the leaf area index (LAI) varies according to species. The LAI results were 5.4 for *F. nitida*, 5.8 for *D. regia*, and 2.9 for *P. dactylifera*. These results can be placed in comparison with those of Liu et al., [89], who found a mean LAI of value 4.73 ± 0.40 for *D. regia* and 5.00 ± 0.47 for *F. nitida*, whereas Lin et al. [90] found an LAI of 6.11 for *Ficus macrocarpa* and 5.05 for *Ficus elastica*, and Awal et al. found an LAI of 1.7 for *P. dactylifera* [91]. A higher leaf area index means higher photosynthesis and efficient use of light, which indicates higher carbon capturing ability and stocks [92].

4.3. Ecosystem Services of Trees in Arid Cities

Urban trees provide ecosystem services [93,94], which can significantly improve the climate in cities [95]. The study estimated above-ground biomass carbon storage and shading potential. Results showed that tree height, dbh, and crown diameter have a strong relationship with above-ground biomass carbon storage in *D. regia* and *F. nitida*. Similarly, tree height, crown diameter, and age have a significant relationship with above-ground biomass carbon storage in *P. dactylifera*, consistent with prior research, (e.g., Yoon et al. [37]). Issa et al. [96] found that the amount of CO_2 absorbed is proportional to the tree component, above-ground biomass can be highly estimated by the green plant component (e.g., canopy area) and tree stems as variables measured in the field. Further, Betemariyam et al. [97], found that *P. dactylifera* trees older than 20 years had a mean above-ground biomass carbon stock of 159.50 kg/plant, in date palm on a farm in north-eastern Ethiopia. Issa et al. [40] found that trunk height and crown diameter are strongly correlated with the age of date palm trees and reported an average carbon storage of 225 kg C of the palm trees in Abu Dhabi, United Arab Emirates, for trees older than 20 years. The results show the average carbon storage of *P. dactylifera* is higher than the averages of *D. regia* and *F. nitida*; these findings support that *P. dactylifera* trees in this study contribute to emission reduction and carbon sink enhancement.

Higher above-ground biomass carbon storage averages for *D. regia* were found in public places and parking lots compared with street trees, whereas *P. dactylifera* trees provided similar rates at all sites. The second ecosystem service is shade potential. The results showed a statistically significant and strong relationship between tree height, diameter at breast height and crown diameter with the shaded area of *D. regia* and *F. nitida*, and a nonsignificant relationship with the shaded area of the *P. dactylifera* tree. This could be due to its monocotyledon nature. The results of the *P. dactylifera* shade area show a smaller value for older trees, where the most likely reason could be leaf senescence due to age. With age, trees may lose some leaf area due to leaf senescence. Another reason could be leaf pruning each year.

Different shaded areas and shading densities exist among urban tree species. *F. nitida* and *D. regia* have the highest shading potential compared to *P. dactylifera*. Shade density is particularly important for lowering surface temperatures and improving thermal comfort [90,98,99]. *F. nitida* and *D. regia* have higher shading potential compared to *P.*

dactylifera, as produced by their crown canopies. A possible reason is that the sampled trees are mixed between taller trees with narrow canopies and shorter trees with wider canopies. This result is in line with Armson et al., [6], with the outcome about tree morphology and shade for five different street tree species in Manchester (UK) exhibiting a significant difference between the species' canopy sizes but nonsignificant differences between tree canopies' shaded areas. Rahman et al. [100] demonstrated that urban trees can mitigate temperatures underneath canopy surfaces during the day through shading.

The potential cooling effect of tropical trees is higher than that of other species, e.g., *Ficus retusa* trees can reduce the temperature values during the summer by 4 °C, while the cooling effect for date palm trees (*P. dactylifera*) is only 1.5 °C, which is characterized by a small canopy [101]. Reflecting the weakness of *P. dactylifera* as a tree for shading benefits and the higher shading potentials of *F. nitida* and *D. regia*, which are characterized as tropical trees in arid cities, shading measures have special importance, where the sun has intense solar radiation, leading to higher air temperatures that can negatively affect most human daily activities [102]. Based on previous studies, the importance of cooling by shading in an arid city is particularly important where solar radiation is intense, leading to higher temperatures that can negatively impact human activities. Overall, this study provides valuable insights into the ecosystem services provided by urban trees, specifically carbon storage and shading potential, and their correlation with structural variables.

5. Conclusions

In conclusion, this study analyzed the growth, ecosystem services, and tree structural characteristics of three common urban trees (i.e., *D. regia*, *F. nitida*, and *P. dactylifera*) in the arid city of Jericho using a numerical approach of a City Tree model. The results showed that tree structural variables (i.e., tree height, crown volume, and crown diameter) have a strong to moderately significant relationship with dbh for *D. regia* and *F. nitida*. The results also show no relationships between leaf area index and tree structure for all tree species' structural variables, while showing a statistically moderate relationship for tree structure with age for *F. nitida*, and no relationship for all other tree species. The tree pit surface area also showed weak significant relationship with tree structure for *F. nitida* and *P. dactylifera*, but not for *D. regia*. Different urban plant growth location also induced various influences among the three species; the results show a significant influence on tree structure for *D. regia* and *F. nitida*, while the influence was not significant on *P. dactylifera*.

D. regia has higher shading potential and above ground biomass carbon storage, compared to *F. nitida* and *P. dactylifera*, respectively, as common urban trees in the city. The results may vary based on species and site conditions. Our results are similar to research from other climates; for example, Moser et al. (2015) carried out similar research in Germany (temperate region) and found strong to moderate relationships between crown dimensions and stem diameter, which is identical to our results except for *P. dactylifera*. Although results can vary based on species and site conditions, overall patterns are comparable, which indicates that similar results are also applicable to other climate regions. However, species functionality should be considered.

Based on these findings for the selected ecosystem services (above-ground biomass carbon storage and shade potential), it is recommended that *D. regia*, *F. nitida* and *P. dactylifera* be considered for future urban greening in arid cities, with *D. regia* outperforming the others. However, further research in other non-arid regions and climate-sensitive growth models are needed to better understand the growth and adaptation capacity of these trees in changing climates. We recommend conducting further research on the relationship between tree species' dimensions and the ecosystem services they provide, with a specific focus on urban areas in Mediterranean and/or arid climates.

Supplementary Materials: The following supporting information can be downloaded at https://www.mdpi.com/article/10.3390/f14040671/s1. Table S1: Means and standard deviation [SD] of all measured and calculated trees' structural data *D. regia*, *F. nitida*, and *P. dactylifera*. Where *n*—number of samples, dbh—diameter at breast height, h–tree height, cl—crown length, cd—crown diameter, cpa—crown projection area, and cv—crown volume, respectively. Table S2: Results of the regression analysis of LAI, the predictor variables, and the tree variables (dbh, h, cd, and cv) as a response and the regression equation (y) = a + b × ln(x). The table below lists the determination of R^2, standard error, and *p*-values. Table S3: Mean of the tree pit and related standard error to the growth site for *D. regia*, *F. nitida*, and *P. dactylifera*, as well as the *p*-values for each ANOVA. Mean values in the same column differ significantly when followed by different letters. Table S4: Results of the summary of the linear mixed model regression analysis of a carbon fixation and the predictor variables, and the tree variables (h, dbh, cd, and age) as a response, and the regression equation $\ln(y) = a + b1 \times \ln(x1) + b2 \times \ln(x2) + b3 \times \ln(x3) + \varepsilon$. The table below lists the determination of R^2, τ00: variance of random intercept, N site refers to the number of distinct groups or sites in the data, where each group may have multiple observations, N T.pit refers to the number of total observations or data points in all the sites, which is equal to the sum of the number of observations in each site, σ^2 refers to the residual variance, and *p*-values. Table S5: Results of the summary of linear mixed model regression analysis of a shaded area and the predictor variables, and the tree variables (h, dbh, cd, and age) as a response and the regression equation $n(y) = a + b1 \times \ln(x1) + b2 \times \ln(x2) + \varepsilon$. The table below lists the determination of R^2, τ00: variance of random intercept, N site refers to the number of distinct groups or sites in the data, where each group may have multiple observations, N T.pit refers to the number of total observations or data points in all the sites, which is equal to the sum of the number of observations in each site, σ^2 refers to the residual variance, and *p*-values.

Author Contributions: Conceptualization, A.A. and T.R.; Methodology, A.A., T.R. and M.A.R.; Formal Analysis, A.A.; Writing—Original Draft Preparation, A.A.; Writing—Review and Editing, A.A., T.R., A.M.-R., M.A.R., E.F., J.H.S., A.H., M.T., H.P. and S.P.; Visualization, A.A.; Supervision, T.R.; Project Administration, T.R., H.P. and S.P.; Funding Acquisition, T.R., H.P., M.A.R. and S.P. All authors have read and agreed to the published version of the manuscript.

Funding: This research was funded by the German Science Foundation (Deutsche Forschungsgemeinschaft), grant number PR 292/21−1 and PA 2626/3−1, and by the Bavarian State Ministry of the Environment and Consumer Protection, grant numbers TUF01UF−64971 and TLK01UFuE69397.

Data Availability Statement: The data are available on request from the corresponding author.

Acknowledgments: We thank the German Research Foundation—DFG (grant PR292/21-1.), which funded this research as part of the Middle East Collaboration Program. In addition, the authors wish to thank the Jericho Municipality for further help and for facilitating the measurements in the city.

Conflicts of Interest: The authors declare no conflict of interest.

References

1. Nowak, D.J.; Greenfield, E.J.; Hoehn, R.E.; Lapoint, E. Carbon Storage and Sequestration by Trees in Urban and Community Areas of the United States. *Environ. Pollut.* **2013**, *178*, 229–236. [CrossRef] [PubMed]
2. Amoatey, P.; Sulaiman, H.; Kwarteng, A.; Al-Reasi, H.A. Above-Ground Carbon Dynamics in Different Arid Urban Green Spaces. *Environ. Earth Sci.* **2018**, *77*, 431. [CrossRef]
3. Troxel, B.; Piana, M.; Ashton, M.S.; Murphy-Dunning, C. Relationships between Bole and Crown Size for Young Urban Trees in the Northeastern USA. *Urban For. Urban Green.* **2013**, *12*, 144–153. [CrossRef]
4. Oke, T.R. The Micrometeorology of the Urban Forest. *Philos. Trans. R. Soc. Lond. B* **1989**, *324*, 335–349. [CrossRef]
5. TEEB Manual for Cities: Ecosystem Services in Urban Management. The Economics of Ecosystems and Biodiversity (TEEB): Geneva. 2011. Available online: www.teebweb.org/ (accessed on 10 February 2023).
6. Armson, D.; Stringer, P.; Ennos, A.R.R. The Effect of Street Trees and Amenity Grass on Urban Surface Water Runoff in Manchester, UK. *Urban For. Urban Green.* **2013**, *12*, 282–286. [CrossRef]
7. Bowler, D.E.; Buyung-Ali, L.; Knight, T.M.; Pullin, A.S. Urban Greening to Cool Towns and Cities: A Systematic Review of the Empirical Evidence. *Landsc. Urban Plan.* **2010**, *97*, 147–155. [CrossRef]
8. Gillner, S.; Vogt, J.; Tharang, A.; Dettmann, S.; Roloff, A. Role of Street Trees in Mitigating Effects of Heat and Drought at Highly Sealed Urban Sites. *Landsc. Urban Plan.* **2015**, *143*, 33–42. [CrossRef]
9. Yu, Z.; Yang, G.; Zuo, S.; Jørgensen, G.; Koga, M.; Vejre, H. Critical Review on the Cooling Effect of Urban Blue-Green Space: A Threshold-Size Perspective. *Urban For. Urban Green.* **2020**, *49*, 126630. [CrossRef]

10. Bourbia, F.; Boucheriba, F. Impact of Street Design on Urban Microclimate for Semi Arid Climate (Constantine). *Renew. Energy* **2010**, *35*, 343–347. [CrossRef]
11. Huang, J.; Yu, H.; Dai, A.; Wei, Y.; Kang, L. Drylands Face Potential Threat under 2 °C Global Warming Target. *Nat. Clim. Chang.* **2017**, *7*, 417–422. [CrossRef]
12. Rötzer, T.; Rahman, M.A.; Moser-Reischl, A.; Pauleit, S.; Pretzsch, H. Process Based Simulation of Tree Growth and Ecosystem Services of Urban Trees under Present and Future Climate Conditions. *Sci. Total Environ.* **2019**, *676*, 651–664. [CrossRef] [PubMed]
13. Foster, R.S.; Blaine, J. Urban Tree Survival: Trees in the Sidewalk. *J. Arboric.* **1978**, *4*, 14–17. [CrossRef]
14. Beatty, R.A.; Heckman, C.T. Survey of Urban Tree Programs in the United States. *Urban Ecol.* **1981**, *5*, 81–102. [CrossRef]
15. Jim, C.Y. Soil Compaction at Tree-Planting Sites in Urban Hong Kong. In *Proceedings of an International Workshop on Tree Root Development in Urban Soils*; International Society of Arboriculture: Hong Kong, 1998; pp. 166–178.
16. Kjelgren, R.K.; Clark, J.R. Microclimates and Tree Growth in Three Urban Spaces. *J. Environ. Hortic.* **1992**, *10*, 139–145. [CrossRef]
17. Day, S.D.; Bassuk, N.L.; Van Es, H. Effects of Four Compaction Remediation Methods for Landscape Trees on Soil Aeration, Mechanical Impedance and Tree Establishment. *J. Environ. Hortic.* **1995**, *13*, 64–71. [CrossRef]
18. Sayad, B.; Alkama, D.; Rebhi, R.; Menni, Y.; Ahmad, H.; Inc, M.; Sharifpur, M.; Lorenzini, G.; Azab, E.; Elnaggar, A.Y.; et al. High-Frequency Densitometry—A New Method for the Rapid Evaluation of Wood Density Variations. *Urban For. Urban Green.* **2018**, *7*, 1–7. [CrossRef]
19. Morgenroth, J.; Buchan, G.D. Soil Moisture and Aeration beneath Pervious and Impervious Pavements. *J. Arboric.* **2009**, *35*, 135. [CrossRef]
20. Rahman, M.A.; Stringer, P.; Ennos, A.R. Effect of Pit Design and Soil Composition on Performance of Pyrus Calleryana Street Trees in the Establishment Period. *Arboric. Urban For.* **2013**, *39*, 256–266. [CrossRef]
21. Rötzer, T.; Moser-Reischl, A.; Rahman, M.A.; Grote, R.; Pauleit, S.; Pretzsch, H. Modelling Urban Tree Growth and Ecosystem Services: Review and Perspectives. *Prog. Bot.* **2021**, *82*, 405–464.
22. Sanders, J.; Grabosky, J.; Cowie, P. Establishing Maximum Size Expectations for Urban Trees with Regard to Designed Space. *Arboric. Urban For.* **2013**, *39*, 68–73. [CrossRef]
23. Clark, J.R.; Kjelgren, R. Water as a Limiting Factor in the Development of Urban Trees. *J. Arboric.* **1990**, *16*, 203–208. [CrossRef]
24. Allen, C.D.; Macalady, A.K.; Chenchouni, H.; Bachelet, D.; McDowell, N.; Vennetier, M.; Kitzberger, T.; Rigling, A.; Breshears, D.D.; Hogg, E.H.T. A Global Overview of Drought and Heat-Induced Tree Mortality Reveals Emerging Climate Change Risks for Forests. *For. Ecol. Manag.* **2010**, *259*, 660–684. [CrossRef]
25. Chen, Z.; He, X.; Cui, M.; Davi, N.; Zhang, X.; Chen, W.; Sun, Y. The Effect of Anthropogenic Activities on the Reduction of Urban Tree Sensitivity to Climatic Change: Dendrochronological Evidence from Chinese Pine in Shenyang City. *Trees* **2011**, *25*, 393–405.
26. Akbari, H.; Pomerantz, M.; Taha, H. Cool Surfaces and Shade Trees to Reduce Energy Use and Improve Air Quality in Urban Areas. *Sol. Energy* **2001**, *70*, 295–310. [CrossRef]
27. Brune, M. Urban Trees under Climate Change. *Clim. Serv. Cent. Ger.* **2016**, 123.
28. Farrell, C.; Szota, C.; Arndt, S.K. Urban Plantings: "Living Laboratories" for Climate Change Response. *Trends Plant Sci.* **2015**, *20*, 597–599. [CrossRef]
29. Roetzer, T.; Wittenzeller, M.; Haeckel, H.; Nekovar, J. Phenology in Central Europe—Differences and Trends of Spring Phenophases in Urban and Rural Areas. *Int. J. Biometeorol.* **2000**, *44*, 60–66. [CrossRef] [PubMed]
30. Gregg, J.W.; Jones, C.G.; Dawson, T.E. Urban Ozone Depletion: Why a Tree Grows Better in New York City. *Nature* **2003**, *424*, 183–187. [CrossRef]
31. Kaye, J.P.; Groffman, P.M.; Grimm, N.B.; Baker, L.A.; Pouyat, R.V. A Distinct Urban Biogeochemistry? *Trends Ecol. Evol.* **2006**, *21*, 192–199. [CrossRef]
32. Jochner, S.; Alves-Eigenheer, M.; Menzel, A.; Morellato, L.P.C. Using Phenology to Assess Urban Heat Islands in Tropical and Temperate Regions. *Int. J. Climatol.* **2013**, *33*, 3141–3151. [CrossRef]
33. George, K.; Ziska, L.H.; Bunce, J.A.; Quebedeaux, B.; Hom, J.L.; Wolf, J.; Teasdale, J.R. Macroclimate Associated with Urbanization Increases the Rate of Secondary Succession from Fallow Soil. *Oecologia* **2009**, *159*, 637–647. [CrossRef] [PubMed]
34. McPherson, E.G. *Tree Guidelines for Coastal Southern California Communities*; Local Government Commission: Sacramento, CA, USA, 2000.
35. Chreptun, C. Kronenstruktureigenschaften von Linden Und Robinien in München: Anwendungen Des Terrestrischen Laserscannings. Doctoral Dissertation, Technische Universität München, München, Germany, 2015.
36. Franceschi, E.; Moser-Reischl, A.; Rahman, M.A.; Pauleit, S.; Pretzsch, H.; Rötzer, T. Crown Shapes of Urban Trees-Their Dependences on Tree Species, Tree Age and Local Environment, and Effects on Ecosystem Services. *Forests* **2022**, *13*, 748. [CrossRef]
37. Yoon, T.K.; Park, C.-W.; Lee, S.J.; Ko, S.; Kim, K.N.; Son, Y.; Lee, K.H.; Oh, S.; Lee, W.-K.; Son, Y. Allometric Equations for Estimating the Aboveground Volume of Five Common Urban Street Tree Species in Daegu, Korea. *Urban For. urban Green.* **2013**, *12*, 344–349. [CrossRef]
38. Nowak, D.J.; Crane, D.E. Carbon Storage and Sequestration by Urban Trees in the USA. *Environ. Pollut.* **2002**, *116*, 381–389. [CrossRef] [PubMed]
39. Moser, A.; Rötzer, T.; Pauleit, S.; Pretzsch, H. Structure and Ecosystem Services of Small-Leaved Lime (Tilia Cordata Mill.) and Black Locust (*Robinia pseudoacacia* L.) in Urban Environments. *Urban For. Urban Green.* **2015**, *14*, 1110–1121. [CrossRef]

40. Issa, S.; Dahy, B.; Ksiksi, T.; Saleous, N. Allometric Equations Coupled with Remotely Sensed Variables to Estimate Carbon Stocks in Date Palms. *J. Arid Environ.* **2020**, *182*, 104264. [CrossRef]
41. Hemery, G.E.; Savill, P.S.; Pryor, S.N. Applications of the Crown Diameter-Stem Diameter Relationship for Different Species of Broadleaved Trees. *For. Ecol. Manag.* **2005**, *215*, 285–294. [CrossRef]
42. Temesgen, H.; Gadow, K.V. Generalized Height-Dimater Models—An Application for Major Tree Species in Complex Stands of Interior British Columbia. *Eur. J. For. Res.* **2004**, *123*, 45–51. [CrossRef]
43. Shinozaki, K.; Yoda, K.; Hozumi, K.; Kira, T. A Quantitative Analysis of Plant Form—The Pipe Model Theory: I. Basic Analyses. *Jpn. J. Ecol.* **1964**, *14*, 97–105.
44. Mäkelä, A. Modeling Structural-Functional Relationships in Whole-Tree Growth: Resource Allocation. *Process Model. For. Growth Responses Environ. Stress* **1990**, *7*, 86–95.
45. Chiba, Y. Architectural Analysis of Relationship between Biomass and Basal Area Based on Pipe Model Theory. *Ecol. Modell.* **1998**, *108*, 219–225. [CrossRef]
46. Pretzsch, H.; Matthew, C.; Dieler, J. Allometry of Tree Crown Structure. Relevance for Space Occupation at the Individual Plant Level and for Self-Thinning at the Stand Level. In *Growth and Defence in Plants*; Springer: Berlin/Heidelberg, Germany, 2012; pp. 287–310.
47. Watt, M.S.; Kirschbaum, M.U.F. Moving beyond Simple Linear Allometric Relationships between Tree Height and Diameter. *Ecol. Modell.* **2011**, *222*, 3910–3916. [CrossRef]
48. Ngo, K.M.; Lum, S. Aboveground Biomass Estimation of Tropical Street Trees. *J. Urban Ecol.* **2018**, *4*, jux020. [CrossRef]
49. Peper, P.J.; Alzate, C.P.; McNeil, J.W.; Hashemi, J. Allometric Equations for Urban Ash Trees (*Fraxinus* spp.) in Oakville, Southern Ontario, Canada. *Urban For. urban Green.* **2014**, *13*, 175–183. [CrossRef]
50. Arzai, A.; Aliyu, B. The Relationship between Canopy Width, Height and Trunk Size in Some Tree Species Growing in the Savana Zone of Nigeria. *Bayero J. Pure Appl. Sci.* **2010**, *3*. [CrossRef]
51. Freedman, D.N.; Myers, A.C. *Eerdmans Dictionary of the Bible*; Amsterdam University Press: Amsterdam, The Netherlands, 2000; ISBN 9053565035.
52. Tuqan, N.; Haie, N.; Ahmad, M.T. Assessment of the Agricultural Water Use in Jericho Governorate Using Sefficiency. *Sustainability* **2020**, *12*, 3634. [CrossRef]
53. Roloff, A. *Baumkronen: Verständnis Und Praktische Bedeutung Eines Komplexen Naturphänomens*; Ulmer: Stuttgart, Germany, 2001.
54. Pretzsch, H.; Biber, P.; Uhl, E.; Dahlhausen, J.; Rötzer, T.; Caldentey, J.; Koike, T.; Van Con, T.; Chavanne, A.; Seifert, T. Crown Size and Growing Space Requirement of Common Tree Species in Urban Centres, Parks, and Forests. *Urban For. Urban Green.* **2015**, *14*, 466–479. [CrossRef]
55. Adeoluwa, O.O.; Akinkunmi, O.Y.; Akintoye, H.A.; Shokalu, A.O. Rooting, Growth and Sustainability of Yellow Ficus (Ficus Retusa 'Nitida') as Affected by Growth Media under Nursery Conditions. *Int. J. Biol. Chem. Sci.* **2014**, *8*, 2071–2080. [CrossRef]
56. Hora, F.B. *The Oxford Encyclopedia of Trees of the World*; Oxford University Press: Oxford, UK, 1981.
57. Riffle, R.L. *The Tropical Look. Portland, Oregon*; Timber Press. Inc.: Portland, OR, USA, 1998.
58. Vogt, K.A.; Vogt, D.J.; Brown, S.; Tilley, J.P.; Edmonds, R.L.; Silver, W.L.; Siccama, T.G.; Vogt, K.A.; Vogt, D.J.; Brown, S.; et al. Dynamics of Forest Floor and Soil Organic Matter Accumulation in Boreal, Temperate, and Tropical Forests. In *Soil Management and Greenhouse Effect*; CRC Press: New York, NY, USA, 1995; pp. 159–178.
59. Chaudhary, L.B.; Sudhakar, J.V.; Kumar, A.; Bajpai, O.; Tiwari, R.; Murthy, G.V.S. Synopsis of the *Genus ficus* L.(Moraceae) in India. *Taiwania* **2012**, *57*, 193–216.
60. Tan, H.T.W.; Yeo, C.K.; Ng, A.B.C. *Native and Naturalised Biodiversity for Singapore Waterways and Water Bodies No. 1 Ficus Microcarpa, Malayan Banyan*; National University of Singapore: Singapore, 2009.
61. Wee, Y.C.C. The Occurrence of Ficus Spp. on High-Rise Buildings in Singapore. *Int. Biodeterior. Biodegrad.* **1992**, *29*, 53–59. [CrossRef]
62. Ankrah, N.; Nyarko, A.K.; Addo, P.G.A.; Ofosuhene, M.; Dzokoto, C.; Marley, E.; Addae, M.M.; Ekuban, F.A. Evaluation of Efficacy and Safety of a Herbal Medicine Used for the Treatment of Malaria. *Phyther. Res.* **2003**, *17*, 697–701. [CrossRef] [PubMed]
63. Modi, A.; Mishra, V.; Bhatt, A.; Jain, A.; Mansoori, M.H.; Gurnany, E.; Kumar, V. Delonix Regia: Historic Perspectives and Modern Phytochemical and Pharmacological Researches. *Chin. J. Nat. Med.* **2016**, *14*, 31–39. [PubMed]
64. Orwa, C. Agroforestree Database: A Tree Reference and Selection Guide, Version 4.0. Available online: http//www.worldagroforestry.org/sites/treedbs/treedatabases.asp (accessed on 11 March 2023).
65. Lib, I.; Webb, D.B.; Wood, P.J. *A Guide to Species Selection for Tropical and Sub-Tropical Plantations*; Commenwealth Forestry Institute, University of Oxford: Oxford, UK, 1984.
66. Singh, S.; Kumar, S.N. A Review: Introduction to Genus Delonix. *World J. Pharm. Pharm. Sci.* **2014**, *3*, 2042–2055.
67. Barrow, S.C. A Monograph of Phoenix L. (Palmae: Coryphoideae). In *Kew Bulletin*; Springer: London, UK, 1998; pp. 513–575.
68. Zohary, D.; Hopf, M.; Weiss, E. *Domestication of Plants in the Old World*; Oxford University Press: Oxford, UK, 2012; ISBN 9780199549061.
69. Shamsi, M.; Mazloumzadeh, S.M. Some Physical and Mechanical Properties of Date Palm Trees Related to Cultural Operations Industry Mechanization. *J. Agric. Technol.* **2009**, *5*, 17–31.
70. Al-Shayeb, S.M.; Seaward, M.R.D. The Date Palm (*Phoenix dactylifera* L.) Fibre as a Biomonitor of Lead and Other Elements in Arid Environments. *Sci. Total Environ.* **1995**, *168*, 1–10. [CrossRef]

71. Manickavasagan, A.; Essa, M.M.; Sukumar, E. (Eds.) *Dates: Production, Processing, Food, and Medicinal Values*; CRC Press: Boca Raton, FL, USA, 2012.
72. Issa, S.; Dahy, B.; Ksiksi, T.; Saleous, N. Development of a New Allometric Equation Correlated WTH RS Variables for the Assessment of Date Palm Biomass. In *Conference Paper*; UAE University, College of Science: Al Ain, United Arab Emirates, 2018.
73. El-Khatib, A.A.; Youssef, N.A.; Barakat, N.A.; Samir, N.A. Responses of Eucalyptus Globulus and Ficus Nitida to Different Potential of Heavy Metal Air Pollution. *Int. J. Phytoremediat.* **2020**, *22*, 986–999. [CrossRef]
74. Agrawal, M. *Relative Susceptibility of Plants in a Dry Tropical Urban Environment*; Springer: Berlin/Heidelberg, Germany, 2001; p. 606. ISBN 9783642624759.
75. Manikandan, S.; Udaykumar, M.; Sekar, T. Woody Stem Density and Above-Ground Biomass in Pachaimalai Hills of Southern Eastern Ghats, Tamil Nadu, India. *Int. J. Res. Appl. Sci. Eng. Tech.* **2019**, *7*, 151–158. [CrossRef]
76. Tollefson, M. Downloading R and RStudio and Setting Up a File System. In *R 4 Quick Syntax Reference*; Springer: Berlin/Heidelberg, Germany, 2022; pp. 3–14.
77. Shapiro, S.S.; Wilk, M.B. An Analysis of Variance Test for Normality (Complete Samples). *Biometrika* **1965**, *52*, 591–611. [CrossRef]
78. Peper, P.J.; McPherson, E.G.; Mori, S.M. Equations for Predicting Diameter, Height, Crown Width, and Leaf Area of San Joaquin Valley Street Trees. *J. Arboric.* **2001**, *27*, 306–317. [CrossRef]
79. Stoffberg, G.H.; Van Rooyen, M.W.; Van der Linde, M.J.; Groeneveld, H.T. Predicting the Growth in Tree Height and Crown Size of Three Street Tree Species in the City of Tshwane, South Africa. *Urban For. Urban Green.* **2008**, *7*, 259–264. [CrossRef]
80. Niklas, K.J. *Plant Allometry: The Scaling of Form and Process*; University of Chicago Press: Chicago, IL, USA, 1994; ISBN 0226580806.
81. Jim, C.Y. Managing Urban Trees and Their Soil Envelopes in a Contiguously Developed City Environment. *Environ. Manag.* **2001**, *28*, 819–832. [CrossRef] [PubMed]
82. Harja, D.; Vincent, G.; Mulia, R.; van Noordwijk, M. Tree Shape Plasticity in Relation to Crown Exposure. *Trees* **2012**, *26*, 1275–1285. [CrossRef]
83. Ali, S.I.A.; Szalay, Z. Towards Developing a Building Typology for Sudan. In Proceedings of the IOP Conference Series: Earth and Environmental Science; IOP Publishing: Bristol, UK, 2019; Volume 323, p. 12012.
84. Dixon, G.R.; Aldous, D.E. *Horticulture: Plants for People and Places, Volume 1*; Springer: Dordrecht, The Netherlands, 2014; ISBN 9401785775.
85. Gilman, E. *An Illustrated Guide to Pruning*, 3rd ed.; Cengage Learning: Belmont, CA, USA, 2012; ISBN 1133715877.
86. Kuser, J.E. (Ed.) *Urban and Community Forestry in the Northeast*; Springer Science & Business Media: Berlin/Heidelberg, Germany, 2006.
87. Coombes, A.; Martin, J.; Slater, D. Defining the Allometry of Stem and Crown Diameter of Urban Trees. *Urban For. Urban Green.* **2019**, *44*, 126421. [CrossRef]
88. Özbayram, A.K.; Cicek, E.; Yılmaz, F. Relationships between Leaf Area Index (LAI) and Some Stand Properties in Turkish Red Pine and Black Pine Stands. *Kastamonu Üniversitesi Orman Fakültesi Derg.* **2015**, *15*, 78–85.
89. Liu, S.; Li, J.P.; Xu, M.F.; Sun, Y.D.; Li, W. Bin Understory Light Environment of Canopy Tree Species in Urban Green Land. In Proceedings of the Advanced Materials Research, Lulea, Sweden, 2–22 March 2013; Trans Tech Publ.: Stafa-Zurich, Switzerland, 2013; Volume 671, pp. 2715–2721.
90. Lin, B.-S.; Lin, Y.-J. Cooling Effect of Shade Trees with Different Characteristics in a Subtropical Urban Park. *HortScience* **2010**, *45*, 83–86. [CrossRef]
91. Awal, M.A.; Ishak, W.I.W.; Bockari-Gevao, S.M. Determination of Leaf Area Index for Oil Palm Plantation Using Hemispherical Photography Technique. *J. Sci. Technol* **2010**, *18*, 23–32.
92. Luo, T.; Pan, Y.; Ouyang, H.; Shi, P.; Luo, J.; Yu, Z.; Lu, Q. Leaf Area Index and Net Primary Productivity along Subtropical to Alpine Gradients in the Tibetan Plateau. *Glob. Ecol. Biogeogr.* **2004**, *13*, 345–358. [CrossRef]
93. Roy, S.; Byrne, J.; Pickering, C. A Systematic Quantitative Review of Urban Tree Benefits, Costs, and Assessment Methods across Cities in Different Climatic Zones. *Urban For. Urban Green.* **2012**, *11*, 351–363. [CrossRef]
94. Pataki, D.E.; Alberti, M.; Cadenasso, M.L.; Felson, A.J.; McDonnell, M.J.; Pincetl, S.; Pouyat, R.V.; Setälä, H.; Whitlow, T.H. The Benefits and Limits of Urban Tree Planting for Environmental and Human Health. *Front. Ecol. Evol.* **2021**, *9*, 603757. [CrossRef]
95. Esperon-Rodriguez, M.; Tjoelker, M.G.; Lenoir, J.; Baumgartner, J.B.; Beaumont, L.J.; Nipperess, D.A.; Power, S.A.; Richard, B.; Rymer, P.D.; Gallagher, R.V. Climate Change Increases Global Risk to Urban Forests. *Nat. Clim. Chang.* **2022**, *12*, 950–955. [CrossRef]
96. Issa, S.; Dahy, B.; Saleous, N.; Ksiksi, T. Carbon Stock Assessment of Date Palm Using Remote Sensing Coupled with Field-Based Measurements in Abu Dhabi (United Arab Emirates). *Int. J. Remote Sens.* **2019**, *40*, 7561–7580. [CrossRef]
97. Betemariyam, M.; Kefalew, T. Carbon Stock Estimation of Mixed-Age Date Palm (*Phoenix dactylifera* L.) Farms in Northeastern Ethiopia. *Heliyon* **2022**, *8*, e08844. [CrossRef] [PubMed]
98. Rahman, M.A.; Stratopoulos, L.M.F.; Moser-Reischl, A.; Zölch, T.; Häberle, K.H.; Rötzer, T.; Pretzsch, H.; Pauleit, S. Traits of Trees for Cooling Urban Heat Islands: A Meta-Analysis. *Build. Environ.* **2020**, *170*, 106606. [CrossRef]
99. Rahman, M.A.; Moser, A.; Rötzer, T.; Pauleit, S. Within Canopy Temperature Differences and Cooling Ability of Tilia Cordata Trees Grown in Urban Conditions. *Build. Environ.* **2017**, *114*, 118–128. [CrossRef]
100. Rahman, M.A.; Moser, A.; Gold, A.; Rötzer, T.; Pauleit, S. Vertical Air Temperature Gradients under the Shade of Two Contrasting Urban Tree Species during Different Types of Summer Days. *Sci. Total Environ.* **2018**, *633*, 100–111. [CrossRef]

101. Potchter, O.; Shashua-Bar, L. Urban Greenery as a Tool for City Cooling: The Israeli Experience in a Variety of Climatic Zones. In Proceedings of the PLEA, Edinburgh, Scotland, 3–5 July 2017.
102. Shashua-Bar, L.; Pearlmutter, D.; Erell, E. The Cooling Efficiency of Urban Landscape Strategies in a Hot Dry Climate. *Landsc. Urban Plan.* **2009**, *92*, 179–186. [CrossRef]

forests

Article

The Effects of Tree Canopy Structure and Tree Coverage Ratios on Urban Air Temperature Based on ENVI-Met

Haihua Wang [1,2,3,4,†], Yue Cai [1,2,3,4,†], Weifen Deng [5], Chong Li [1,2,3,4], Ya Dong [1,2,3,4], Lv Zhou [6], Jingyi Sun [1,2,3,4], Chen Li [1,2,3,4], Bingzheng Song [1,2,3,4], Fangfang Zhang [1,2,3,4] and Guomo Zhou [1,2,3,4,*]

1 State Key Laboratory of Subtropical Silviculture, Zhejiang A&F University, Lin'an 311300, China
2 Zhejiang Provincial Collaborative Innovation Centre for Bamboo Resources and High-Efficiency Utilization, Zhejiang A&F University, Lin'an 311300, China
3 Key Laboratory of Carbon Cycling in Forest Ecosystems and Carbon Sequestration of Zhejiang Province, Zhejiang A&F University, Lin'an 311300, China
4 School of Environmental and Resources Science, Zhejiang A&F University, Lin'an 311300, China
5 Forestry Bureau of Longquan City, Longquan 323700, China
6 College of Forestry, Beijing Forestry University, Beijing 100083, China
* Correspondence: wanghaihua@stu.zafu.edu.cn
† These authors contributed equally to this work.

Abstract: Vegetation configuration in residential districts improves human comfort by effectively moderating the thermal environment. Herein, the reliability of ENVI-met is verified by comparing the field measured with simulated data, including air temperature and relative humidity. The cooling effect of trees gradually increased with increasing tree coverage. Under the same coverage, trees with a tree crown diameter (TCD) of 3 m have the strongest cooling capacity, followed by trees with a TCD of 7 m, and trees with a TCD of 5 m have the weakest cooling capacity. The cooling capacity of a TCD of 3 m is considerably higher than that a TCD of 5 m and a TCD of 7 m. When the tree coverage ratio is 50%, the difference among the three TCDs is the largest. When the tree coverage is 50% or 70%, the cooling effect of TCD at 7 m is considerably higher than that at 5 m. For different canopy sizes and shapes under the same degree of tree coverage, only when the tree coverage is more than 50% and TCD is 3 m, the cooling capacity of a cylindrical shape is 0.2 to 0.3 °C higher than that of conical and ellipsoidal shapes. However, the difference between conical and ellipsoidal shapes when TCD is 5 or 7 m is not significant ($\Delta Ta < 0.1$ °C). Our results suggest that small canopy trees have a better cooling effect than large canopy trees for the same level of coverage.

Keywords: ENVI-met; residential district; canopy structure; tree coverage; cooling effect

Citation: Wang, H.; Cai, Y.; Deng, W.; Li, C.; Dong, Y.; Zhou, L.; Sun, J.; Li, C.; Song, B.; Zhang, F.; et al. The Effects of Tree Canopy Structure and Tree Coverage Ratios on Urban Air Temperature Based on ENVI-Met. *Forests* **2023**, *14*, 80. https://doi.org/10.3390/f14010080

Academic Editors: Thomas Rötzer, Stephan Pauleit, Mohammad A Rahman and Astrid Reischl

Received: 9 November 2022
Revised: 25 December 2022
Accepted: 29 December 2022
Published: 1 January 2023

1. Introduction

Over the past few decades, with rapid population growth, rapid urban expansion, and global climate warming, the phenomenon of urban heat islands has become a common issue in urban cities around the globe [1], significantly impacting human health and daily living, and increasing the frequency of extreme heat stress. Furthermore, the urban thermal environments and the thermal comfort of residents have become challenging. When temperatures become too hot, lives are threatened, and studies have found that temperature is an important factor affecting morbidity and mortality [2,3]; indeed, heat and heat waves are an additional contributing factor to increased mortality in summer. Exposure to heat can impair the body's ability to regulate its internal temperature, and can result in temperature-related deaths [4–6]. Urban cooling can help to mitigate the impact of extreme heat conditions on lives and livelihoods.

In most developed cities, urban green space is extremely limited, hence maximizing the cooling effect of green space is critical. Previous studies have also found that vegetation is one of the effective measures to mitigate the heat island effect and improve human comfort [7–9], since it affects air temperature (AT), relative humidity (RH), rainfall, and other conditions in urban areas, and the cooling capacity of trees is much more pronounced than that of grass. The physiological characteristics of trees provide a higher cooling capacity than other vegetation since trees can block the sunlight and absorb some heat and transpiration. Therefore, trees can effectively moderate the urban thermal environment [10,11].

Studies have confirmed that the key factor for trees to exert a cooling effect resides in the canopy [12–14], which blocks direct sunlight on the surface, while the leaves of the canopy can perform photosynthesis and transpiration [15,16]. Different canopies have different three-dimensional characteristics, leading to differences in the cooling capacity of trees.

The surface temperature of pileated canopy shapes had the strongest cooling and humidification effects, while the shape of the columnar canopy had the weakest [17,18]. This suggests that canopy shape should be considered when investigating the cooling effect of trees. The size of trees can be described by tree crown diameter (TCD) and tree height (TH), both of which have different effects on the urban thermal environment. At the same time, increasing the diameter of the crown will have better cooling and humidification effects than increasing the height of the trunk [18,19]. A larger TCD is more conducive to improving human thermal comfort [19] since increasing TCD will reduce the average sky view factor of the city [20,21]. TCD can indicate tree coverage; a larger canopy indicates a higher percentage of tree coverage and thus the more shade it provides [19,22].

Tree coverage and number are important factors in urban planning. In addition to crown size and canopy shape, tree coverage is also an important factor in the thermal environment of urban open green spaces [23,24]. Urban vegetation coverage is closely related to urban thermal environment and thermal comfort conditions [25]. In the hot summer months, areas with higher vegetation coverage are cooler than those with lower, as vegetation coverage effectively reduces air and radiation temperatures. Areas with higher vegetation coverage provide greater shade and transpiration, thus effectively improving the urban thermal environment [26]. When the canopy diameter of the trees is fixed, tree coverage is a key factor influencing the number of trees in a green space. The amount of cooling provided by green space and the distance over which that cooling extends depend on factors such as green space size, but this relationship is not linear [27,28].

Previous studies have confirmed that crown size, canopy shape, and tree coverage are important factors affecting AT and RH [25,27,29], although only few studies explained how the canopy affects the surrounding environment. To quantify the effect of crown size, canopy shape, and tree coverage on AT and RH, numerical simulation was used in this study to analyze and predict the urban thermal environment. ENVI-met, RayMan, and SOLWEIG are software packages frequently used in microclimate simulation studies. Among them, RayMan simulation has the fastest operation speed, yet only outputs few radiation parameters and cannot simulate reflected radiation well. SOLWEIG is suitable for radiation simulation of enclosed spaces, while ENVI-met software can calculate more radiation parameters and long-wave and short-wave radiation between buildings and surfaces [30].

ENVI-met is a 3D numerical simulation software based on fluid mechanics and thermodynamic equations [31]. According to the principle of iterative calculation, ENVI-met can simulate the "surface-plant-atmosphere" interaction in the city; then, the influence characteristics of urban planning form and landscape elements on complex space microclimate are quantified, which can be used to assess the influence of architecture and urban planning on environmental variables. ENVI-met can also simulate AT and RH, soil temperature and humidity, solar radiation, surface temperature, and wind speed and direction scenarios; hence, it is widely used in urban green space microclimate regulation effects, especially to simulate the heat island effect [32–34], improve human thermal comfort [35–37], and reduce air pollutants [38,39].

Previous studies using ENVI-met software found branching position height, planting orientation, crown size, tree shape, and the number of trees differently impact the surrounding microclimate [40,41]. However, these studies are based on the vegetation database in the system, while there are various kinds of vegetation that vary regionally. Therefore, in this study, we created a vegetation model consistent with the study area with the help of a 3D laser scanner. To ensure the reliability of the model's simulation output, it is often necessary to calibrate the simulated values with the use of measured values and to assess the accuracy of the model using statistical values such as root mean square error (RMSE) and index of agreement (d) [42,43].

Although ENVI-met's database contains various types of vegetation, in reality, the size, shape, and types of trees are more complicated. To truly restore the information of trees and improve the simulation accuracy of the model, we used the terrestrial laser scanning (TLS) to reconstruct the trees in the sample plot. TLS is a measuring tool that can accurately provide detailed stand and individual tree information without destroying trees, including crown size and shape, tree height, and diameter at breast height (DBH) [44,45]. The tree crown is reconstructed by dividing the point cloud acquired by the TLS into a three-dimensional grid of volume elements (voxel) [46].

In this study, we used the ENVI-met software to simulate a realistic scenario and used the measured data to verify the accuracy of the simulation. Firstly, five groups of different coverage of trees with three tree crown sizes were simulated during the summer. Secondly, three groups of different TCD and three groups of different canopy shapes of trees were simulated in five tree coverage scenarios. Our study aims are to (1) verify the simulation accuracy of the ENVI-met model; (2) study the effects of different crown sizes and tree coverage on the urban thermal environment in the summer; (3) quantify the cooling effect of three canopy shapes on the urban thermal environment in the summer.

2. Materials and Methods

2.1. Study Area

The study area was located in the Lin'an district (118°51′–119°52′ E, 29°56′–30°23′ N), Hangzhou City, Zhejiang Province, China (Figure 1). It has a typical subtropical climate with sufficient light and abundant rainfall. The weather forms of the four seasons are quite different: rainy in spring, hot and humid in summer, cool in autumn, and low temperature with little rain in winter. The annual rainfall is 1628.6 mm, 158 days of precipitation, a frost-free period averaging 235 days per year, with an average temperature of 16.4 °C throughout the year [47].

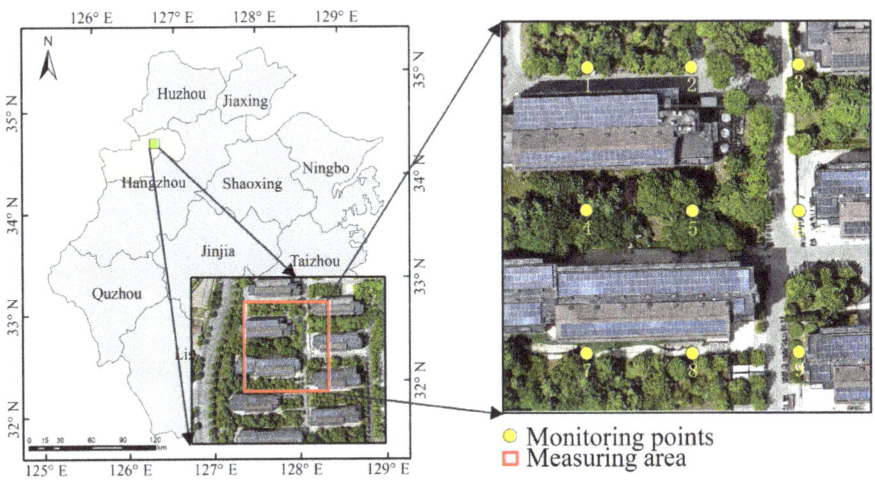

Figure 1. Location of the study area, and arrangement of wireless sensors. Wireless sensors at points 1, 2, 4, 5, and 7 were placed in the shade of trees, and wireless sensors at points 3, 6, 8, and 9 were placed in unobstructed places.

We selected a 100 m × 100 m size as the study area, and the features included buildings, roads, concrete roads, green space, and vegetation (Table 1). The experiment was conducted from 15 to 17 July 2021. The weather conditions in these three days were similar, with stable weather, cloudless and low wind speed.

Table 1. Area and percentage of land cover types in the study area.

Open Space Elements	Area (m^2)	Percent (%)
Student apartment	3014	30.14
Green space	4142	41.12
Roads	2844	28.44

2.2. Meteorological Data and Measured Data Collection

In this study, GHHB-001-485 wireless sensor equipment was used to monitor environmental factors. To ensure the accuracy of experimental results and reduce systematic errors among devices, all wireless sensor equipment were calibrated prior to the start of the experiments. All data were calibrated after being collected via sensors and used in this study. The AT error after calibration was <0.2 °C, while the RH error was <0.5%. Detailed data specifications for wireless sensor equipment are listed in Table 2.

Table 2. Specifications data of wireless sensor.

Parameter	Sensor Type	Measuring Range	Accuracy	Resolution	Response Time
Air Temperature	GHHB-001-485	−40–80 °C	0.1 °C	0.1 °C	60 s
Relative Humidity	GHHB-001-485	0–100%	0.3%	0.1%	60 s

We divided the study area into nine equal-sized cells, and arranged a wireless sensor in the center of each cell. As the center points of some cells fall on the roof, we moved these center points to be distributed on the roof horizontally along the north to the green space, and arranged sensors. Since the air temperature was relatively stable, and the composition and distribution of features in the study area were similar to those outside the study area, we think that the data difference caused by this small-scale movement can be ignored. Finally, nine wireless sensors were arranged in total (Figure 1), and some wireless sensors were under trees, some were in grassland, and some were next to buildings. The sensors were positioned at 1.5 m above the ground (Figure 2a), and recorded data at a frequency of 1 per min. In addition, we obtained three days' weather data consistent with the experimental time from a weather station located on an open asphalt surface in the campus and 2 m above the ground (Figure 2b). The collection frequency of the weather station was 1 per min. Furthermore, we obtained the wind speed and direction information at 10 m in the experimental day from the nearby Lin'an Meteorological Station.

2.3. Tree Information Collectiona

In this study, we used the DJI M300 UAV equipped with a visible light camera to take pictures of the study area to obtain images of the study site, including the composition of the underlying surface and the location of trees, and set the flight altitude at 60 m. We carried out a per-tree survey of the trees in the study area to obtain detailed information about each tree, including TH and DBH. There are 167 trees in the study area, primarily including Koelreuteria paniculata, Prunus cerasifera "Atropurpurea", Osmanthus fragrans, Elaeocarpus decipiens Hemsl, and Firmiana simplex (Linnaeus) W. Wight. We used a TLS device to obtain the parameters such as TH, TCD, and trunk height. Detailed parameters of the TLS equipment are listed in Table 3.

Figure 2. (**a**) Pictures of wireless sensor monitoring in sample plots. (**b**) Image of the reference weather station in the campus and at 2 m above the ground.

Table 3. System specifications for Leica ScanStation C5 equipment.

Parameter	Value
Field of view	270° × 360°
Range size	35 m @ ≥ 5% albedo
Scan rate	25,000 pts/s
Accuracy of position	6 mm
Accuracy of distance	4 mm
Minimum point spacing	<1 mm
Operating temperature	0–40 °C

2.4. Individual Tree Creation Process

The point cloud data collected using the TLS equipment were imported into the Cyclone software for processing and noise point removal near the point cloud of a single tree. Then, the separated point cloud of a single tree was voxelized using the Python 3.6 software, and the voxelization operation was performed on the point cloud data to convert the geometric form of the object into the closest representation form. In the sub-module Albero of ENVI-met software, the voxelized point cloud data were used to create a realistic single-tree model. We have created a total of 14 vegetation models. Cloud point voxelization and single tree creation are shown in Figure 3.

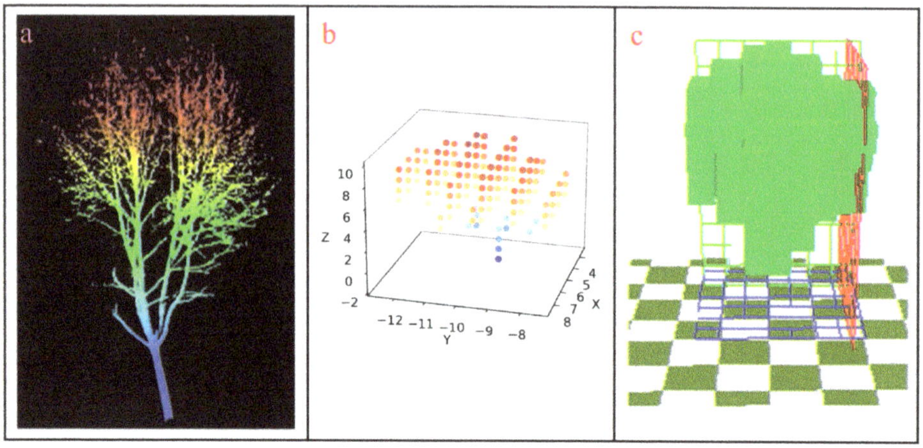

Figure 3. Creation of a single tree model. (**a**) Single tree point cloud data obtained using the TLS device. (**b**) Single point cloud data after voxel processing. (**c**) Single tree model created using ENVI-met.

2.5. ENVI-Met Scenario Setting and Simulation

The input parameter settings of the model include area input, configuration, and database files. The area input file included location and size of the study site, categories of ground objects, and areas of various types of ground objects. The related software Context Capture was used to stitch the photos taken via a drone to obtain a complete image map of the study areas, following which ArcGIS 10.4 was used to vectorize the study area photos obtained via the UAV to generate an image interface that can be input into the ENVI-met software. The area input file was mainly the setting of the simulated area file. The realistic scenario creation is illustrated in Figure 4.

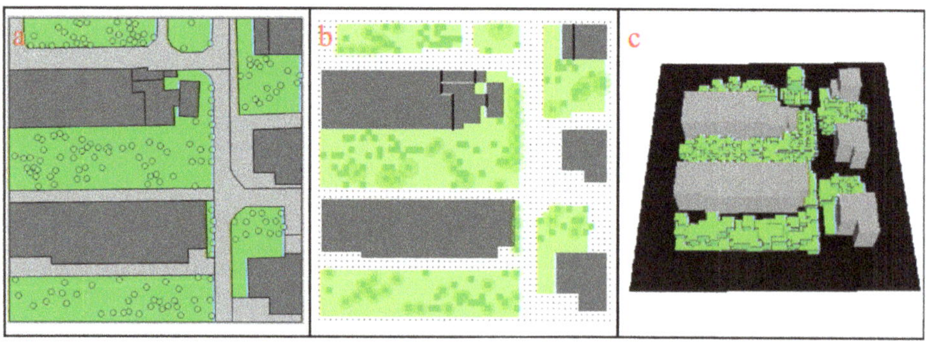

Figure 4. Realistic scenario creation process: (**a**) plane figure of the real scene, (**b**) the 2D display diagram of the real scene in ENVI-met, and (**c**) the 3D display diagram of the real scene in ENVI-met.

The database file includes parameter settings such as vegetation, green space, buildings, and soil. When setting vegetation, users can set parameters such as tree height, tree height width, and leaf shape according to their actual needs. The parameter settings of the configuration file include the AT and RH, wind speed and direction at 10 m, ground roughness, building material, and height. The hourly AT and RH obtained from the weather station were used as meteorological boundary conditions. Our actual field size was 100 m × 100 m. However, in order to minimize the edge effect and enhance the numerical stability, we added 10 empty cells: for each lateral boundary we expanded each edge outwards by 20 m in the simulation to reduce the influence of the model boundary on the simulation results. The final simulation area was a 70 × 70 × 20 3D cell grid, with a grid resolution of 2 × 2 ×3 (dx = 2 m, dy = 2 m, dz = 3 m), and the vertical grid cells of the near-ground was divided into five sub-grid cells. The simulation results were saved at hourly intervals. The specific parameter settings are listed in Table 4.

Table 4. ENVI-met model parameter settings.

Parameters	Value/Source
Maximum air temperature at 2 m (°C)	38.2
Minimum air temperature at 2 m (°C)	28.2
Maximum relative humidity at 2 m (%)	87.5
Minimum relative humidity at 2 m (%)	48.2
Wind speed at 10 m (m/s)	1.1
Wind direction at 10 m	SW
Grid cell size (Δx, Δy, Δz)	2, 2, 3
Number of grid cells (Δx, Δy, Δz)	70, 70, 20
Boundary condition	Simple Forcing
Simulation duration	24 h
Roughness length	0.01
Albedo of road	0.2
Emissivity of road	0.9
Albedo of glass	0.2
Transmittance of glass	0.3

2.6. *Numerical Simulation*

2.6.1. Simulation under Different TCDs and Tree Coverage

In this study, 16 July 2021 was selected as the simulation date. To study the cooling effect of different cover and TCD in summer, five different steps of tree cover (10, 30, 50, 70, and 90%) and three different groups of TCD (3, 5, and 7 m) were set. A total of 15 scenes were created, in which we did not consider the influence of the crown shape; hence, we set the crown shape as a cylinder. In this study, the tree height was set to 7 m, except for the trees in the real scene. The cover ratio used in the study is the ratio of vegetation to the overall green space.

Base model: The basic scenario was the part left after all the trees in the realistic scenario were removed. In other words, the difference between the realistic scenario and the basic scene is only whether there are trees or not. In this paper, the analysis of the cooling effect of different scenes was based on the basic scenes.

Case 1 (C1): Tree coverage of 10% with 46 trees (17, 9) based on the base model, and TCD = 3(5, 7).

Case 2 (C2): Tree coverage of 30% with 138 trees (50, 25) based on the base model, and TCD = 3 (5, 7).

Case 3 (C3): Tree coverage of 50% with 230 trees (83, 42) based on the base model, and TCD = 3 (5, 7).

Case 4 (C4): Tree coverage of 70% with 322 trees (116, 59) based on the base model, and TCD = 3 (5, 7).

Case 5 (C5): Tree coverage of 90% with 414 trees (149, 76) based on the base model, and TCD = 3 (5, 7).

2.6.2. Simulation under Different TCDs, Tree Coverage, and Crown Shapes

Based on the above 15 scenes, we set three canopy shapes, namely cylindrical, conical, and ellipsoidal, for each group, creating a total of 45 scenes. According to three groups of different TCDs sizes and three different crown shapes, a total of nine single tree models (Table 5) were created in the vegetation database.

Table 5. Nine individual trees of different crown sizes and canopy shapes.

Shape	TCD = 3 m	TCD = 5 m	TCD = 7 m
Cylindrical shape		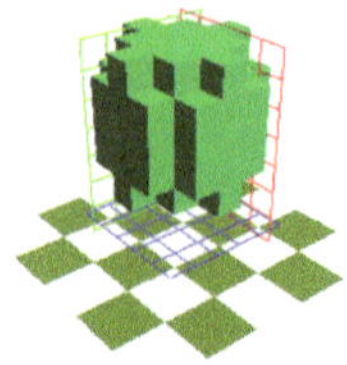	
Ellipsoidal shape	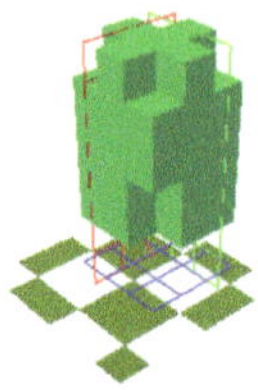		
Conical shape	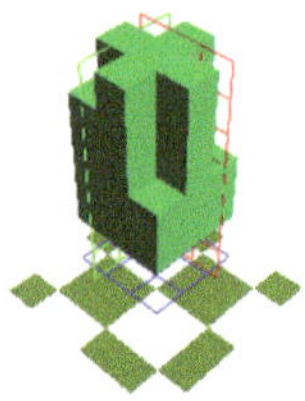		

The single tree model used in the scenario was constructed in the submodule Albero in the ENVI-met software. The specific parameter settings of scenes were set in the same way as in the real scenario. In this study, a total of 5 h (09:00–14:00) were simulated with the first 2 h of data being used to stabilize the model, and the average of the simulated data for the 3 h (12:00–14:00) was selected to study the cooling and humidification effects of different scenarios.

Except for the tree configuration, all other settings of scenes, including weather boundary, forcing conditions, and category settings of ground objects, were consistent with the basic model, which can avoid the influence of the model itself on the simulation results. The average temperature with vegetation were compared with that without vegetation, and the cooling effect of trees under different configuration methods was analyzed.

2.7. Statistical Analysis

Python 3.6 and Sigmaplot 14.0 were adopted for statistical analysis. We also used IBM SPSS Statistics 26 for variance analysis to compare the differences in cooling capacity between different scenes. The data acquired by the sensors and those exported using the ENVI-met software were processed using Python software; we used the software to calculate the correlation coefficient R2 between simulated and measured data, as well as the RMSE and index of agreement (d). Scatter plots, histograms, and line plots were produced using Sigmaplot software to compare the cooling and humidification effects of the different tree coverage, TCD, and canopy shapes in summer.

3. Results

3.1. ENVI-Met Accuracy Verification

To ensure the reliability of the model output, the model was validated to ensure the authenticity of the simulation results of the later scenarios and facilitate further analysis and comparison of the results. In this study, we selected the data from 16 July to verify the model, because it was the hottest day in the three-day experiment: the maximum

AT was 38.2 °C and average AT was 32.8 °C. Since the height of the sensor was set to 1.5 m, the simulation results should also be at 1.5 m for the AT and RH. The output results following simulation were analyzed, viewed, and exported through LEONARDO in the ENVI-met module.

Figure 5 shows the changes of the measured AT data of nine wireless sensors and the simulated AT data of the corresponding points of the wireless sensors during the whole day of 16 July. In the daytime, the AT data of wireless sensors which were placed in the unobstructed positions were overestimated, while the AT data of wireless sensors which were placed under the tree were underestimated. All measuring points were overestimated in the nighttime. Although there were differences between the simulated data and the measured data, the simulated data were in good agreement with the measured data of the corresponding sensor location, and the maximum temperature of each point in the simulated data appeared at the same time as the measured data of the corresponding sensor location. From the simulation results, we know that the temperature of the uncovered monitoring points was always higher than that of the trees in the shade, which was completely consistent with the actual situation, which also indicates that ENVI-met model can simulate the spatial distribution of temperature.

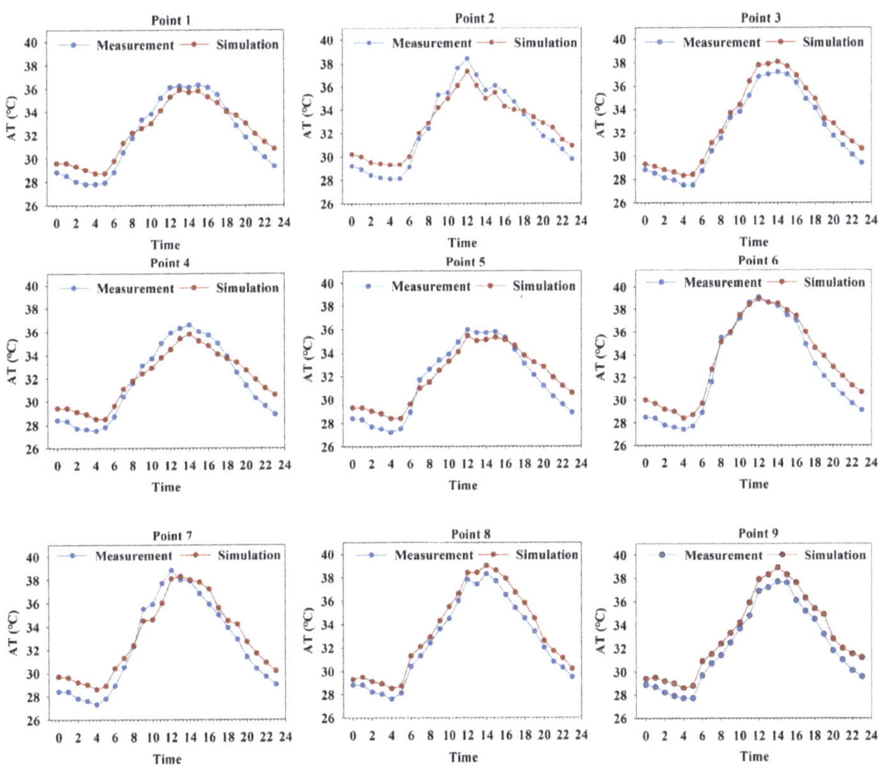

Figure 5. Simulated and measured air temperature data of nine points on 16 July 2021.

Figure 6 shows the 24-h simulated humidity data were consistent with measured humidity data. In the daytime, the measured humidity data in the uncovered place were lower than those from around the grass and trees, because trees provide shade and increase humidity, and the simulated humidity data also conform to this phenomenon. Humidity was generally underestimated in the nighttime.

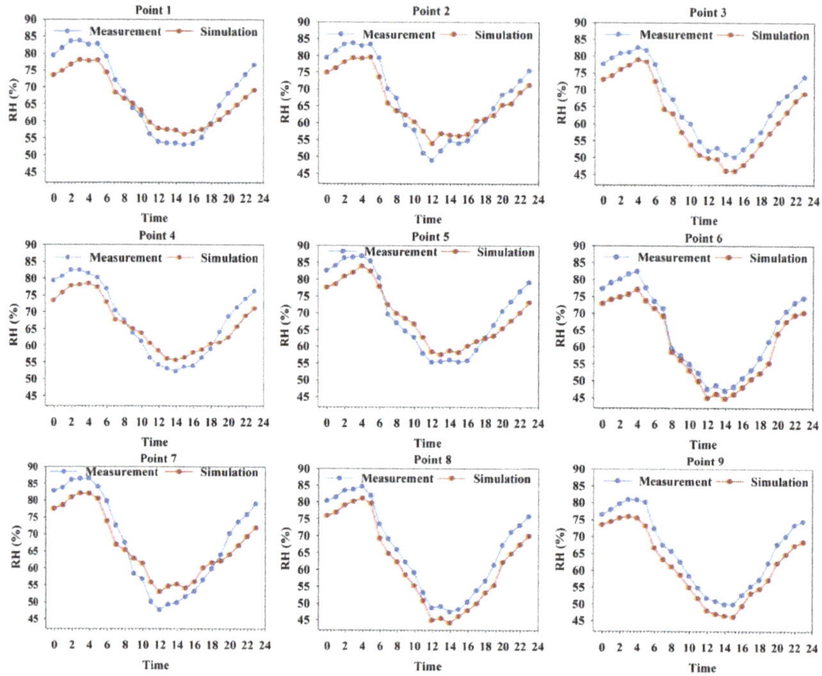

Figure 6. Simulated and measured relative humidity data of nine points on 16 July 2021.

Although the simulated data of nine points were in good agreement with the measured data, we also needed to evaluate the overall accuracy of the model. We extracted the hourly data of the nine wireless sensors corresponding to the monitoring point at 1.5 m above the ground from the simulation results, with a total of 216 simulate results. At the same time, we also analyzed the correlation between the 24-h measured data collected by nine sensors and the 24-h simulated data of the corresponding nine points. Finally, a total of 216 pairs of simulated and measured data were collected. Figure 7 shows that the R^2 of AT was 0.9307, indicating that the simulated values were strongly correlated with the measured values. Root mean square error (RMSE) and index of agreement (d) were used to assess the difference between simulated and measured data. The RMSE of AT and RMSE of RH were 1.13 °C and 4.35%, respectively, and the d of AT and d of RH were 0.945 and 0.94, respectively.

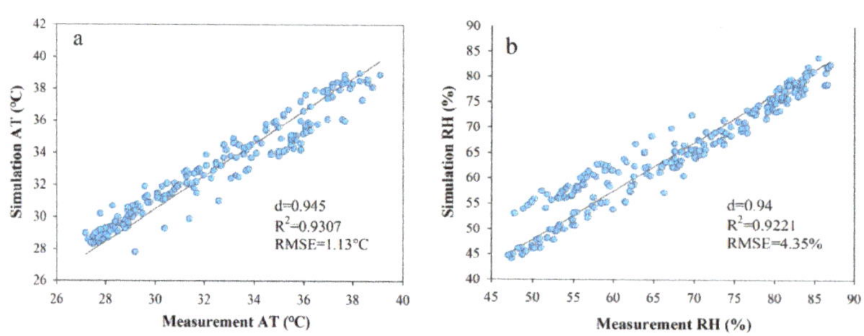

Figure 7. The relationship between the measured and simulated data of AT (**a**) and RH (**b**) on 16 July 2021.

3.2. Effects of TCD on AT and RH under Different Coverages

To better compare the cooling and humidifying effects of each created scene, we used the scene without trees as the base scene for comparison and selected 12:00–14:00 as the period for analysis, with the AT and RH values at 1.5 m. The variation range of AT of the basic model was 37.79–38.46 °C, and the average temperature was 38.14 ± 0.34 °C. The variation range of RH was 48.33–51.27%, and the average AT was 50.07 ± 1.54%. The maximum temperature occurs at 14:00 (38.46 °C), which is very hot in summer. Figure 8 illustrates the AT and RH plan of the basic model at a height of 1.5 m at 14:00.

$$\Delta AT_{ij} = AT_{base\ model} - AT_{ij} \tag{1}$$

where i could be 1, 2, or 3, which corresponded to TCD of 3 m, 5 m, and 7 m, respectively, and j could be 1, 2, 3, 4, or 5 which corresponded to coverage of 10%, 30%, 50%, 70%, and 90%, respectively. ΔAT_{ij} indicates the average cooling capacity of a scenario with TCD of i and coverage of j. $AT_{base\ model}$ indicates the average air temperature of the base scenario. AT_{ij} indicates the average air temperature of the TCD of i and coverage of j.

$$\Delta RH_{ij} = RH_{ij} - RH_{base\ model} \tag{2}$$

where i could be 1, 2, 3, which corresponded to TCD of 3 m, 5 m, and 7 m, respectively, and j could be 1, 2, 3, 4, 5 which corresponded to coverage of 10%, 30%, 50%, 70%, 90%, respectively. ΔRH_{ij} indicates the average humidity capabilities of a scenario with TCD of i and coverage of j. $RH_{base\ model}$ indicates the average relative humidity of the base scenario. RH_{ij} indicates the average relative humidity of the TCD of i and coverage of j.

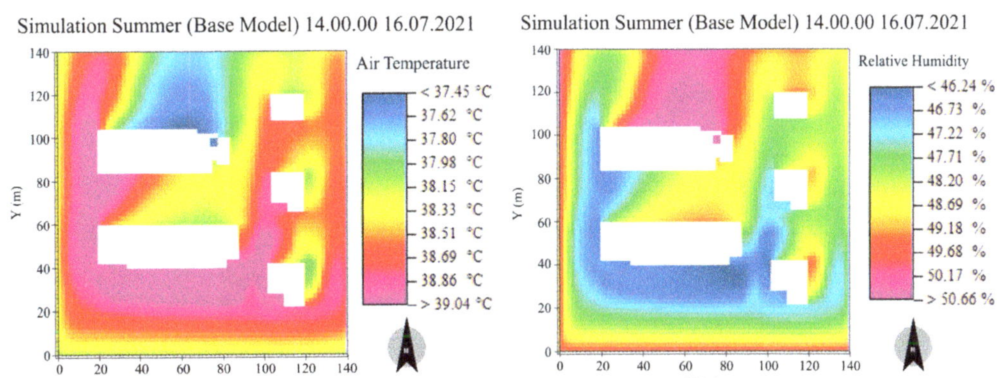

Figure 8. Distribution of simulated AT and RH at 1.5 m in the simulation domain at 14:00 in the summer (16 July 2021) base model.

Figure 9 demonstrates the increasing magnitude of the cooling capacity of trees with an increasing tree coverage ratio. At C1, all three scenes with different TCD showed the weakest cooling effect, the ΔAT_{11}, ΔAT_{21}, ΔAT_{31} is 0.34, 0.04, and 0.10 °C, respectively; meanwhile, at C5, all three scenes with different TCD showed the strongest cooling effect, with an average cooling of 1.80, 1.39, and 1.54 °C, respectively. The cooling effect of TCD at the 3 m scene was significantly higher than that at 5 and 7 m ($p < 0.05$), and the biggest difference was among the three scenarios in C3; the ΔAT_{13}, ΔAT_{23}, ΔAT_{33} is 1.59, 0.76, and 1.03 °C, respectively. Only in scene C3 or C4 was the average cooling effect of TCD at 7 m significantly higher than that at 5 m ($p < 0.05$). Under the same tree coverage ratio, the cooling effect of TCD at 7 m scene was consistently stronger than that at 5 m.

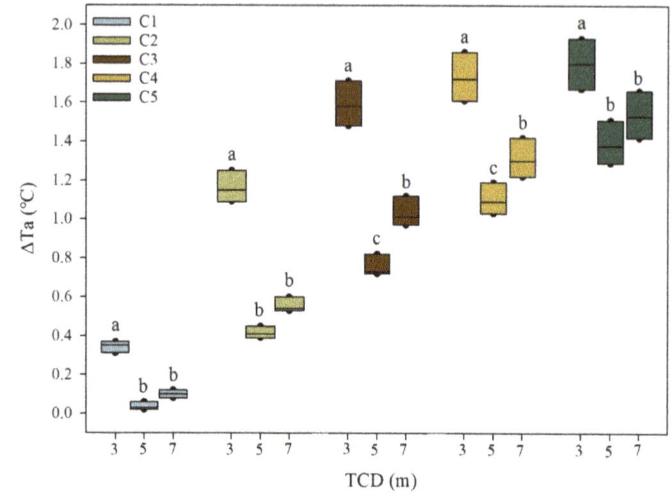

Figure 9. Comparison of cooling effects of three different crown diameters (TCD = 3, 5, and 7 m) under five groups of coverage, the tree coverage of C1, C2, C3, C4, and C5 was 10, 30, 50, 70, and 90%, respectively. Lowercase letter (a, b, c) above the boxes indicate significant differences among three different crown diameters (TCD = 3, 5, and 7 m) for five groups of coverage (C1, C2, C3, C4, C5) at $p < 0.05$.

3.3. Effects of Canopy Shapes on AT and RH under Different Coverage and TCD

Crown shape also affected the cooling effect of trees. The influence of the same crown shape differed under different crown diameters and coverage. The histogram shown in Figure 10 illustrates the effect of changing in canopy shape of the trees on the temperature of the study area for five different tree coverage ratios and three different TCDs.

The cooling effect of all scenarios increases with increasing tree coverage. When the tree coverage was at 10%, the difference between the three different canopy shapes at the same TCD was minimal and did not exceed 0.1 °C. The maximum cooling of the three canopy shapes with a TCD of 3 m could reach 0.3–0.4 °C. However, the effect of the three canopy shapes with a TCD of 5 and 7 m on the air temperature was minimal. The maxi-mum cooling value of the scene was reached when the tree coverage was 90%, where the average ΔAT of the three crown shapes, cylindrical shape, conical shape, and ellipsoidal shape, was 2.11 ± 0.15 °C, 1.82 ± 0.13 °C, and 1.86 ± 0.14 °C, respectively; at TCD of 3 and 5 m, the average ΔAT was 1.31 ± 0.10 °C, 1.29 ± 0.12 °C, and 1.28 ± 0.12 °C, respectively, and 1.34 ± 0.09 °C, 1.24 ± 0.09 °C, and 1.20 ± 0.10 °C, respectively, at 7 m. When the tree coverage ratio was 50%, the difference between three TCDs was the biggest, the difference between different TCDs of the same canopy shape was also the biggest, and the cooling effects of trees with cylindrical shapes under three TCDs (3, 5, and 7 m) were 1.86 ± 0.14 °C, 0.71 ± 0.04 °C, and 0.82 ± 0.05 °C, respectively; those with conical and ellipsoidal shapes both were 1.60 ± 0.12 °C, 0.70 ± 0.06 °C, and 0.73 ± 0.04 °C, respectively.

Figure 10 also shows no significant difference among the three crown shapes with a TCD at 5 or 7 m, with a difference of <0.1 °C, regardless of coverage.

The three crown shapes with a TCD of 3 m differed under different coverage. Under the same coverage, the cylindrical shape showed the strongest cooling capacity, while conical and ellipsoidal shapes showed the same cooling capacity, with a maximum difference of <0.1 °C. In addition, there was no significant difference among the three crown shapes when the tree coverage was <50%, while with a TCD of 3 m, the difference between them was <0.2 °C. At TCD of 3 m and tree coverage ratio of >50%, the cooling capacity of the tree with a cylindrical shape was significantly higher than that of the other two crown shapes ($p < 0.05$).

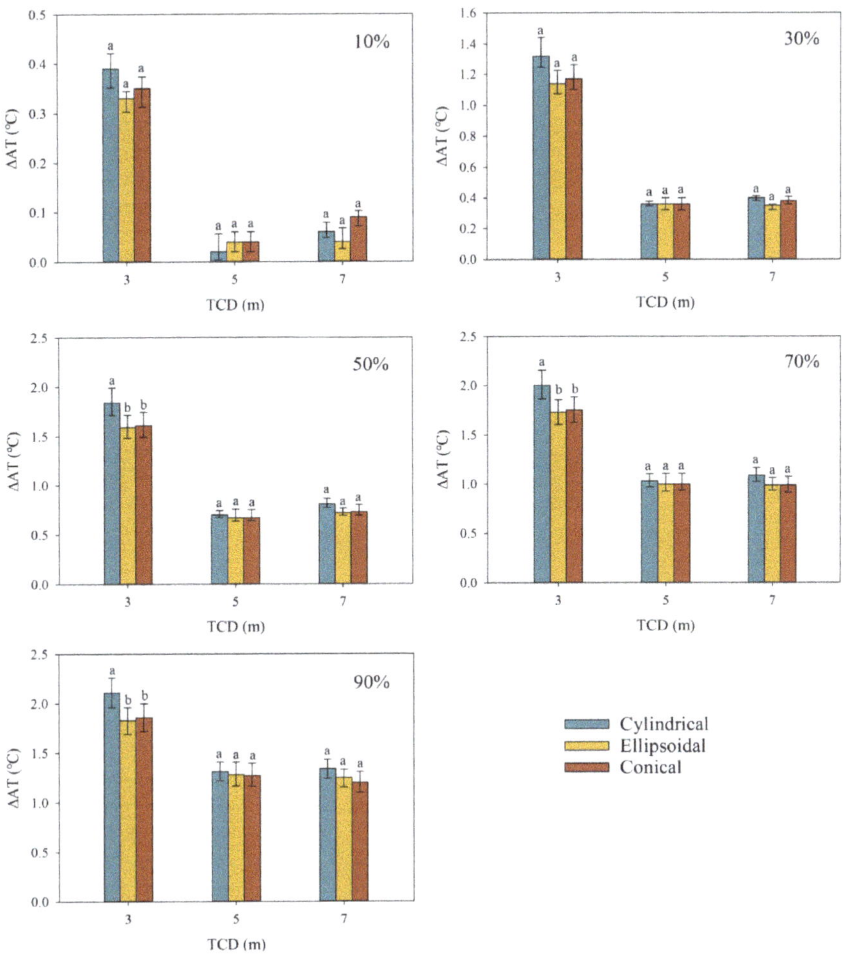

Figure 10. Effects of the three canopy shapes with three crown diameters on AT and RH under five groups of tree coverage during summer. Lowercase letter (a, b) above the boxes indicate significant differences among three different canopy shapes with same tree crown diameter for five groups of coverage (10%, 30%, 50%, 70%, 90%) at $p < 0.05$.

4. Discussion

4.1. Differences between the Simulated and Measured Values

The AT correlation between the simulated and measured values was 0.9307, the RH correlation was 0.9221, and the d between the simulated and measured values was >0.90, indicating that TLS can be used as an important means of tree reconstruction and be applied to the creation of vegetation in ENVI-met model. The RMSE was 1.13 °C for AT and 4.35% for RH in the simulated and measured values. The results were similar to some studies [40,48], with an underestimation during the day and an overestimation at night. The reasons for the discrepancy between simulated and measured data are attributed to three reasons, namely the limitations of the model itself, the complexity of the realistic scenario, and the data of processing method.

Firstly, the ENVI-met model is limited in the solar radiation input and does not allow users to input by the hour, which may cause the AT in the uncovered place to be overestimated. The inaccuracy in the calculation of energy consumption of simulated buildings accounts for the difference between the simulated and measured values [49]. The model also does not consider radiation between buildings and leaves. These factors impact the surrounding environment [50].

Secondly, the vegetation used in the ENVI-met model was a simplified tree model, while the actual tree geometry is irregular and difficult to digitize [51]. The climate of a residential area is affected by several factors, such as the thermal insulation and radiation characteristics of buildings, concrete, and soil. Human factors also lead to greater differences when the flow of people and vehicles increases, as vehicle fuel combustion and human metabolism release a lot of waste heat [52]. Weather condition is another factor since the wind speed and direction change in real time, while the wind speed and direction in the simulation always remain constant [53]. These factors contribute to the differences between the measured and simulated values. ENVI-met could not accurately simulate the shaded areas of the tree, hence the temperature in the shade of trees is easily underestimated in the daytime and overestimated in the nighttime [54,55].

However, there was a limitation in our study that the sensors we used were not verified for radiation prior to use. Both natural shading and incorrect shading methods can lead to bias in the sensor data, and the sensor bias may be more obvious when the solar radiation was too high or the wind speed was too low; this bias was particularly apparent during the day and has little effect at night [56–58]. This may result in a greater data collection from sensors placed in exposed environments than the truthful data.

4.2. Effect of TCD and Tree Coverage Ratio on the Cooling Effect

Vegetation is one of the factors that alleviate urban thermal environment problems; the canopy structure is the main factor for vegetation to exert its cooling effects. In the limited urban open space, we must select the appropriate tree size and tree coverage to maximize the cooling effect of trees. Previous studies have suggested tree coverage ratio as a factor affecting air and surface temperatures, with higher coverage having a more powerful effect. When air flows through green space, trees can alter the wind characteristics that affect air dispersion. Generally, a higher tree coverage rate can effectively reduce wind speed through the canopy while increasing the time to change the thermal characteristics of the air mass [23]. The cooling effect of trees in this study also generally increased with increasing tree coverage ratio [24]; however, after reaching a certain peak, the increase gradually decreases, which indicates that when the vegetation coverage reaches a certain threshold, the cooling effect is optimum. It has been previously reported that the threshold of optimal vegetation coverage depends on the regional location and climatic conditions. The cooling effect of trees is the greatest when the vegetation coverage is 40% in the Midwest of the United States [27], while for some subtropical seasonal climatic regions, the cooling effect of vegetation tends to be stable when the coverage is 20–30% [59], similar to our results.

When the tree coverage rate exceeded 50%, and the increasing rate of the cooling effect gradually decreased as the tree coverage rate increased. Our study also confirmed that canopy size is an important factor in the cooling effect. Indeed, the cooling effect of trees of different canopy sizes in a scene with equal coverage varied between scenes; the cooling effect was strongest in the TCD of 3 m, followed by 7 m, while TCD at 5 m showed the weakest cooling effect. The scene with a TCD of 3 m was considerably higher than that of the other two crown shapes when the tree coverage ratios were >30%. When considering only a single tree in the thermal environment, the cooling effect of large canopies was stronger than that of small canopies. Still, under the same level of tree coverage, trees with smaller crown diameters indicated more trees.

Multiple small patches can provide more shade than a single large patch when the total area covered is the same; the more small patches a scene has, the more shade the tree will provide [26,60]. For the same level of coverage, scenes with a 7 m canopy are cooler than those with a 5 m canopy, possibly since larger canopy trees are more effective at blocking short-wave radiation from the sun and sky, as well as long-wave radiation from the sky and from inside nearby buildings [61]. The cooling capacity and shading effect of larger canopy diameter is greater than that of a small canopy [29,62]. At a coverage of 30%, the ΔAT between the three groups of different TCDs was the largest. The average ΔAT of the TCD at the 3 m scene was 0.56 and 0.83 °C higher than that of the other two scenes, respectively, while at a tree coverage of 10%, the TCD of the 5 m tree scene had a negligible cooling effect on the area (<0.1 °C), yet markedly impacts surface and building temperatures [63].

4.3. Effect of Different Canopy Shapes on the Thermal Environment

In this study, the tree canopy characteristics were correlated with the surrounding thermal environment, as previously reported [17,29]. The differences in the effect of different canopy shapes on the surrounding environment are mainly due to different canopy shapes having different canopy structures, resulting in different shading and wind protection capacities [64,65]. Further, we observed that the cooling effect of the cylindrical shape was greater than that of the ellipsoidal and conical shapes, consistent with a previous study [18]. When TCD was at 3 m, the ΔAT of cylindrical shape was larger with a value of 0.2–0.3 °C than the other shapes. Although the three crown shapes with a TCD of 3 m are narrow at the same coverage level, they have a higher total canopy density, and are associated with the highest number of trees.

A canopy with multiple layers of leaves can reduce the maximum transmittance, and the photosynthesis and transpiration of the trees are also stronger [66,67], which indirectly leads to a better cooling effect. However, different crown shapes do not show great differences in cooling effect in TCD at 5 or 7 m. At a TCD of 5 m, the influence of canopy shape on the surrounding thermal environment becomes less important, with no significant differences between crown shapes. At a TCD of 7 m, the difference between the three canopy shapes occurred only when there was sufficient tree coverage (>70%), and the cylindrical shape was 0.1 °C higher than the other two. The difference between the three canopy shapes with TCD at 5 and 7 m was the smallest. The total leaf area index of all three canopy shapes was set to a constant value to eliminate the influence of other parameters, in which leaf area index was an important factor affecting the cooling capacity of the trees [13,29]. Furthermore, the three canopy shapes have the same short-wave radiation transmittance and dense crowns, hence the solar radiation can hardly penetrate the crown and only a very small amount of light beams and radiation reach the ground [16].

5. Conclusions

In this study, we reconstructed trees in the real scene with TLS and used the reconstructed trees in the real scene simulation. The simulation results were verified, and the simulation data were in good agreement with the measured data. Therefore, when ENVI-met is used to simulate the real scene, TLS can be considered as an important means of tree reconstruction, which can restore the real trees and improve the simulation accuracy of the model. Under the five coverage gradients, the cooling effect of trees increases with the increase in coverage. Tree crown diameter can produce significant cooling effect under the same coverage; the cooling effect of trees with TCD of 3 m was significantly higher than that of the other two crown diameters ($p < 0.05$). We recommend planting trees with small crown diameter since they provide a better cooling effect. Moreover, at a coverage of 50%, the cooling effect among the three tree crowns showed the biggest difference. The crown shape was also an important factor affecting the cooling effect of several trees, as the tree coverage exceeded 50%, the cooling effect of the cylindrical shapes was significantly higher than that of the other two crowns under the same coverage ($p < 0.05$), however, the difference of the cooling effect between TCD at 5 and 7 m remained small (<0.1 °C). In future urban planning and construction projects, we recommend selecting appropriate tree coverage and tree crown diameter and making use of the limited available space to achieve maximum cooling effect. More attention should be paid to the use of sensors in future studies, especially in open sites, and radiation shields need to be repeatedly verified before use or higher quality radiation shields and sensors need to be used to reduce errors; when this error is minimized, the study can be reliable.

Author Contributions: Methodology, H.W. and Y.C.; software, H.W., W.D. and F.Z.; validation, H.W. and Y.C.; formal analysis, H.W. and Y.C.; investigation, H.W., L.Z., Y.D., J.S., C.L. (Chen Li) and B.S.; resources, H.W., Y.C., Y.D., J.S., C.L. (Chen Li) and B.S.; data curation, H.W.; writing—original draft preparation, H.W.; writing—review and editing, H.W. and Y.C.; visualization, H.W., C.L. (Chong Li) and G.Z.; supervision, C.L. (Chong Li) and G.Z.; project administration, H.W, Y.C. and G.Z. All authors have read and agreed to the published version of the manuscript.

Funding: This research was funded by the National Natural Science Foundation of China (grant number: U1809208).

Data Availability Statement: Not applicable.

Conflicts of Interest: The authors declare that they have no known competing financial interests or personal relationships that could have appeared to influence the work reported in this paper.

References

1. Qaid, A.; Bin Lamit, H.; Ossen, D.R.; Raja Shahminan, R.N. Urban Heat Island and Thermal Comfort Conditions at Micro-Climate Scale in a Tropical Planned City. *Energy Build.* **2016**, *133*, 577–595. [CrossRef]
2. Ha, J.; Kim, H. Changes in the Association between Summer Temperature and Mortality in Seoul, South Korea. *Int. J. Biometeorol.* **2013**, *57*, 535–544. [CrossRef] [PubMed]
3. Xinchuang, X.; Quansheng, G.; Shanfeng, H.; Xuezhen, Z.; Xunliang, X.; Guang Xu, L. Impact of High Temperature on the Mortality in Summer of Wuhan, China. *Environ. Earth Sci.* **2016**, *75*, 543. [CrossRef]
4. Allen, M.J.; Sheridan, S.C. Mortality Risks during Extreme Temperature Events (ETEs) Using a Distributed Lag Non-Linear Model. *Int. J. Biometeorol.* **2018**, *62*, 57–67. [CrossRef] [PubMed]
5. Chen, K.; Bi, J.; Chen, J.; Chen, X.; Huang, L.; Zhou, L. Influence of Heat Wave Definitions to the Added Effect of Heat Waves on Daily Mortality in Nanjing, China. *Sci. Total Environ.* **2015**, *506–507*, 18–25. [CrossRef] [PubMed]
6. Shindell, D.; Zhang, Y.; Scott, M.; Ru, M.; Stark, K.; Ebi, K.L. The Effects of Heat Exposure on Human Mortality Throughout the United States. *GeoHealth* **2020**, *4*, e2019GH000234. [CrossRef]
7. Avissar, R. Potential Effects of Vegetation on the Urban Thermal Environment. *Atmos. Environ.* **1996**, *30*, 437–448. [CrossRef]
8. Liao, J.; Tan, X.; Li, J. Evaluating the Vertical Cooling Performances of Urban Vegetation Scenarios in a Residential Environment. *J. Build. Eng.* **2021**, *39*, 102313. [CrossRef]
9. Barradas, V.L.; Miranda, J.A.; Esperón-Rodríguez, M.; Ballinas, M. (Re)Designing Urban Parks to Maximize Urban Heat Island Mitigation by Natural Means. *Forests* **2022**, *13*, 1143. [CrossRef]
10. Roth, M.; Lim, V.H. Evaluation of Canopy-Layer Air and Mean Radiant Temperature Simulations by a Microclimate Model over a Tropical Residential Neighbourhood. *Build. Environ.* **2017**, *112*, 177–189. [CrossRef]
11. Gkatsopoulos, P. A Methodology for Calculating Cooling from Vegetation Evapotranspiration for Use in Urban Space Microclimate Simulations. *Procedia Environ. Sci.* **2017**, *38*, 477–484. [CrossRef]
12. Deng, J.; Pickles, B.J.; Kavakopoulos, A.; Blanusa, T.; Halios, C.H.; Smith, S.T.; Shao, L. Concept and Methodology of Characterising Infrared Radiative Performance of Urban Trees Using Tree Crown Spectroscopy. *Build. Environ.* **2019**, *157*, 380–390. [CrossRef]
13. Lin, B.S.; Lin, Y.J. Cooling Effect of Shade Trees with Different Characteristics in a Subtropical Urban Park. *HortScience* **2010**, *45*, 83–86. [CrossRef]
14. Cheung, P.K.; Jim, C.Y.; Hung, P.L. Preliminary Study on the Temperature Relationship at Remotely-Sensed Tree Canopy and below-Canopy Air and Ground Surface. *Build. Environ.* **2021**, *204*, 108169. [CrossRef]
15. Kántor, N.; Kovács, A.; Takács, Á. Small-Scale Human-Biometeorological Impacts of Shading by a Large Tree. *Open Geosci.* **2016**, *8*, 231–245. [CrossRef]
16. Konarska, J.; Lindberg, F.; Larsson, A.; Thorsson, S.; Holmer, B. Transmissivity of Solar Radiation through Crowns of Single Urban Trees-Application for Outdoor Thermal Comfort Modelling. *Theor. Appl. Climatol.* **2014**, *117*, 363–376. [CrossRef]
17. Cai, Y.; Li, C.; Ye, L.; Xiao, L.; Gao, X.; Mo, L.; Du, H.; Zhou, Y.; Zhou, G. Effect of the Roadside Tree Canopy Structure and the Surrounding on the Daytime Urban Air Temperature in Summer. *Agric. For. Meteorol.* **2022**, *316*, 108850. [CrossRef]
18. Speak, A.; Montagnani, L.; Wellstein, C.; Zerbe, S. The Influence of Tree Traits on Urban Ground Surface Shade Cooling. *Landsc. Urban Plan.* **2020**, *197*, 103748. [CrossRef]
19. Wang, Y.; Akbari, H. The Effects of Street Tree Planting on Urban Heat Island Mitigation in Montreal. *Sustain. Cities Soc.* **2016**, *27*, 122–128.
20. Oshio, H.; Kiyono, T.; Asawa, T. Numerical Simulation of the Nocturnal Cooling Effect of Urban Trees Considering the Leaf Area Density Distribution. *Urban For. Urban Green.* **2021**, *66*, 127391. [CrossRef]
21. Karimi, A.; Sanaieian, H.; Farhadi, H.; Norouzian-Maleki, S. Evaluation of the Thermal Indices and Thermal Comfort Improvement by Different Vegetation Species and Materials in a Medium-Sized Urban Park. *Energy Rep.* **2020**, *6*, 1670–1684. [CrossRef]
22. Middel, A.; Chhetri, N.; Quay, R. Urban Forestry and Cool Roofs: Assessment of Heat Mitigation Strategies in Phoenix Residential Neighborhoods. *Urban For. Urban Green.* **2015**, *14*, 178–186. [CrossRef]

23. Wu, Z.; Man, W.; Ren, Y. Influence of Tree Coverage and Micro-Topography on the Thermal Environment within and beyond a Green Space. *Agric. For. Meteorol.* **2022**, *316*, 108846. [CrossRef]
24. Ouyang, W.; Morakinyo, T.E.; Ren, C.; Ng, E. The Cooling Efficiency of Variable Greenery Coverage Ratios in Different Urban Densities: A Study in a Subtropical Climate. *Build. Environ.* **2020**, *174*, 106772. [CrossRef]
25. Aminipouri, M.; Knudby, A.J.; Krayenhoff, E.S.; Zickfeld, K.; Middel, A. Modelling the Impact of Increased Street Tree Cover on Mean Radiant Temperature across Vancouver's Local Climate Zones. *Urban For. Urban Green.* **2019**, *39*, 9–17. [CrossRef]
26. Jiao, M.; Zhou, W.; Zheng, Z.; Wang, J.; Qian, Y. Patch Size of Trees Affects Its Cooling Effectiveness: A Perspective from Shading and Transpiration Processes. *Agric. For. Meteorol.* **2017**, *247*, 293–299. [CrossRef]
27. Ziter, C.D.; Pedersen, E.J.; Kucharik, C.J.; Turner, M.G. Scale-Dependent Interactions between Tree Canopy Cover and Impervious Surfaces Reduce Daytime Urban Heat during Summer. *Proc. Natl. Acad. Sci. USA* **2019**, *116*, 7575–7580. [CrossRef]
28. Wu, Z.; Dou, P.; Chen, L. Comparative and Combinative Cooling Effects of Different Spatial Arrangements of Buildings and Trees on Microclimate. *Sustain. Cities Soc.* **2019**, *51*, 101711. [CrossRef]
29. Zhang, J.; Gou, Z. Tree Crowns and Their Associated Summertime Microclimatic Adjustment and Thermal Comfort Improvement in Urban Parks in a Subtropical City of China. *Urban For. Urban Green.* **2021**, *59*, 126912. [CrossRef]
30. Liu, D.; Hu, S.; Liu, J. Contrasting the Performance Capabilities of Urban Radiation Field between Three Microclimate Simulation Tools. *Build. Environ.* **2020**, *175*, 106789. [CrossRef]
31. Bruse, M.; Fleer, H. Simulating Surface-Plant-Air Interactions inside Urban Environments with a Three Dimensional Numerical Model. *Environ. Model. Softw.* **1998**, *13*, 373–384. [CrossRef]
32. Fahed, J.; Kinab, E.; Ginestet, S.; Adolphe, L. Impact of Urban Heat Island Mitigation Measures on Microclimate and Pedestrian Comfort in a Dense Urban District of Lebanon. *Sustain. Cities Soc.* **2020**, *61*, 102375. [CrossRef]
33. Faragallah, R.N.; Ragheb, R.A. Evaluation of Thermal Comfort and Urban Heat Island through Cool Paving Materials Using ENVI-Met. *Ain Shams Eng. J.* **2022**, *13*, 101609. [CrossRef]
34. Tsoka, S.; Tsikaloudaki, A.; Theodosiou, T. Analyzing the ENVI-Met Microclimate Model's Performance and Assessing Cool Materials and Urban Vegetation Applications–A Review. *Sustain. Cities Soc.* **2018**, *43*, 55–76. [CrossRef]
35. Abdallah, A.S.H.; Hussein, S.W.; Nayel, M. The Impact of Outdoor Shading Strategies on Student Thermal Comfort in Open Spaces between Education Building. *Sustain. Cities Soc.* **2020**, *58*, 102124. [CrossRef]
36. Rui, L.; Buccolieri, R.; Gao, Z.; Gatto, E.; Ding, W. Study of the Effect of Green Quantity and Structure on Thermal Comfort and Air Quality in an Urban-like Residential District by ENVI-Met Modelling. *Build. Simul.* **2019**, *12*, 183–194. [CrossRef]
37. Salata, F.; Golasi, I.; de Lieto Vollaro, R.; de Lieto Vollaro, A. Urban Microclimate and Outdoor Thermal Comfort. A Proper Procedure to Fit ENVI-Met Simulation Outputs to Experimental Data. *Sustain. Cities Soc.* **2016**, *26*, 318–343. [CrossRef]
38. Vos, P.E.J.; Maiheu, B.; Vankerkom, J.; Janssen, S. Improving Local Air Quality in Cities: To Tree or Not to Tree? *Environ. Pollut.* **2013**, *183*, 113–122. [CrossRef]
39. Jing, L.; Liang, Y. The Impact of Tree Clusters on Air Circulation and Pollutant Diffusion-Urban Micro Scale Environmental Simulation Based on ENVI-Met. *IOP Conf. Ser. Earth Environ. Sci.* **2021**, *657*, 12008. [CrossRef]
40. Cortes, A.; Rejuso, A.J.; Santos, J.A.; Blanco, A. Evaluating Mitigation Strategies for Urban Heat Island in Mandaue City Using ENVI-Met. *J. Urban Manag.* **2022**, *11*, 97–106. [CrossRef]
41. Yang, J.; Hu, X.; Feng, H.; Marvin, S. Verifying an ENVI-Met Simulation of the Thermal Environment of Yanzhong Square Park in Shanghai. *Urban For. Urban Green.* **2021**, *66*, 127384. [CrossRef]
42. Simon, H.; Lindén, J.; Hoffmann, D.; Braun, P.; Bruse, M.; Esper, J. Modeling Transpiration and Leaf Temperature of Urban Trees—A Case Study Evaluating the Microclimate Model ENVI-Met against Measurement Data. *Landsc. Urban Plan.* **2018**, *174*, 33–40. [CrossRef]
43. Šimek, M.; Elhottová, D.; Klimeš, F.; Hopkins, D.W. Emissions of N2O and CO2, Denitrification Measurements and Soil Properties in Red Clover and Ryegrass Stands. *Soil Biol. Biochem.* **2004**, *36*, 9–21. [CrossRef]
44. Trochta, J.; Kr\uuček, M.; Vrška, T.; Král, K. 3D Forest: An Application for Descriptions of Three-Dimensional Forest Structures Using Terrestrial LiDAR. *PLoS ONE* **2017**, *12*, e0176871. [CrossRef] [PubMed]
45. Jacobs, M.; Rais, A.; Pretzsch, H. Analysis of Stand Density Effects on the Stem Form of Norway Spruce Trees and Volume Miscalculation by Traditional Form Factor Equations Using Terrestrial Laser Scanning (TLS). *Can. J. For. Res.* **2020**, *50*, 51–64. [CrossRef]
46. Weiser, H.; Winiwarter, L.; Anders, K.; Fassnacht, F.E.; Höfle, B. Opaque Voxel-Based Tree Models for Virtual Laser Scanning in Forestry Applications. *Remote Sens. Environ.* **2021**, *265*, 112641. [CrossRef]
47. Gao, X.; Li, C.; Cai, Y.; Ye, L.; Xiao, L.; Zhou, G.; Zhou, Y. Influence of Scale Effect of Canopy Projection on Understory Microclimate in Three Subtropical Urban Broad-Leaved Forests. *Remote Sens.* **2021**, *13*, 3786. [CrossRef]
48. Baloloy, A.; Cruz, J.A.; Sta Ana, R.R.; Blanco, A.; Lubrica, N.V.; Valdez, C.J.; Bernardo, J.J. Modelling and Simulation of Potential Future Urbanization Scenarios and Its Effect on the Microclimate of Lower Session Road, Baguio City. *ISPRS Ann. Photogramm. Remote Sens. Spat. Inf. Sci.* **2020**, *5*, 187–194. [CrossRef]
49. López-Cabeza, V.P.; Galán-Marín, C.; Rivera-Gómez, C.; Roa-Fernández, J. Courtyard Microclimate ENVI-Met Outputs Deviation from the Experimental Data. *Build. Environ.* **2018**, *144*, 129–141. [CrossRef]
50. Krayenhoff, E.S.; Christen, A.; Martilli, A.; Oke, T.R. A Multi-Layer Radiation Model for Urban Neighbourhoods with Trees. *Boundary-Layer Meteorol.* **2014**, *151*, 139–178. [CrossRef]

51. Liu, Z.; Zheng, S.; Zhao, L. Evaluation of the ENVI-Met Vegetation Model of Four Common Tree Species in a Subtropical Hot-Humid Area. *Atmosphere* **2018**, *9*, 198. [CrossRef]
52. Hu, Z.; Yu, B.; Chen, Z.; Li, T.; Liu, M. Numerical Investigation on the Urban Heat Island in an Entire City with an Urban Porous Media Model. *Atmos. Environ.* **2012**, *47*, 509–518. [CrossRef]
53. Acero, J.A.; Arrizabalaga, J. Evaluating the Performance of ENVI-Met Model in Diurnal Cycles for Different Meteorological Conditions. *Theor. Appl. Climatol.* **2018**, *131*, 455–469. [CrossRef]
54. Li, X.; Zhou, W.; Ouyang, Z. Relationship between Land Surface Temperature and Spatial Pattern of Greenspace: What Are the Effects of Spatial Resolution? *Landsc. Urban Plan.* **2013**, *114*, 1–8. [CrossRef]
55. Rahman, M.A.; Hartmann, C.; Moser-Reischl, A.; von Strachwitz, M.F.; Paeth, H.; Pretzsch, H.; Pauleit, S.; Rötzer, T. Tree Cooling Effects and Human Thermal Comfort under Contrasting Species and Sites. *Agric. For. Meteorol.* **2020**, *287*, 107947. [CrossRef]
56. Terando, A.J.; Youngsteadt, E.; Meineke, E.K.; Prado, S.G. Ad Hoc Instrumentation Methods in Ecological Studies Produce Highly Biased Temperature Measurements. *Ecol. Evol.* **2017**, *7*, 9890–9904. [CrossRef]
57. Nakamura, R.; Mahrt, L. Air Temperature Measurement Errors in Naturally Ventilated Radiation Shields. *J. Atmos. Ocean. Technol.* **2005**, *22*, 1046–1058. [CrossRef]
58. Erell, E.; Leal, V.; Maldonado, E. Measurement of Air Temperature in the Presence of a Large Radiant Flux: An Assessment of Assively Ventilated Thermometer Screens. *Boundary-Layer Meteorol.* **2005**, *114*, 205–231. [CrossRef]
59. Grimmond, C.S.B.; Oke, T.R. Turbulent Heat Fluxes in Urban Areas: Observations and a Local-Scale Urban Meteorological Parameterization Scheme (LUMPS). *J. Appl. Meteorol.* **2002**, *41*, 792–810. [CrossRef]
60. He, C.; Zhou, L.; Yao, Y.; Ma, W.; Kinney, P.L. Cooling Effect of Urban Trees and Its Spatiotemporal Characteristics: A Comparative Study. *Build. Environ.* **2021**, *204*, 108103. [CrossRef]
61. Liu, H.; Lim, J.Y.; Wint Hnin Thet, B.; Lai, P.Y.; Koh, W.S. Evaluating the Impact of Tree Morphologies and Planting Densities on Outdoor Thermal Comfort in Tropical Residential Precincts in Singapore. *Build. Environ.* **2022**, *221*, 109268. [CrossRef]
62. Skelhorn, C.; Lindley, S.; Levermore, G. The Impact of Vegetation Types on Air and Surface Temperatures in a Temperate City: A Fine Scale Assessment in Manchester, UK. *Landsc. Urban Plan.* **2014**, *121*, 129–140. [CrossRef]
63. Petri, A.C.; Wilson, B.; Koeser, A. Planning the Urban Forest: Adding Microclimate Simulation to the Planner's Toolkit. *Land use policy* **2019**, *88*, 104117. [CrossRef]
64. Kong, F.; Yan, W.; Zheng, G.; Yin, H.; Cavan, G.; Zhan, W.; Zhang, N.; Cheng, L. Retrieval of Three-Dimensional Tree Canopy and Shade Using Terrestrial Laser Scanning (TLS) Data to Analyze the Cooling Effect of Vegetation. *Agric. For. Meteorol.* **2016**, *217*, 22–34. [CrossRef]
65. Wang, J.; Guo, W.; Wang, C.; Yao, Y.; Kou, K.; Xian, D.; Zhang, Y. Tree Crown Geometry and Its Performances on Human Thermal Comfort Adjustment. *J. Urban Manag.* **2021**, *10*, 16–26. [CrossRef]
66. de Abreu-Harbich, L.V.; Labaki, L.C.; Matzarakis, A. Effect of Tree Planting Design and Tree Species on Human Thermal Comfort in the Tropics. *Landsc. Urban Plan.* **2015**, *138*, 99–109. [CrossRef]
67. Shahidan, M.F.; Shariff, M.K.M.; Jones, P.; Salleh, E.; Abdullah, A.M. A Comparison of *Mesua ferrea* L. and *Hura crepitans* L. for Shade Creation and Radiation Modification in Improving Thermal Comfort. *Landsc. Urban Plan.* **2010**, *97*, 168–181. [CrossRef]

forests

MDPI

Article

Daily Dynamics of Soil Heat Flux and Its Relationship with Net Radiation in Different Urban Riparian Woodlands

Anze Liang [1], Changkun Xie [2,*], Jing Wang [2] and Shengquan Che [2,3]

1 School of Agriculture and Biology, Shanghai Jiao Tong University, Shanghai 200240, China
2 School of Design, Shanghai Jiao Tong University, Shanghai 200240, China
3 Shanghai Engineering Research Center of Sustainable Plant Innovation, Shanghai 200231, China
* Correspondence: xiechangkun@sjtu.edu.cn

Abstract: Soil heat flux (*G*) not only affects the Earth's surface energy balance but also models of calculating soil evaporation. A better understanding on the effect of timing, soil and vegetation on riparian *G* helps to improve energy balance closure and *G* simulation in riparian areas with various woodlands. This paper examined diurnal and seasonal variation patterns of soil heat flux in urban riparian areas, together with its relationship with net radiation (*Rn*) including midday *G/Rn* and the hysteresis phenomenon under the mutual influence of the timing, soil wetness and vegetation conditions. Study sites lie in the riparian areas of Shanghai with seven vegetation-covered conditions—grassland (C_H), broadleaf evergreen woodlands with shrubs (C_{CO}), broadleaf evergreen woodlands (C_{CH}), broadleaf deciduous woodlands with shrubs (C_{UO}), broadleaf deciduous woodlands (C_{UH}), conifer with shrubs (C_{MO}) and conifer (C_{MH}). Hourly data of *Rn* and *G* on typical days in four seasons starting from 11/2020 to 10/2021 were obtained with automated data-logging sensors. Diurnal variations in soil heat flux were characterized as two patterns depending on leaf area index (*LAI*)—unimodal curves followed cycles of *Rn* in woodlands with low *LAI* (C_{CH}, C_{CO}, C_H and C_{UO}) and sinusoidal ones in woodlands with high *LAI* (C_{MO}, C_{MH} and C_{UH}). Midday *G*/*Rn* was generally no more than 10% with slight variations in most woodlands across the four seasons, but upward trends in the grass and C_{UO} were observed in the afternoon. They were found significantly correlated with *SWC*. For sparse-canopied riparian sites, hourly *G* was found to be significantly correlated with *Rn* and *SWC* in summer, whereas, for dense sites, the role of canopy characteristics overwhelmed soil properties. Equations were derived to estimate diurnal *G* from *Rn*, *SWC* and *LAI*. The *G* of all riparian sites was subject to hysteresis problems to *Rn*. Phase shifts ranged from one to eight hours in riparian sites and were positively related with *LAI* and *SWC*, mainly accounting for the second diurnal pattern of *G*.

Keywords: urban riparian woodlands; soil heat flux; phase shift; urban microclimate

Citation: Liang, A.; Xie, C.; Wang, J.; Che, S. Daily Dynamics of Soil Heat Flux and Its Relationship with Net Radiation in Different Urban Riparian Woodlands. *Forests* **2022**, *13*, 2062. https://doi.org/10.3390/f13122062

Academic Editors: Thomas Rötzer, Stephan Pauleit, Mohammad A Rahman and Astrid Reischl

Received: 17 October 2022
Accepted: 29 November 2022
Published: 4 December 2022

1. Introduction

The microclimatic characteristics of one ecosystem depends on a series of energy exchanging and partitioning from net radiation (*Rn*) into sensible and latent heat, as well as soil heat [1]. Soil heat flux (*G*) indicates energy exchanges between the soil surface and subsoil during a given time period [2]. It determines how fast soil temperature changes and influences the rates of chemical and biological processes in the soil, which are essential to plant growth. As a key component of the surface energy balance, *G*, despite being typically smaller than sensible and latent heat flux (*H* and *LE*), not only plays an important role in energy balance closure [3] but also interacts energy transfer processes at the surface (surface energy balance) with energy transfer processes in the soil (soil thermal regime). In that case, information on soil heat variations exerts significant implications for microclimate maintenance and micro-habitats construction, which, in turn, affects the performance and management strategies of plant communities [4–6].

G was parameterized as a constant proportion of *Rn* that is fixed for a period of interest in some studies [2,7]. However, it is neither constant nor negligible on diurnal timescales, particularly for sites of sparse vegetation [8]. The achievement of accurate variations in *G* is important for measurements and analysis with the Bowen ratio energy balance approach, since it depends on the accurate value of available energy (*Rn*-*G*). A number of studies have resulted in a wealth of data concerning *G* on different temporal scales, either as a parameter alone or as one of the energy balance components. The direction and magnitude of *G* and its relationship with *Rn* are proven to be closely related to aboveground vegetation properties [9–11], water availability (e.g., soil moisture), topographic features and macroclimate conditions (e.g., the time of a day). For example, *G* in the mosses of high northern latitudes was found to be 57% lower in summer [12]. Studies on annual crops prove changes in vegetation covers can greatly influence the relative magnitude of energy flux components as a response to differences in morphological canopy attributes of various plant species [8,13]. Payero et al. proposed that relationships between *G* and *Rn* in grass and alfalfa depended on plant canopy height [8]. Canopy characteristics (e.g., leaf area index) have a great influence on the amount of radiation that soil surface can receive. In addition, a previous study observed differences in *G* values due to soil moisture that can be affected by differences in vegetation physiological activities (i.e., evapotranspiration and root water uptake) [14]. Traditionally, the canopied woodland was regarded as a whole to make comparisons of vegetation types, mainly between forests and grassland [15,16]. However, although the role of vegetation characteristics in the soil heat flux and its relationship with net radiation have been increasingly noticed [17], relatively few efforts have been made to characterize the impact of their specific morphologies and structures on soil heat and relationships of soil heat flux to net radiation.

Riparian areas are characterized by reciprocal exchanges on energy occurring between terrestrial and aquatic systems. Different from other ecosystems, the presence of rivers in these areas substantially affects the energy partitioning into soil heat [18], and the relationship between soil heat and net radiation. Concerning soil heat, a wide variety of homogeneously covered surfaces, including deserts, grasslands, crops and orchards, were researched [6], but little attention is given to forests, especially riparian forests, and their inter-site differences under different vegetation indices and soil moisture. It is well acknowledged that the cool and moist microclimate that riparian areas maintain makes these areas function as valuable habitats for distinctive flora and fauna communities. Considering this, an enhanced understanding on the spatial–temporal variabilities of *G* and its relationships with *Rn* in vegetation-covered riparian areas may theoretically provide insights for energy balance in riparian areas and practically contribute to urban riparian forestry management, such as species selection, plant community configuration, pruning and thinning strategies of woodlands to maintain the riparian microclimate and soil thermal regimes.

G influences soil temperature regime and evapotranspiration at a daily scale. In addition, it had a negligible impact on the annual energy balance [19], so only soil heat flux at the daily scale are discussed. This work was conducted in the riparian areas of Shanghai, a typical metropolis located within a riverine network. Through approximately one-year field monitoring and measurement of soil heat flux in seven different riparian sites, the objective of this study was to quantify the diurnal variabilities in riparian *G* and analyze its relationship with *Rn* (midday *G/Rn* and phase shifts of *G* to *Rn*) over a range of vegetation indices and to identify the effect of timing, vegetation and soil wetness. We mainly addressed three questions: (1) What are the diurnal variability patterns of *Rn* and *G* in each riparian site during different seasons? (2) What are relationships between *G* and *Rn* (midday *G*/*Rn* and phase shifts of *G* to *Rn*) in each site? (3) What is the potential influencing mechanism for variations in *G* and its relationships with *Rn*? This work ought to be of value for the theoretical understanding of energy partitioning into soil heat under different vegetation-covered riparian sites. It also provides practical and technical guidance for the planning and management strategies of urban riparian woodlands.

2. Methods

2.1. Site Description

Experimental sites (121°44′ N, 31°03′ W) were chosen along the north–south-running Danshui River (Figure 1), an approximately 20–30 m wide river in Minhang District of Shanghai. The climate of Shanghai is classified as north subtropical monsoon climate with four distinct seasons, with moist and cool winter. Annual air temperature is 17.6 °C and rainfall is approximately 1100 mm. Seven sites (20 m × 20 m) were simultaneously sampled in the western reaches, with consistent open surrounding environment and uncompressed traffic. All revetments are impervious though; they are lower than the ground surface of terrestrial land.

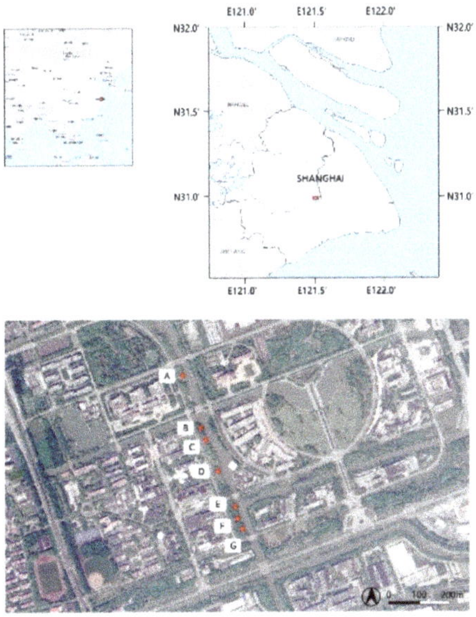

Figure 1. Location of 7 riparian experimental sites along Danshui River in Minhang District, Shanghai—C_{CO} (A), C_{MO} (B), C_H (C), C_{CH} (D), C_{UO} (E), C_{MH} (F) and C_{UH} (G).

Here, vertical structural characteristics of the forest floor are categorized into three types, regarding their layers—grass, arbor-grass and arbor–shrub–grass. Horizontally, we mainly categorized into three groups for arbored communities—evergreen broadleaves, deciduous broadleaves and conifers. In addition, an open grassland under regular mowing and maintenance was chosen to be a contrast. Based on the frequency of arbor species planted in urban Shanghai, seven riparian sites were chosen, and they are plant communities of (1) Camphor (*Cinnamomum camphora*), Osmanthus (*Osmanthus fragrans* (Thunb.) Lour.) and Creeping Woodsorrel (*Oxalis corniculate* L.), labelled as C_{CO}; (2) Dawn Redwood (*Metasequoia glyptostroboides*), Osmanthus and Dwarf Lilyturf (*Ophiopogon japonicus*), labelled as C_{MO}; (3) Japanese Lawn Grass (*Zoysia japonica*), labelled as C_H; (4) Camphor and Creeping Woodsorrel, labelled as C_{CH}; (5) Elm (*Ulmus parvifolia Jacq*), Osmanthus and Dwarf Lilyturf, labelled as C_{UO}; (6) Dawn Redwood and Japanese Lawn Grass, labelled as C_{MH} and (7) Elm and Creeping Woodsorrel, labelled as C_{UH}, respectively (Figure 1 and Table 1). Canopied communities are all artificially planted and aged more than 15 years. All plants in sites generally have a good growing conditions and maintenance.

Table 1. Plant communities in each riparian site along the Danshui River, Shanghai.

No.	1	2	3	4	5	6	7
Plant type	Evergreen broadleaf woodland +shrubland	Coniferous woodland +shrubland	Grassland	Evergreen broadleaf woodland +grassland	Deciduous woodland +shrubland	Coniferous woodland +grassland	Deciduous woodland +grassland
Plant type	C_{CO}	C_{MO}	C_{H}	C_{CH}	C_{UO}	C_{MH}	C_{UH}
DBH (cm)	26 ± 2.43	18.80 ± 2.95	/	26 ± 4.79	21 ± 7.23	17.50 ± 4.34	19 ± 1.87
Height (m)	9.76	12.80	0.20	9.80	9.60	13.00	9.04
Density (/hm^2)	375	1050	/	325	400	850	400
Crown ratio	59.6%	72.68%	/	49.45%	52.20%	74.91%	64.85%
Summer LAI	1.80	2.99	/	2.10	2.45	3.95	3.22
Site photos							

2.2. Experimental Design

Microclimatic variables, soil temperature (*Ts*), volumetric soil water content (*SWC*), net radiation (*Rn*) and soil heat flux (*G*), were monitored with automated data-logging sensors during the period from November, 2020 to October, 2021. *Ts*, *SWC* and *G* values were measured at three points (A1–3 in Figure 2) at each site, approximately 1, 6 and 11 m distance from the river, and averaged to be fully representative of the whole site (Figure 2, Table 2). Since *Rn* shows little spatial variability above complex and patchy surfaces [20], one net radiation radiometer was set up above each riparian plant community at 6 m to the river (B in Figure 2). Soil parameters were measured 3 cm under the surface. All measurements were taken at 10 min intervals. Instruments were inter-calibrated before the experiment and installed in the same batch. Note that net radiation towards the soil surface is defined as positive, and opposite for soil heat flux here. In addition, vegetation structure indexes, including plant height, density, crown ratio and summer leaf area index (*LAI*), were measured.

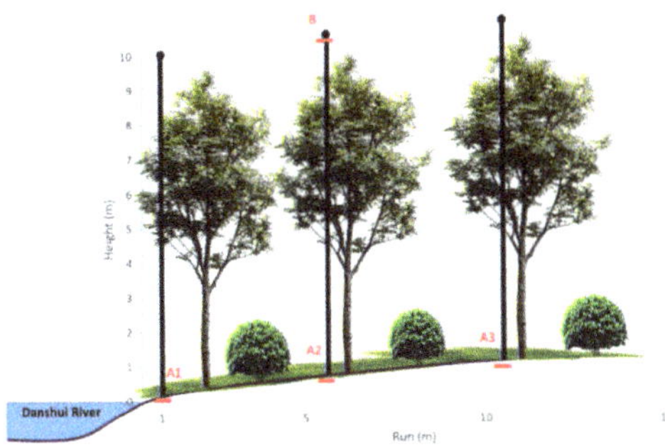

Figure 2. The schematic diagram of the south-facing slope of the study transect in the arbor–shrub–herbaceous community along the Danshui River. Approximate sensor locations are shown with red boxes. A1–A3 are three measuring points of soil heat flux, respectively, and B is the measuring point of net radiation.

Table 2. Environmental variables monitored and instruments applied during data collection.

Parameter	Instrument	Model	Accuracy	Measuring Height [1]
Net radiation	Net radiometer	QT-1	±5%	10/12 m
Soil temperature	Soil temperature and humidity probes	JXBS-3001	±0.5 °C	−0.2 m
Soil humidity			±3%	−0.2 m
Soil heat flux	Soil heat flux plate	HFP01	<0.1%/°C	−0.2 m

[1] Measuring height is 10 m for broadleaf woodlands and 12 m for conifers.

2.3. Data Analysis

2.3.1. Data Control

Data of three typical sunny days (three days before which no rain was observed) in spring (from March to May, 2021), summer (from June to August, 2021), autumn (from September to November, 2020) and winter (from December, 2020 to February, 2021) were chosen, respectively. All data were applied by data quality control, including performing a de-spiking process to screen outlying measurements [21]. Missing and screened meteorological data during observation periods took account of 3.00%. For gap filling, each gap was a single ten-minute gap and was replaced with the average of two observation values with two adjacent time points. The remaining data were examined to ensure it followed typical daily patterns. After that, all data were compiled into 1 h average to smooth the random errors and higher-frequency fluctuations [22,23].

2.3.2. Data Analysis

Daily variation patterns of *Rn* and *G* and midday *G/Rn* in four seasons were exhibited to identify the dynamic variations in time–energy relationships in line charts. The contour plot was used to illustrate a three-dimensional surface by plotting summer *Ts* and *SWC* against *G* as the independent variable. Relationship between *SWC* and midday *G/Rn* was analyzed with linear regression. Linear and polynomial regression models were fitted for *Ts*, *SWC* and *G*, respectively. These analyses were visualized by Origin (Origin Lab, 2019). For the analysis of the relationship between *SWC* and midday *G*/*Rn* and phase shift of *G* to *Rn*, data were normalized first with Equation (1), before linear fitting was performed. Pearson correlation analysis was performed to correlate *Rn* and *G* at different lagged period conditions, and to correlate lagged time with daily *SWC* and *LAI* for each riparian site, respectively. The normality of all data was checked using Kolmogorov–Smirnov's test prior to the variance analysis. Additionally, hierarchical cluster analysis was applied to identify possible groups of plant communities regarding daily changing amplitude of *Rn* and *G* in four seasons, respectively. All above analysis were performed with SPSS 24.0 (IBM SPSS, Chicago, IL, USA).

$$x^{*}=\frac{x-x_{min}}{x_{max}-x_{min}}, \tag{1}$$

3. Results

3.1. Diurnal and Seasonal Variations in Rn

Above-canopy net radiation is the determinant of the natural energy processes in riparian areas. Its diurnal variation patterns at sunny days in four seasons are shown in Figure 3. They varied greatly during the whole day but were mostly concentrated on the daytime (approximately starting from 0800 h to 1500 h local time). Regardless of season changes and vegetation types, all variations exhibited as unimodal curves, with peak values occurring around the noon. Daytime peak Rn ranged from 383.01 W/m^2 to 1001.59 W/m^2 in spring, from 223.61 W/m^2 to 919.6 W/m^2 in summer, from 167.64 W/m^2 to 1109.16 W/m^2 in autumn and from 241.77 W/m^2 to 1033.72 W/m^2 in winter, respectively. All nighttime Rn fluctuated around 0 W/m^2 approximately.

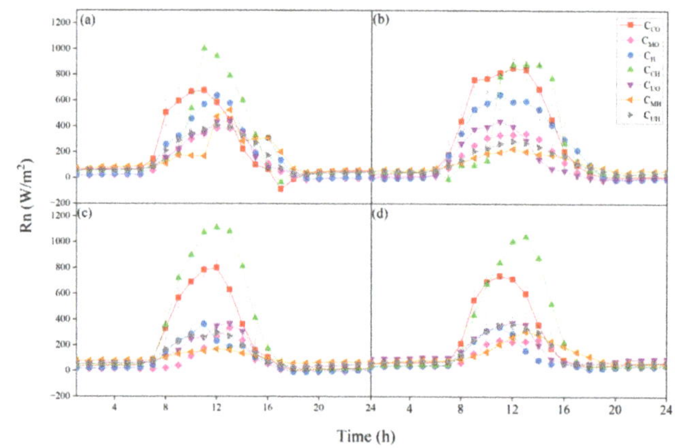

Figure 3. Diurnal variations in net radiation (*Rn*) above seven riparian communities in spring (**a**), summer (**b**), autumn (**c**) and winter (**d**). The bars show the standard deviations of *Rn*.

The inter-site distinctions of daytime *Rn* were observed, mainly in two aspects—daytime peak values and the duration of which net radiation was positive. For peaks of *Rn*, evergreen broadleaf woodlands always received the highest daily net radiation in each season, with maximum values ranging from 652.29 W/m^2 to 1109.16 W/m^2, followed by the open grassland (from 359.75 W/m^2 and 645.62 W/m^2), and then conifers and deciduous broadleaf woodlands. No significant differences were found between coniferous and deciduous evergreen woodlands. For the time period of positive net radiation, all-day *Rn* of C_{CH}, C_{UO}, C_{MH} and C_{UH} was always positive in four seasons. However, for C_{CO}, C_{MO} and C_{H}, their Rn occurred as negative after sunset.

Considering the consistent daily variations among seven riparian plant communities, hierarchical cluster analysis was performed to recognize the groups of plant communities, regarding changing ranges of *Rn* values (ΔRn, Rn_{max}-Rn_{min}, Table 3), indicating the stability of the change in energy that the ecosystem receives. Seven riparian plant communities could be categorized into three groups in spring and summer-I: C_{CH}, II: C_{CO} and C_{H}, III: C_{MO}, C_{UO}, C_{MH} and C_{UH}; three groups in autumn-I: C_{CH} and C_{CO}, II: C_{H}, III: C_{MO}, C_{UO}, C_{MH} and C_{UH}; two groups in winter-I: C_{CH} and C_{CO}, II: C_{H}, C_{MO}, C_{UO}, C_{MH} and C_{UH}.

Table 3. Daily ranges of *Rn* (ΔRn, W/m^2) of seven riparian plant communities during four seasons.

ΔRn (W/m^2)	C_{CH}	C_{CO}	C_{H}	C_{UO}	C_{MO}	C_{UH}	C_{MH}
Spring	1035.72	767.45	642.6	420.08	400.82	377.25	484.88
Summer	898.40	881.83	662.93	462.3	359.05	257.38	181.17
Autumn	1099.93	796.17	372.32	348.17	342.85	266.17	111.77
Winter	981.98	708.63	328.73	318.54	228.68	312.75	258.84

Note: The grid color indicates the groups of plant communities—pink for Group I, grey for Group II and blue for Group III.

In addition, each site showed similar seasonal variation trends in *Rn* but seasonal differences existed in its ranges, mainly in the maximum *Rn*, rather than the minimum one. The *Rn* achieved by the grassland in spring and summer (636.7 and 645.62 W/m^2, respectively) was approximately twice that in autumn and winter (363.98 and 359.75 W/m^2, respectively). For woodlands, *Rn* in spring was higher than that in summer. Downwards shortwave radiation was far higher than upwards longwave radiation due to low ground surface temperature, while for summer, higher upwards longwave radiation could offset downwards shortwave radiation because of high ground surface temperature. For evergreen broad-leaved woodlands (C_{CH} and C_{CO}), litterfall of Camphor mostly occurs in spring, in which *Rn* tended to be lower than those in other seasons. Small differences between *Rn* results were observed in summer, autumn and winter for evergreen woodlands. On the contrary, for deciduous woodlands, *Rn* was higher in the spring than in other seasons.

3.2. Diurnal and Seasonal Variations in G

Daily *G* variations in different riparian sites at sunny days of each season are shown in Figure 4. The 24 h *G* values in each season for all riparian woodlands were positive, indicating that the soil absorbing heat throughout the whole daily cycle during four seasons. Additionally, soil surfaces in all canopied communities function as heat sink. However, owing to radiative cooling at the ground surface, the release of heat from the soil resulted into negative *G* values at night in the C_{H}, and soil worked as a heat source during this period.

Daily variations could be mainly grouped into two patterns. The first one is that fluctuations in *G* varied diurnally in the same pattern as corresponding *Rn* but with several slight phase differences in each season, exhibiting as unimodal curves (Figure 4a–d). They generally increased during the morning, reaching maximum values before they decreased in the afternoon. This pattern was applied to riparian woodlands with lower *LAI*-C_{CH}, C_{CO}, C_{H} and C_{UO} (Table 1), while in those with higher *LAI*- C_{MO}, C_{MH} and C_{UH}, the second variation pattern was observed. Their diurnal variations were assumed to be sinusoidal curves, with a decrease in the morning towards minimum values at 1000 local time and then gradually increasing to maximum values in the late afternoon before a second decrease (Figure 4e–h). For both patterns, soil heat flux and net radiation were observed to vary in terms of synchronization with the 24 h daily cycle at all sites. However, the phase shifts vary among riparian sites, resulting in different daily variation curves (details discussed later).

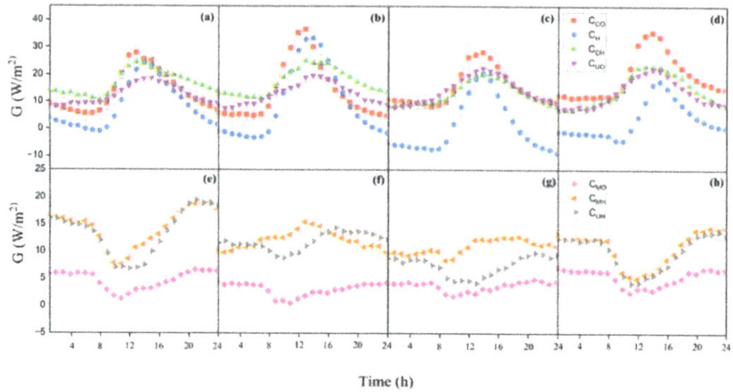

Figure 4. Diurnal variations in soil heat fluxes (G) in seven riparian communities, in spring (**a**,**e**), summer (**b**,**f**), autumn (**c**,**g**) and winter (**d**,**h**). The bars show the standard deviations.

Higher maximum values of G (G_{max}) and wider amplitudes of G variations (ΔG) were found in the first pattern than those in the second one. G_{max} in the C_{CO} were the highest all year round, with maximum values of 41.50, 36.56, 28.33 and 35.68 W/m^2 in spring, summer, autumn and winter, respectively, while G_{max} in the C_{MO} were lowest at 6.63, 4.30, 4.95 and 7.12 W/m^2, respectively. Similarly, hourly G for a moist soil beneath plant canopies was often found to be less than 20 W/m^2 [5]. G values in evergreen broadleaf communities were the highest, followed by deciduous communities and conifers. For broadleaf communities with shrubs in the understory layer, both G_{max} and ΔG were found larger than those in corresponding communities without shrubs—$C_{CO} > C_{CH}$ and $C_{UO} > C_{UH}$. However, in coniferous communities, the roles of shrublands were not obvious in influencing the magnitude of G. For G_{min}, G values in all canopied communities were positive, except that the nighttime G in the grassland dropped below zero (especially during autumn and winter). G values in the C_H could go towards −0.90, −3.12, −8.73 and −4.26 W/m^2 at night in spring, summer, autumn and winter, respectively.

G in the open grassland exhibited seasonal differences. It was approximately double in summer of what in winter. Similarly, findings were reported in a grassland area of the Netherlands by Jacobs et al. [24]. For canopied woodlands with shrubs (C_{CO}, C_{MO} and C_{UO}), no significant seasonal differences were found, while for those without shrubs (C_{CH}, C_{MH} and C_{UH}), G in spring and summer was found to be higher than that in autumn and winter.

Although diurnal variations in conifers were found to be different from some studies on grassland and crops, their sinusoidal G variations, with minimum values occurring in the morning and maximum in the late afternoon, were similar to the "S-shape" of diurnal variations in G found in the *Robinia Pseudoacacia* plantation in the Yellow River Delta and tropical forests in Guangdong, China [25,26]. Less amplitudes found in coniferous woodlands than those in the grassland and broadleaf woodlands indicate less intensity of soil heat energy, mainly attributing to morphological features of conifers.

Correlation analysis between summer G and SWC showed that G was highly related with soil moisture in sparse-canopied riparian woodlands and grassland ($LAI < 2.45$). Since they were found with consistent variation patterns, C_{CH} was chosen here to show their specific relationships as an example (Figure 5). Most soil heat transmissions occurred under relatively high soil water content, whereas higher peaks were found in Ts higher than 28.5 °C and a medium range of SWC (Figure 5a). The linear regression model in Figure 5b indicated a significant positive relationship between Ts and G ($r^2 = 0.91$, $p < 0.05$). A non-linear curve between SWC and G was obtained by the polynomial regression model (Figure 5c). G increased first and then decreased with the increase in water content in soil ($r^2 = 0.90$, $p < 0.05$). The optimal range of SWC was 32%–33% at the site of C_{CH}.

However, for dense-canopied (*LAI* > 2.99 in this study) woodlands—C_{MO}, C_{MH} and C_{UH}—no significant relationship between *SWC* and *G* was observed.

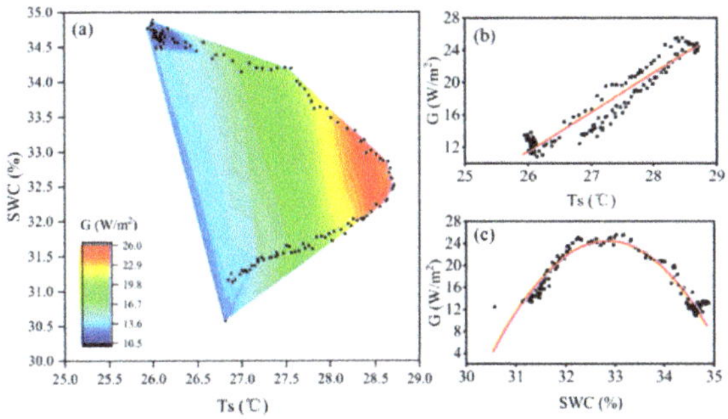

Figure 5. Relationships between soil temperature (*Ts*), soil water content (*SWC*) and soil heat flux (*G*) in the C_{CH} over a 24 h cycle in summer. (**a**) Contour plot showing relationships between *Ts*, *SWC* and *G* (n = 144). (**b**) Regression curve of *Ts* vs. *G*. Linear fitted regression of *Ts* and *G* (r^2 = 0.91, $p < 0.05$, n = 144). (**c**) Regression curve of *SWC* vs. *G*. Polynomial fitted regression of *SWC* and *G* (r^2 = 0.90, $p < 0.05$, n = 144).

3.3. Relationships between G and Rn

3.3.1. Midday Variations in G/Rn Ratios

The ratio of soil heat to net radiation (*G*/*Rn*) is indicative of the relative importance of soil heat to net radiation. It is important not only for simple modeling parameterizations but also for microclimate regulation and productivity promotion [27]. Although midday values of *G*/*Rn* are not representative of the entire diurnal cycle, considering variations in heat fluxes were mainly focused during the daytime, differences among variations in ratios of soil heat to net radiation in seven riparian sites were only evaluated for the midday figure (1000–1500 h, local time), as shown in Figure 6. Field observations in our study show that *G*/*Rn* is not always constant for some vegetated surfaces at midday. On the one hand, *G* was found to be accounted for a little proportion of *Rn* in all riparian sites, especially for those with high *LAI*. Midday *G*/*Rn* ranged from 0.07% to 31.60%, lower than that found in the desert ecosystems in which midday *G* typically ranged from 20% to 40% in terms of net radiation [28]. On the other hand, regardless of its daytime variation trends, maximum of *G*/*Rn* occurs at 1400 h local time or later, which is later than the maximum of *Rn* (see Figure 3).

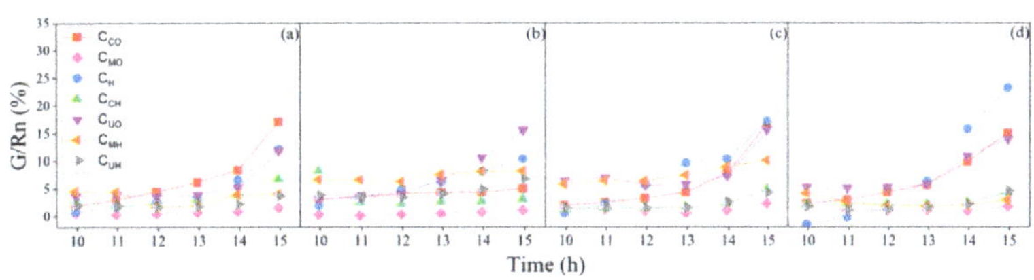

Figure 6. Midday (10–15 h) variations in ratios of soil heat flux to net radiation (*G*/*Rn*) in seven riparian communities in spring (**a**), summer (**b**), autumn (**c**) and winter (**d**).

For conifers (C_{MO} and C_{MH}) with dense canopies intercepting heat transmitting towards the ground, and broadleaf woodlands without shrubs (C_{UH} and C_{CH}) with high net radiation, soil heat accounts for little of *Rn* and, undoubtedly, variations in *G/Rn* were not obvious in each season. *G* accounted for <8.36% of *Rn* in these sites. Similarly, no more than 10% with slight diurnal variations for full canopied surface and <5% in forests were observed [29–31].

For grassland and broadleaf woodlands with shrubs (C_{CO}, C_H and C_{UO}), upward trends were observed. They generally maintained less than 10% until 1400 h local time and then increased quickly afterwards. This is because *Rn* in these sites decreased dramatically after 1400 h local time. However, G/Rn in the C_{CO} generally maintained at a stable rate in the summer, ranging from 3.06% to 5.03%. This could be ascribed to the notion that net radiation in the summer was too high to observe obvious variations. Furthermore, it was higher in winter and autumn (23.26% and 17.28%, respectively) than those in spring and summer (12.18% and 10.40%, respectively), within ranges of 10% to 50% that was confirmed in bare and sparsely covered soils [28]. Similarly, Meyers found nearly 25% of midday *Rn* was partitioned into *G* in a water-stressed watershed, while that figure was 15% for a non-stressed summer [32]. Midday G/Rn variabilities could partially ascribe to soil wetness. The relationship between normalized *SWC* and midday G/Rn of seven riparian sites in summer in this study indicated their well correlations (Figure 7, $r^2 = 0.64$, $p < 0.05$).

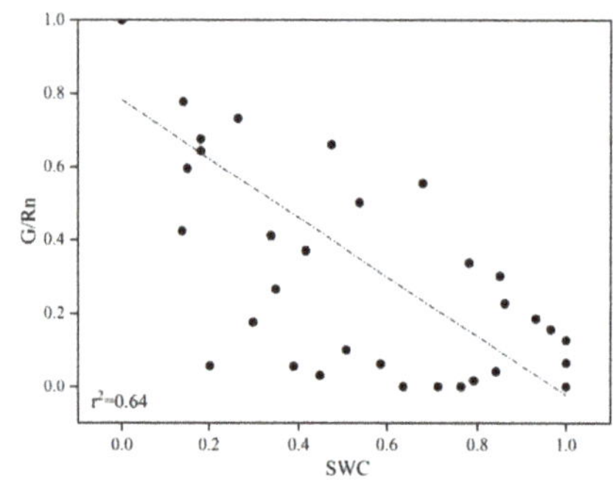

Figure 7. Relationships between normalized soil water content (*SWC*) and the normalized ratio of midday G/Rn in summer. Linear fitted regression of *SWC* and G/Rn ($r^2 = 0.64$, $p < 0.05$, $n = 42$).

3.3.2. Phase Shift of G to Rn

The linear analysis between the normalized data of hourly G and *Rn* values for different riparian communities in summer was conducted on a daily scale (Figure 8). It showed soil heat flux varied out of synchronization with net radiation, indicating that soil heat flux on a daily basis, to some extent, suffers from a hysteresis problem to varying degrees. For a given *Rn* in any riparian site, different G values occurred during the midnight to noon hours, compared to noon to midnight hours, since the diurnal G and *Rn* waves did not reach their peaks at the same time. This hysteresis phenomenon essentially comes from the existence of phase difference between the diurnal variations in net radiation and soil heat flux [33].

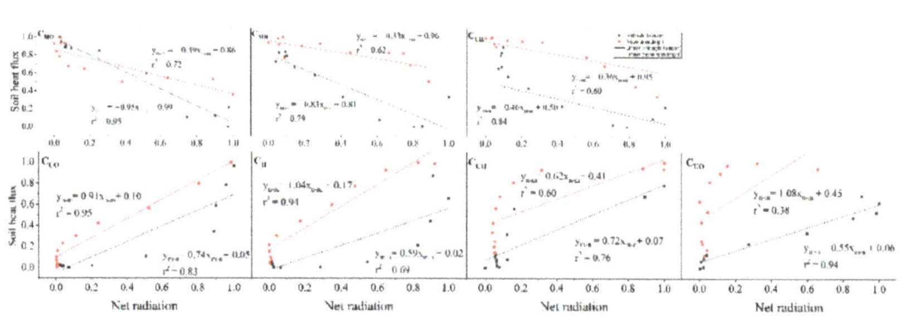

Figure 8. Relationship between normalized net radiation and soil heat flux for different riparian plant communities during sunny days in summer. Each point represents a half-hour average. The relationships between normalized net radiation and soil heat flux were statistically significant ($p < 0.05$) for all sites. G_{m-n} is G from midnight to noon, and G_{n-m} is G from noon to midnight.

To evaluate the extent of phase shift, a linear regression between net radiation and soil heat flux within different lagged durations, such as real-time G (G_0) and one-hour-lagged G (G_{1h}) until eight-hour-lagged G (G_{8h}), was conducted for each riparian site during summertime (Table 4). The r^2 values could be interpreted as indicative of the degree of hysteresis. The relationship between Rn and G_{nh} (n = 0, 1, 2, … 8) with the highest r^2 values represents the lagging duration of soil heat in this site.

Table 4. Correlations between summertime net radiation and soil heat flux at different lagged conditions for all riparian sites.

	G_0		G_{1h}		G_{2h}		G_{3h}		G_{4h}		G_{5h}		G_{6h}		G_{7h}		G_{8h}	
	r^2	p	r^2	p	r^2	p	r^2	p	r^2	p	r^2	p	r^2	p	r^2	p	r^2	p
C_{CO}	0.888	0.01	0.967	0.01	0.935	0.01	0.797	0.01	0.563	0.01	/		/		/		/	
C_H	0.759	0.01	0.913	0.01	0.986	0.01	0.957	0.01	0.817	0.01	0.567	0.05	/		/		/	
C_{CH}	0.75	0.01	0.823	0.01	0.817	0.01	0.705	0.01	0.552	0.01	/		/		/		/	
C_{UO}	0.393	0.01	0.58	0.01	0.747	0.01	0.874	0.01	0.923	0.01	0.86	0.01	0.674	0.01	/		/	
C_{MO}	−0.908	0.01	−0.846	0.01	−0.695	0.01	−0.476	0.05	/		/		/		0.604	0.05	0.797	0.01
C_{MH}	−0.671	0.01	−0.448	0.05	/		/		/		0.685	0.01	0.831	0.01	0.876	0.01	0.834	0.01
C_{UH}	−0.787	0.01	−0.564	0.01	/		/		/		0.6	0.01	0.796	0.01	0.894	0.01	0.906	0.01

Results showed differences in phase shift among different plant communities (Table 4). The shortest phase shift occurred in evergreen broadleaf woodlands (C_{CO} and C_{CH}), which lagged one hour (r^2 = 0.967 and 0.823, respectively, $p < 0.01$). Soil heat flux in the open grassland lagged two hours to net radiation ($r^2 = 0.986$, $p < 0.01$) and four hours in the C_{UO} ($r^2 = 0.923$, $p < 0.01$). However, it was found that conifers with shrubs lagged 8 h ($r^2 = 0.797$, $p < 0.01$) and 7 h in those without shrubs ($r^2 = 0.876$, $p < 0.01$). At the site of C_{MH}, soil heat flux also lagged 8 h ($r^2 = 0.906$, $p < 0.01$).

To evaluate the cause of phase shift, lagged time in each riparian site was correlated with LAI and daily SWC, respectively, via Pearson correlation analysis. Results indicated a positive relationship between lagged time and LAI ($r^2 = 0.857$, $p < 0.05$) and significantly positive relationship with daily SWC ($r^2 = 0.842$, $p \leq 0.01$), suggesting that the hysteresis problem was related to both soil wetness and canopy cover conditions.

4. Discussion

4.1. Factors Influencing Rn

Net radiation is influenced by the transmission of solar radiation and long-wave radiation in the layers of the plant canopy, residue, and soil [34]. In this study, the one-peaked daily behaviors of Rn in sunny days were dominated by the sun's daily path, quantified as solar zenith angle [35]. The observed diurnal variation curves above some sites were asymmetrical, probably because probes were not set completely horizontal. Diurnal variations in Rn for seven different riparian plant communities during four seasons showed all communities received energy as a result of the incoming shortwave radiation during the day, and a counterbalance reflected longwave radiation from the underlying and incoming longwave radiation from the sky at night.

Inter-site differences in Rn can be attributed to characteristics of the ground surface–soil structural features (soil color and porosity) and physical features (soil temperature and moisture). Evergreen woodland had the highest net radiation all-year-round, while C_{MH} got least. On the one hand, extensive and evergreen Camphor canopies form structural complexity, increasing aerodynamical roughness [36], forming a small surface albedo (α), and thus leading to a high gain of Rn. On the other hand, bare soil under plant communities characterized by an absence of foliage, is almost completely exposed, leading to high α, and thus low net radiation [18]. Heavy litters and dead biomass under deciduous woodlands and conifers, especially during autumn and winter, also have different spectral characteristics, resulting in high α and low Rn received [37,38].

4.2. Factors Influencing G and Its Relationships with Rn

For riparian grassland and sparse-canopied woodlands (C_{CO}, C_H, C_{CH} and C_{UO}), daily G varied with variations in Rn, consistent with many previous studies conducted in other ecosystems. Positive linear regression between normalized G and Rn could explain this variation pattern. For dense-canopied riparian woodlands (C_{MO}, C_{MH} and C_{UH}), the hysteresis phenomenon bringing phase shifts of G caused sinusoidal variations in G and non-linearly relationship between G and Rn.

Apart from *Rn*, soil water content was proven to influence soil heat flux, since it determines the hydraulic and thermal properties for heat transfer. Our study shows that in sparse-canopied riparian sites, soil wetness has dual effects on soil heat flux. *G* increased when soil moisture was lower and decreased as soil moisture continued to increase. However, in dense-canopied woodlands, insignificant relationships between them indicated that, in this case, the roles of vegetation characteristics overwhelm that of soil moisture. Similarly, Santanello and Friedl [33] proposed that a single relationship of *G* and *Rn* is sufficient for dense-covered surfaces ($LAI > 2.5$), regardless of soil conditions. Potential reasons might lie in species-dependent parameters, such as the height and size of the canopy, and specific crown attributes of trees. They also affect the amount, temporal or spatial variability of light [38], and radiation taken to reach the soil surface. For instance, hemispherical canopies of broadleaf trees favor soil receiving heat, and conic canopies together with a large canopy height of conifers weaken and delay solar radiation hitting the ground, explaining the second variation pattern of soil heat. Meanwhile, the phenological stages for deciduous trees, including foliage emergence and senescence, can alter the exchange properties between vegetation, the atmosphere and soil [20]. Litter layer under conifers, as a mulch at the soil surface, could be regarded as a heat reservoir [38]. Conifers and deciduous woodlands without shrub were covered with thick litter layers, even in summer, where a part of the heat storage term was not included in this work.

Based on two types of hysteresis, multiple regression analysis was performed to derive equations to estimate *G* from *Rn*, *SWC* and *LAI*. For grassland and sparse-canopied woodlands (C_{CO}, C_H, C_{CH} and C_{UO}), the analysis resulted in the following equation ($n = 96$, $r^2 = 0.735$):

$$G = 123.2 + 0.048Rn - 5.780SWC - 34.69LAI - 9.10\ LAI^2 - 0.016Rn * LAI + 2.794SWC * LAI \quad (2)$$

For dense-canopied woodlands (C_{MO}, C_{MH} and C_{UH}), the analysis resulted in the following equation ($n = 72$, $r^2 = 0.974$):

$$G = -853.4 + 0.2665Rn + 2.98SWC + 482.3LAI - 64.30LAI^2 - 0.006Rn * SWC - 0.035Rn * LAI - 1.119SWC * LAI \quad (3)$$

All terms included in these two equations were statistically significant ($p < 0.01$).

Concerning midday G/Rn ratio, it depends on the time of the day/year, soil properties, vegetation amount and height [33]. On the one hand, part of the G/Rn variability on hourly timescale arises from soil moisture. For less-canopied surfaces, such as deciduous woodlands and grassland, a significant fraction of the soil surface is exposed to radiation, resulting in the midday G/Rn taking on a larger range than conifers and evergreen woodlands when considering their growth cycle. This was due to phase shifts in G/Rn at diurnal timescales in our riparian sites. Increased soil heat and decreased net radiation during the late midday period led to an upward pattern of G/Rn ratio in this work. On the other hand, the roles of vegetation characteristics in influencing the magnitude of G/Rn were developed in previous studies with remote sensing [39]. The functions of G and *Rn* show that near-noon G/Rn ratio decreased with plant canopy, and the cover increased [8]. However, the compounding influences of soil moisture, soil types and phase shift in diurnal variations in G should be included in future work. In our study of full canopied riparian sites, the midday of G/Rn was less than 0.1 and works reasonably well as a constant value.

The relationship between *Rn* and G in each riparian site suffered with a hysteresis problem, with different extents ranging from 1 h lagged to 8 h lagged. A similar phenomenon was reported before either theoretical assumption [40] or by monitoring measurements. We found that soil wetness partially accounted for this phenomenon. Phase lag was calculated as $\pi/4$ for dry soil surfaces, but for moist soil it varied with soil wetness [40,41]. In addition, canopy characteristics were related to the hysteresis. Similarly, it is found that hysteresis changed during the growing cycle of Alfalfa [8]. Horizontally, *LAI* positively influenced the hysteresis. However, the role of the vertical layering of plant communities was not found in riparian sites. Lagged time was found 0.5 h in an alpine meadow and 1.5–2 h in paddy fields [6]. Additionally, the mutual influence of soil moisture and vegetated conditions could probably explain why phase shifts for each riparian woodland type were larger than those found in other ecosystems. Covered canopies brought hysteresis, but it was one-hour lagged in the grass, compared to the evergreen broadleaf woodlands. One possible reason might lie in the occasional shading from one conifer with a height of over 12 m next to the grass.

5. Conclusions

This study explored diurnal and seasonal variations in *G* and *Rn*; their relationships (midday *G/Rn* ratios and phase shifts of *G* to *Rn*); and the effect of timing, vegetation and soil wetness. In general, hourly soil heat flux on a daily scale changes with vegetation characteristics and soil moisture. It was categorized into two patterns—sparse-canopied plant communities with *LAI* < 2.45 (C_{CH}, C_{CO}, C_H and C_{UO}) followed the diel cycles of *Rn* exhibiting as unimodal curves, and dense-canopied woodlands with an *LAI* > 2.99 exhibited as sine curves, including the C_{MO}, C_{MH} and C_{UH}. A clear dependence of summertime *G* on soil water content was found for the first pattern—it increased first and then decreased with *SWC*.

Midday *G/Rn* in C_{MO}, C_{MH}, C_{UH} and C_{CH} were no more than 10% with slight variations. Upward trends in C_{CO}, C_H and C_{UO} were due to the phase shift of *G*, leading to increased *G* but meanwhile decreased *Rn*. One-hour *Rn* and *G* values in summer were linearly related. For all riparian sites, relationships between *G* and *Rn* suffered from hysteresis problems. Phase shifts between riparian *Rn* and *G* ranged from one to eight hours, and they were significantly correlated with *LAI* and soil moisture, mainly accounting for the second pattern of hourly *G* variations. The equations of diurnal changes in *G* as a function of *Rn*, *SWC* and *LAI* were obtained with multiple regression analysis. In summary, our work provides a supplemental understanding on the influence of vegetation indices and soil moisture on riparian soil heat flux. It is instructive for scientifical planning and management of riparian woodlands and provides favorable micro-habitats of riparian organisms.

Author Contributions: Conceptualization, A.L. and C.X.; methodology, A.L. and C.X.; software, A.L.; validation, A.L. and C.X.; formal analysis, A.L. and C.X.; investigation, J.W., A.L. and C.X.; resources, C.X. and S.C.; data curation, A.L.; writing—original draft preparation, A.L.; writing—review and editing, A.L. and C.X.; visualization, A.L.; supervision, C.X. and S.C.; project administration, C.X. and S.C.; funding acquisition, C.X. and S.C. All authors have read and agreed to the published version of the manuscript.

Funding: This research was funded by NATIONAL NATURAL SCIENCE FOUNDATION OF CHINA, grant number 31971712.

Data Availability Statement: Not applicable.

Conflicts of Interest: The authors declare no conflict of interest.

References

1. Sun, R.; Chen, L. How can urban water bodies be designed for climate adaptation? *Landsc. Urban Plan.* **2012**, *105*, 27–33. [CrossRef]
2. Zhu, W.; Wu, B.; Yan, N.; Feng, X.; Xing, Q. A method to estimate diurnal surface soil heat flux from MODIS data for a sparse vegetation and bare soil. *J. Hydrol.* **2014**, *511*, 139–150. [CrossRef]

3. Liu, X.; Yang, S.; Xu, J.; Zhang, J.; Liu, J. Effects of soil heat storage and phase shift correction on energy balance closure of paddy fields. *Atmósfera* **2017**, *30*, 39–52. [CrossRef]
4. Bianca, N.I.E.; Madsen, L.; Hagar, J.C.; Temesgen, H. Estimating Riparian Understory Vegetation Cover with Beta Regression and Copula Models. *For. Sci.* **2011**, *57*, 212.
5. Yue, P.; Zhang, Q.; Zhang, L.; Li, H.; Yang, Y.; Zeng, J.; Wang, S. Long-term variations in energy partitioning and evapotranspiration in a semiarid grassland in the Loess Plateau of China. *Agric. For. Meteorol.* **2019**, *278*, 107671. [CrossRef]
6. Hatfield, J.L.; Baker, J.M. *Micrometeorology in Agricultural Systems*; American Society of Agronomy: Madison, WI, USA, 2005; pp. 131–143.
7. Tanguy, M.; Baille, A.; González-Real, M.M.; Lloyd, C.; Cappelaere, B.; Kergoat, L.; Cohard, J.M. A new parameterisation scheme of ground heat flux for land surface flux retrieval from remote sensing information. *J. Hydrol.* **2012**, *454–455*, 113–122. [CrossRef]
8. Payero, J.O.; Neale, C.M.U.; Wright, J.L. Estimating soil heat flux for alfalfa and clipped tall fescue grass. *Appl. Eng. Agric.* **2005**, *3*, 401–409. [CrossRef]
9. Krishnan, P.; Meyers, T.P.; Scott, R.L.; Kennedy, L.; Heuer, M. Energy exchange and evapotranspiration over two temperate semi-arid grasslands in North America. *Agric. For. Meteorol.* **2012**, *153*, 31–44. [CrossRef]
10. Bryś, K.; Bryś, T.; Ojrzyńska, H.; Sayegh, M.A.; Głogowski, A. Variability and role of long-wave radiation fluxes in the formation of net radiation and thermal features of grassy and bare soil active surfaces in Wrocław. *Sci. Total Environ.* **2020**, *747*, 141192. [CrossRef]
11. Zhang, Y.; Zhao, W.; He, J.; Zhang, K. Energy exchange and evapotranspiration over irrigated seed maize agroecosystems in a desert-oasis region, northwest China. *Agric. For. Meteorol.* **2016**, *223*, 48–59. [CrossRef]
12. Jason Beringer, A.H.L.F. The Representation of Arctic Soils in the Land Surface Model: The Importance of Mosses. *J. Clim.* **2001**, *14*, 3324–3335. [CrossRef]
13. Hernandez-Ramirez, G.; Hatfield, J.L.; Prueger, J.H.; Sauer, T.J. Energy balance and turbulent flux partitioning in a corn–soybean rotation in the Midwestern US. *Theor. Appl. Climatol.* **2010**, *100*, 79–92. [CrossRef]
14. Tianjiao, F.; Dong, W.; Ruoshui, W.; Yixin, W.; Zhiming, X.; Fengmin, L.; Yuan, M.; Xing, L.; Huijie, X.; Caballero-Calvo, A.; et al. Spatial-temporal heterogeneity of environmental factors and ecosystem functions in farmland shelterbelt systems in desert oasis ecotones. *Agric. Water Manag.* **2022**, *271*, 107790. [CrossRef]
15. Drexler, J.Z.; Anderson, F.E.; Snyder, R.L. Evapotranspiration rates and crop coefficients for a restored marsh in the Sacramento–San Joaquin Delta, California, USA. *Hydrol. Process.* **2008**, *22*, 725–735. [CrossRef]
16. Welsh, J.H.H.; Hodgson, G.R.; Karraker, N.E. Influences of the vegetation mosaic on riparian and stream environments in a mixed forest-grassland landscape in "Mediterranean" northwestern California. *Ecography* **2005**, *28*, 537–551. [CrossRef]
17. Acharya, R.H.; Sigdel, M.; Ma, Y.; Wang, B. Diurnal and seasonal variation of heat fluxes over an agricultural field in southeastern Nepal. *Theor. Appl. Clim.* **2019**, *137*, 2949–2960. [CrossRef]
18. Masseroni, D.; Facchi, A.; Romani, M.; Chiaradia, E.A.; Gharsallah, O.; Gandolfi, C. Surface energy flux measurements in a flooded and an aerobic rice field using a single eddy-covariance system. *Paddy Water Environ.* **2015**, *13*, 405–424. [CrossRef]
19. Ma, J.; Zha, T.; Jia, X.; Tian, Y.; Bourque, C.P.A.; Liu, P.; Bai, Y.; Wu, Y.; Ren, C.; Yu, H.; et al. Energy and water vapor exchange over a young plantation in northern China. *Agric. For. Meteorol.* **2018**, *263*, 334–345. [CrossRef]
20. Wilson, K.B.; Baldocchi, D.D.; Aubinet, M.; Berbigier, P.; Bernhofer, C.; Dolman, H.; Falge, E.; Field, C.; Goldstein, A.; Granier, A.; et al. Energy partitioning between latent and sensible heat flux during the warm season at FLUXNET sites. *Water Resour. Res.* **2002**, *38*, 30–31. [CrossRef]
21. Dupont, S.; Patton, E.G. Influence of stability and seasonal canopy changes on micrometeorology within and above an orchard canopy: The CHATS experiment. *Agric. For. Meteorol.* **2012**, *157*, 11–29. [CrossRef]
22. Leuning, R.; Van Gorsel, E.; Massman, W.J.; Isaac, P.R. Reflections on the surface energy imbalance problem. *Agric. For. Meteorol.* **2012**, *156*, 65–74. [CrossRef]
23. Nelli, N.R.; Temimi, M.; Fonseca, R.M.; Weston, M.J.; Thota, M.S.; Valappil, V.K.; Branch, O.; Wizemann, H.; Wulfmeyer, V.; Wehbe, Y. Micrometeorological measurements in an arid environment: Diurnal characteristics and surface energy balance closure. *Atmos. Res.* **2020**, *234*, 104745. [CrossRef]
24. Scott, R.L.; Edwards, E.A.; Shuttleworth, W.J.; Huxman, T.E.; Watts, C.; Goodrich, D.C. Interannual and seasonal variation in fluxes of water and carbon dioxide from a riparian woodland ecosystem. *Agric. For. Meteorol.* **2004**, *122*, 65–84. [CrossRef]
25. Wang, X.; Zhou, G.; Zhang, D.; Wang, C. Soil heat fluxes of mixed coniferous and broad-leaf forest in the south subtropics in China. *Ecol. Environ.* **2005**, *14*, 260–265. (In Chinese)
26. Wang, X.; Liu, D.; Li, Y.; Wang, Z.; Ma, B. Soil Heat Fluxes of Larix gmelinii in Daxing'anling of Heilongjiang Province. *Prot. For. Sci. Technol.* **2020**, *52*, 88–93. (In Chinese)
27. Hossen, M.S.; Mano, M.; Miyata, A.; Baten, M.A.; Hiyama, T. Surface energy partitioning and evapotranspiration over a double-cropping paddy field in Bangladesh. *Hydrol. Process.* **2012**, *26*, 1311–1320. [CrossRef]
28. Kustas, W.P.; Prueger, J.H.; Hatfield, J.L.; Ramalingam, K.; Hipps, L.E. Variability in soil heat flux from a mesquite dune site. *Agric. For. Meteorol.* **2000**, *103*, 249–264. [CrossRef]
29. Ogée, J.; Lamaud, E.; Brunet, Y.; Berbigier, P.; Bonnefond, J.M. A long-term study of soil heat flux under a forest canopy. *Agric. For. Meteorol.* **2001**, *106*, 173–186. [CrossRef]

30. Shao, C.; Chen, J.; Li, L.; Xu, W.; Chen, S.; Gwen, T.; Xu, J.; Zhang, W. Spatial variability in soil heat flux at three Inner Mongolia steppe ecosystems. *Agric. For. Meteorol.* **2008**, *148*, 1433–1443. [CrossRef]
31. Beringer, J.; Chapin, F.S.; Thompson, C.C.; McGuire, A.D. Surface energy exchanges along a tundra-forest transition and feedbacks to climate. *Agric. For. Meteorol.* **2005**, *131*, 143–161. [CrossRef]
32. Meyers, T.P. A comparison of summertime water and CO2 fluxes over rangeland for well watered and drought conditions. *Agric. For. Meteorol.* **2001**, *106*, 205–214. [CrossRef]
33. Santanello, J.A., Jr.; Friedl, M.A. Diurnal covariation in soil heat flux and net radiation. *Am. Meteorol. Soc.* **2003**, *42*, 851–862. [CrossRef]
34. Guo, D.; Yang, M.; Wang, H. Sensible and latent heat flux response to diurnal variation in soil surface temperature and moisture under different freeze/thaw soil conditions in the seasonal frozen soil region of the central Tibetan Plateau. *Environ. Earth Sci.* **2011**, *63*, 97–107. [CrossRef]
35. Lenters, J.D.; Cutrell, G.J.; Istanbullupglu, E.; Scott, D.T.; Herman, K.S.; Irmak, A.; Eisenhauer, D.E. Water and Energy Balance in Response to the Removal of Invasive Phragmites Australis in a Riparian Wetland. *J. Hydrol.* **2011**, *2011*, 19–34. [CrossRef]
36. Behera, S.K.; Mishra, A.K.; Sahu, N.; Kumar, A.; Singh, N.; Kumar, A.; Bajpai, O.; Chaudhary, L.B.; Khare, P.B.; Tuli, R. The study of microclimate in response to different plant community association in tropical moist deciduous forest from northern India. *Biodivers. Conserv.* **2012**, *21*, 1159–1176. [CrossRef]
37. Zheng, L.; Zhao, G.; Dong, J.; Ge, Q.; Tao, J.; Zhang, X.; Qi, Y.; Doughty, R.B.; Xiao, X. Spatial, temporal, and spectral variations in albedo due to vegetation changes in China's grasslands. *ISPRS J. Photogramm. Remote Sens.* **2019**, *152*, 1–12. [CrossRef]
38. Kovács, B.; Tinya, F.; Ódor, P. Stand structural drivers of microclimate in mature temperate mixed forests. *Agric. For. Meteorol.* **2017**, *234–235*, 11–21. [CrossRef]
39. Friedl, M.A. Forward and inverse modeling of land surface energy balance using surface temperature measurements. *Remote Sens. Environ.* **2002**, *79*, 344–354. [CrossRef]
40. Gao, Z.; Horton, R.; Liu, H.P. Impact of wave phase difference between soil surface heat flux and soil surface temperature on soil surface energy balance closure. *J. Geophys. Res.* **2010**, *115*, D16112. [CrossRef]
41. Sun, T.; Wang, Z.; Ni, G. Revisiting the hysteresis effect in surface energy budgets. *Geophys. Res. Lett.* **2013**, *40*, 1741–1747. [CrossRef]

forests

MDPI

Article

Tree Information Modeling: A Data Exchange Platform for Tree Design and Management

Qiguan Shu [1,*], Thomas Rötzer [2], Andreas Detter [3] and Ferdinand Ludwig [1]

1 School of Engineering and Design, Technical University of Munich, Arcisstr. 21, 80333 Munich, Germany
2 School of Life Sciences, Technical University of Munich, Hans-Carl-v.-Carlowitz-Platz 2, 85354 Freising, Germany
3 Brudi & Partner Treeconsult, Berengariastr. 9, 82131 Gauting, Germany
* Correspondence: qiguan.shu@tum.de

Abstract: Trees integrated into buildings and dense urban settings have become a trend in recent years worldwide. Without a thoughtful design, conflicts between green and gray infrastructures can take place in two aspects: (1) tree crown compete with living space above ground; (2) built underground environment, the other way round, affect tree's health and security. Although various data about urban trees are collected by different professions for multiple purposes, the communication between them is still limited by unmatched scales and formats. To address this, tree information modeling (TIM) is proposed in this study, aiming at a standardized tree description system in a high level of detail (LoD). It serves as a platform to exchange data and share knowledge about tree growth models. From the perspective of architects and landscape designers, urban trees provide ecosystem services (ESS) not only through their overall biomass, shading, and cooling. They are also related to various branching forms and crown density, forming new layers of urban living space. So, detailed stem, branch and even root geometry is the key to interacting with humans, building structures and other facilities. It is illustrated in this paper how these detailed data are collected to initialize a TIM model with the help of multiple tools, how the topological geometry of stem and branches in TIM is interpreted into an L-system (a common syntax to describe tree geometries), allowing implementation of widely established tree simulations from other professions. In a vision, a TIM-assisted design workflow is framed, where trees are regularly monitored and simulated under boundary conditions to approach target parameters by design proposals.

Keywords: tree information modeling; tree engineering; building information modeling; computational design; urban green infrastructure

Citation: Shu, Q.; Rötzer, T.; Detter, A.; Ludwig, F. Tree Information Modeling: A Data Exchange Platform for Tree Design and Management. *Forests* **2022**, *13*, 1955. https://doi.org/10.3390/f13111955

Academic Editor: Tommaso Sitzia

Received: 11 October 2022
Accepted: 12 November 2022
Published: 19 November 2022

1. Introduction

1.1. Aim of This Study

1.1.1. History and Trend of Tree Use Integrated in Human Habitats

Integrating trees in nowadays urban spaces as well as buildings is driven by multiple benefits: people's psychological health [1,2], thermal comfort in the context of the urban heat island (UHI) effect [3,4] and sustainability [5]. In architectural history, trees were already used at an early stage of settlement to provide shelter against flood and beasts. Later historical cases saw trees utilized more multifunctionally: espalier trees are trained in geometrical forms for acquiring structural stability and increasing fruit yield in the city [6] (i.e., an espalier tree in England shown in Figure 1c); through pollarding and coppicing, trees were manipulated to produce firewood and building materials [7] (i.e., pollarded trees were bent into umbrella shapes in the center of Labouheyre, France to shade the public square shown in Figure 1d); Devon hedges were built by injuring and laying down trees on earth banks to protect cattle or crops [8,9]; vite maritata systems used trees as supports for vines, whilst providing wind- and sun-protection for field crops and an ecosystem for diverse dependent species [10]; Hausschutzhecken are trees weaved into stable structures to protect buildings and gardens from wind [11]. Street trees nowadays, however, are mainly kept in freely growing forms to reduce maintenance costs and failure risks [12]. Owing to this situation, studies regarding their ecosystem services (ESS) focus on certain aspects such as cooling, shading, carbon storage and reduction in rainfall-runoff [13,14]. But beyond these, from the perspective of architects and landscape designers, trees can provide further values in ESS with specially trained forms.

Figure 1. Challenges and potentials of trees in dense urban areas. (Image (**b–e**) are retrieved from the internet under Creative Commons licenses. Image (**a**,**f**,**g**) are own images).

Interest in multifunctional trees in public spaces and in the building context has been growing in the architecture and landscape architecture industry: in Figure 1a, a tree is planted close to a building façade in Munich, growing only one side of its canopy; in Figure 1e, ficus trees that take vertical walls as their foundation are preserved in Hong Kong. Besides, more proposals are framing trees as their core concept in recent international design competitions; for example, Madrid's RENAZCA development by MVRDV [15] illustrates a public space surrounded by luxuriant "floating" plants grown on metal scaffolding; Street Tree Pods by Matthew Chamberlain [16] offers single apartment on street trees to alleviate London's housing crisis. Buildings with trees integrated into the envelope appear recently in practice, like Bosco Verticale in Milan, designed by Boeri Studio (see Figure 1b) [17] and Kö-Bogen II in Düsseldorf designed by Ingenhoven Architects [18]. Living Architecture, especially Baubotanik [19] (i.e., the house of future proposed for a museum in Berlin by Ludwig Schönle, shown in Figure 1f), go one step further, exploring trees as load-bearing structure. In Figure 1g, the tree façade forms a vertical open space. In this way, the ESS of urban trees lies not only in their general biomass and canopy volume for increasing biodiversity and thermal comfort but also in configurations of roots and branches as a sustainable material to enclose, support and co-create living spaces.

1.1.2. Conflicts between Gray and Green Infrastructure

This trend to enhance trees' multifunctional use in the building industry is confronted now with conflicts from two sides: on the one hand, trees physically can reach a height of at least 10–25 m [20–22] in cities. This height range is also occupied by multi-layer traffic systems and the pedestrian bridges like those in Hongkong [23] as well as public spaces like the High Line in New York City [24]. Within 25 m height are also common residential buildings of 7 floors. Therefore, free-growing tree canopies are competing with these building structures in space above the ground. Too densely aligned canopy can reduce street ventilation, causing traffic-related pollutant concentrations [25,26]. On the other hand, densely built underground environments led to a high removal rate for urban trees. Although some tree species can live up to 200 years in principle, most of them would not be retained longer than 40–60 years in cities [27] due to damages caused by humans [28] or low mechanical performance. Consequently, the average lifespans of urban trees are shorter than the average operating stage of residential housing (61 and 120 years in the US [29] and Denmark [30], respectively). If trees involved in a building structure should not be "temporary" installations but accompany the whole operational period of the building, thoughtful tree design and management are in urgent demand. Healthy and secure growth of individual tree branches and roots must be wisely integrated with urban gray infrastructures (e.g., building façade, foundations, underground pipelines, and even subways).

1.1.3. A Novel Workflow for Tree Design and Management

In this scenario, project planning and maintenance will play a key role in its success. As tree growth is complex and dynamic, the chance to precisely predict and control this process for every branch is small. If pruning all unwanted branches away constantly, this does not effectively reach the purposed functional use provided by specific configurations of branches. Therefore, tree design and management must take place through the whole life cycle in those tree–closely–integrated structures. As illustrated in Figure 2, a typical contemporary architecture design is a set of definitive solutions based on boundary conditions and clients' requirements. If unforeseen circumstances occur during the construction, designers usually seek minor adjustments in response to the problem. The form of the original building design, at least, will be maximally preserved. This common workflow is seen in most of nowadays building projects. But to design and build with living trees, projects by Ludwig and Schwertfreger [19] followed an iterative design approach. It means designers repetitively check tree growth (every 1–3 years) and accordingly make adaptions to the design proposal.

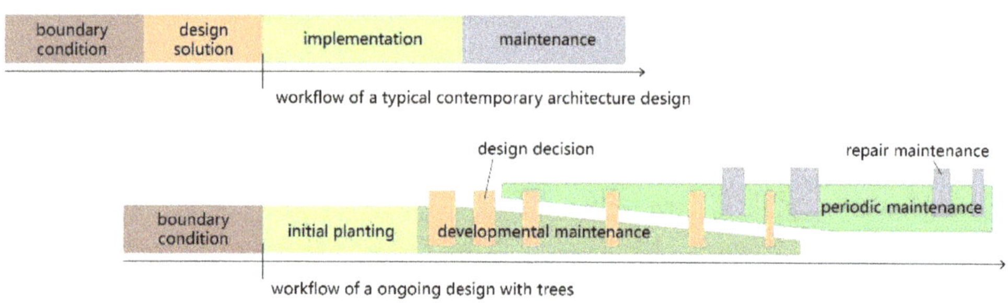

Figure 2. Proposed workflow for a dynamic design to deal with tree growth and death. Unlike a contemporary architecture design (shown on the **top**), the design decision (see dark orange rectangles) in the new workflow will no longer be made only once but multiple times through the tree's life cycle (shown on the **bottom**).

However, traditional tree management is either empirical-based labor-intensive work or standardized periodic pruning [31]. Under current conditions, tree management has a high demand for labor resources. It is not feasible to implement such workflow widely only with human forces, especially when more buildings embedded with trees are being built. To solve this problem in the future without losing the goal of achieving both high ESS values of urban trees as well as their long-term healthy living, efforts must be made in an automated tree management system. With the rapid development of machine vision and automatic robots, urban trees could be taken care of by (swarms of) robots as a low-cost working force in the future [32]. Meanwhile, the public might also be interested in participating in managing trees (including harvesting fruits) in the community as their recreational activity during their free time. Such participation in managing urban green infrastructure by either robots or non-professional personnel requires an appropriate knowledge base about tree management. This knowledge base must incorporate at least the following topics: (1) the impact of the urban environment on tree growth in various LoD; (2) the impact of human intervention (i.e., pruning and bending branches) to tree behaviors; (3) evaluation index for ESS of urban trees in different professions and aspects; (4) efficient and low-cost solution for tree measuring, modeling, and manipulation; (5) solution for measuring, modeling and maintaining buildings and other gray infrastructures in a more dynamic context; (6) design strategy and practice how trees best co-live with buildings and other urban facilities.

1.1.4. This Solution Requires Interdisciplinary Cooperations

These topics are already partly covered or being studied in multiple fields like urban forestry and arboriculture. But different professions, although all have their general interests in trees, aim at completely different goals (see details in Section 1.2). For example, forestry scientists analyze trees' role in an ecosystem; arboriculturists ensure trees' safety and health; biologists simulate the physiological process of tree growth; computer graphic experts contribute to tree geometries and visualizations. For such purposes, they independently collect relevant data about trees to build their own models in various scales, LoD and functions. However, cross-disciplinary communication remains limited owing to unmatched scales and data formats. Cross-platform media, which allows data exchange and model sharing between tree-related professions, is of great importance.

The building industry experienced a similar problem when data from architecture designers, civil engineers, constructors, and equipment suppliers did not communicate. Even on the same industrial chain, data were repeatedly collected and semi-manually transferred from one model to another for different purposes until Building Information Modeling (BIM) was developed. Standards were set for integrating digital models and properties of a building system, such as water, electricity, and gas infrastructure/installations [33]. Such digital copies of building systems also provided an interface for adding dynamic operational data, e.g., via smart meters. These data, referring to the physical entities in the real world, are now becoming the base for the Internet of Things (IoT) [34,35]. In an urban scope, this idea of establishing a data sharing and exchange platform leads to digital twins of cities whose plan, design, construction, and operation can be guided and driven by data to improve working efficiency and quality while bringing down energy consumption and waste. City Information modeling (CIM) is being discussed and developed in several cities around the world [36], for example, Digital Twin Munich [37] and Chinese Xiongan new area [38]. Landscape Information Modelling (LIM) was also proposed for efficiently managing projects in the landscape architecture industry [39]. But until now, a standard media for sharing data and models of urban trees is absent. This standard must reach a certain LoD to enable participation from more professions. If in a low LoD, generic indexes proposed in LIM and CIM, such as tree height, diameter at breast height (DBH), and canopy diameter for describing one tree, cannot guide tree manipulation in arboriculture and cannot integrate the physiological tree growth model from biological studies.

1.1.5. A Starting Point for the Proposed Workflow

This study, therefore, put forward the concept of tree information modeling (TIM), aiming at a data exchange and model sharing platform for gathering cross-disciplinary knowledge about urban trees. For architects and landscape designers, this is a starting point to bring trees and building structures in harmony. To achieve this aim, we investigated interdisciplinary fields (see Section 1.2) to propose a unified tree description system consisting of information tags and geometrical representations of trees. This modeling framework meets the scope of future requirements for tree design and maintenance introduced above. Several example methods for data acquisition (Section 2.2.1) and data interpretation (Section 2.2.2) are described in further detail. With TIM, it is expected to establish a survey-simulation-manipulation workflow for managing future urban tree systems (Section 3.3).

1.2. Related Studies in Interdisciplinary Fields

1.2.1. Tree Management and Risk Assessment in Arboriculture

In current arboricultural practice, urban trees in public are often managed in this manner [40]: for young trees, branches at the lower side of the canopy will be trimmed off to allow for traffic to pass under, which is called lifting or crown raising; before the trees reach a certain height, part of dense branches can be removed to avoid collision and to reduce competition between leading branches for space and light, which is known as crown thinning; any branches growing close to power lines, traffic lanes and private spaces, are often removed or reduced in length, which is called crown reduction. Current best practice includes many more arboricultural measures and specifications for their application (e.g., ATTC [41], Lilly, Gilman [42]).

For all these approaches, pruning stands at the heart of arboriculture [43]. Pruning not only alters the short-term appearance of trees but impacts phytohormone (i.e., auxin-cytokinin) distribution inside them (i.e., apical dominance [44]). Therefore, which branches to be pruned demands craftsmanship. For instance, when a tree fork consists of two sub-branches of comparable size and angle, one of them may be pruned to slow down its growth. After 5–10 years, as a result, the reduced branch on this fork will be clearly subordinated. Otherwise, a fork union with stems of similar diameter may evolve, which forms a weakness in the tree's mechanical stability [45].

For arborists, safety is always a priority in conducting tree management (compared to health and aesthetics listed by Bedker [40]). Root failure and decay inside the trunk are two major causes of tree risks [46]. Due to their high complexity, both risks cannot be precisely analyzed by computational simulations yet. Respectively, to rule out root and trunk failure, practical pulling tests are developed to measure the tree's tilt and deformation using an inclinometer and elastometer under a given force on the main stem [47,48]. This method bypasses the technical obstacle in detecting actual root geometry and decay locations as well as understanding the force transmission from the roots to the soil or within the stem's geometry. To spot weak spots along trunks, an empirical visual inspection could solve the vast majority of cases. The rests require assistance from advanced methods of tree assessment, for example, tomography based on either the time of flight of sound waves or the electrical resistivity of wood. These tomograms display the distribution of resistivity across the stem cross-section, which can indicate featured patterns of decay taking place [49].

Studies have been carried out in recent years about how these data can be collected on a large scale in real-time for monitoring tree tilt angle and sway under natural wind loading [50]. Integrated with GIS, these data can be utilized more fruitful than studying tree risks. The sensors installed on a huge number of trees across the city become a network of mini weather stations, which can be used to precisely model near-earth wind speed and wind-load effect of trees [51]. Data collected and studied in arboriculture do not rely on the detailed geometric representation of tree branches and roots at the current stage. But other professions have different situations.

1.2.2. Urban Tree Models and Databases from Forestry Science

The rise of urbanization since the 1800s has driven various professions to investigate the hybrid system of man-made environment and nature: In architecture, Frederick Law Olmsted and Calvert Vaux [52] for the first time used the term "landscape architecture" in 1863 for a new profession, which extended traditional architecture studies to open space systems. Natural ecosystems, in this way, gained a place in urban-scaled planning. Forestry science developed its scope in the opposite direction. It started by managing natural resources in the natural environment but then faced the challenges of re-evaluating such resources in an urban context for making public policies. To merge this gap, "urban forestry" was first invented in North America in 1965 to integrate a broader group of experts (e.g., psychologists and sociologists) in forestry education [53,54]. Urban forestry is now commonly recognized as the sub-discipline for managing trees, other vegetation and water resources in urban ecosystems for benefits in multiple aspects like sociology, economy, and aesthetics [55]. The integration of all these aspects holds the overall goal: amenity and the promotion of human well-being [56,57]. This is how both urban planning and forestry science meet in this field.

In forestry studies, foliar and woody biomass are the two most important indicators to evaluate a tree's long-term contribution to its ecosystem. Accordingly, several key parameters are widely documented to describe tree stands: trunk diameter at breast height (DBH) and tree height are key factors for estimating the woody biomass; leaf surface area [58], height and diameter of the crown are used for calculating the foliar biomass [59]. Empirical equations are summarized from long-term forestry investigations to predict biomass increment of different species in specific climate zones at different ages [60]. Such equations need to be adapted to urban contexts owing to higher air temperatures and less precipitation in high-dense areas [61]. Therefore, recent development adds further data to describe the surroundings, for example, tree distance from buildings and whether tree crown conflict with overhead wires [62].

The most popularly used urban tree database, including numerical models, is now i-tree, developed by the United States Department of Agriculture (USDA). It was originally the Urban Forest Effects model (UFORE) in the 1990s before the concept of "ecosystem services" was brought out [63]. Ecosystem services (ESS) are defined as the functional components of urban greening that are directly enjoyed, consumed, or used to produce specific, measurable human benefits [14]. Under the demand of quantifying ESS in its subcategories (provision, regulation, support and cultural services), urban tree growth models today include a variety of empirical (e.g., i-tree) and process-based (e.g., CityTree) equations that encompass trees on urban and rural lands for estimating their performance such as cooling, pollution mitigation, stormwater run-off reduction, carbon sequestration and storage [64,65]. Besides this, databases supporting tree selection are also developed [66].

In brief, these models, as well as the databases developed in the field of urban forestry, are global descriptions of the trees, where relatively reliable top-down simulations (see Section 1.2.3) are built but not knowing detailed physiological processes among tree organs and branches.

1.2.3. Functional Structural Plant Models in Varies Scales

Plants are a typical complex system [67]: this system has both biotic- (e.g., leaf, stoma, and cell) and abiotic environmental (e.g., light, water, and nutrients) elements that interact [68,69]; these interactions can be physical, chemical and in other forms, but are often interlinked, resulting in partly deterministic partly stochastic performances [70]; when observing an overall outcome of the system (e.g., growing direction of the shoot), however, certain patterns would be recognized as a result of emergent behaviors [71]. To describe the macroscopic outcome of trees like the total biomass increment, Top-Down models (i.e., empirical equations) are feasible to predict the general growth tendency; but for understanding emergent behaviors (i.e., branches competing for light), Bottom-Up models work with more similar principles as natural phenomena and processes.

While forestry scientists, by studying wood production on a large scale, developed top-down models (already described in Section 1.2.2), botanists, by studying plant physiology on a micro scale, developed bottom-up plant growth simulations. Since the 1990s, the term 'functional structural plant model' (FSPM) has been used to describe such bottom-up models [72], which contain descriptions of metabolic (physiological) processes that are combined in the presentation of the 3D structure of the tree [73]. After more than two decades of development, FSPMs are gradually studied and applied in multiple fields (i.e., biology, animation, forestry and agronomy) in various scales ranging from meristems to plant communities [74]. While studies of metabolic processes are mainly concentrated on several key aspects (like water uptake, transport, photosynthesis, etc.), plant structure is represented variously in different scales.

On a population-to-plant scale, plant structures are summarized as global presentations [75]. An overall geometric shell (mostly a sphere, ellipse, and cylinder) is used to represent the size and volume of the tree canopy and trunk. This is already capable of calculating several fundamental interactions of a tree and its environment, such as a rough light reception [76] and wind force on the canopy [77]. To understand basic carbon distribution [78] or water transport [79] between pools of leaves, roots, fruits and stems, conceptual compartments are built to set individual equations for different pools. But these plant structures [78,79] remain at a global level. In smaller scales, such as the scale of one plant or the scale of plant organs, the applied plant architecture is summarized as a modular representation [75]. The module can be either a spatial cell, geometric cell, or topological cell. The spatial cell is voxel to illustrate the spatial occupation of the objects, i.e., leaf area density [80]; the geometric cell uses a common set of parameters to describe shapes of similar elements, i.e., leaf length, radius and size [81]; topological cell indicates exact connections between plant organs [82]. Topological models of a tree, due to their unique ability to bridge plant organs and the individual plant [83], are framed in most of FSPMs on a plant-organ scale (see also Section 1.2.4).

1.2.4. Quantitative Structure Models (QSMs) of Trees

Topology describes the properties of a geometric object that are preserved under continuous deformations, such as stretching and twisting, but not tearing or gluing [84]. For studying an object with complex geometry (compared to a simple geometric solid such as a cube or sphere), a "thin" version of the shape is commonly used for representing its geometrical and topological properties, i.e., its connectivity, length, and direction, in an easier form. Such an abstraction of the shape that is equidistant to its boundaries is called a topological skeleton [85]. Its mathematical definition varies from distance function [86] and medial axis [87] to morphological operators [88]. Despite these different types, the skeleton, together with the distance of its points to the shape boundary, contains all the necessary information to reconstruct the shape.

The essential format for skeleton data consists of vertices (also called nodes or points) which are connected by edges (also called links or lines). In this perspective, a topological skeleton is also a graph (a mathematical structure [89], which can model pairwise relations between objects [90]. For modeling trees specifically, vertices represent buds, apexes, and nodes for locating other plant organs (i.e., fruits, flowers, and leaves); edges represent internodes, the trunk and branches; a combination of these vertices and edges can form growth module like apical meristem to perform a certain metabolic process such as blossom or elongation [91]. This tree graph has three possible computational data types [75]: a chained list of records that use a single pointer at each child node pointing towards its parent node [92]; an incidence matrix with each vertex in a column, each edge in a row and a number to indicate their relationships [93]; strings of characters that use specific marks to encode graph architecture [82]. Due to its convenience in reading, rewriting, retrieving, and calculating, the plain text string has become the most popular computational data type for FSPM studies. It was a powerful medium to convey topological information when the computational power was limited compared to today [94].

The way strings of characters encode plant structure is called L-system [95]. Based on L-system as the general approach, multiple FSPM platforms have been developed: L-studio coded in C++ provides a library of programs for simulating environmental processes that affect plant development [96]; GroIMP based on the relational growth grammar coded in java enables parallel modification of the geometries while performing the rewriting rules [97]; OpenAlea achieves a graphical programming environment in python offering FSPMs to a larger range of audiences [98]; most recently, L-Py further improved the flexibility of building FSPMs in python and kept compliance with other platforms [99]. Varies plant structures are built, such as Kiwi fruit [100], peach [101] and apple [102,103], in relation to multi-aspects of the metabolic process like photosynthesis (see kiwi), water stress (see peach), pruning (see apple), gravity and light competition [104]. Despite substantial results of these studies using L-system, strings for describing the topology of trees have certain limitations: firstly, interpreting 3D structures of real trees with massive nodes through linear string has low fault-tolerance; secondly, L-system cannot describe re-joint branching networks such as inosculated tree structures [105].

In the past ten years, the popular use of Terrestrial LiDAR Scanning (TLS), structure from motion (SfM) [106] and rapid growth in computing power for 3D graphics enabled detailed documentation of objects with point cloud data. In the field of remote sensing and computer graphics, these technologies were soon applied to tree surveys. 3D geometric primitives of trees can be abstracted from discrete points to represent the structure and topology of their trunk and branches [107]. Multiple approaches were developed for this purpose: Raumonen [108] developed their own method using "cover sets" to reconstruct tree topology; similar to the cover-set idea, PypeTree [109] rebuilt trunk and branches by their "segment" based on skeleton curves in python and then used semi-supervised adjustment to correct the errors; SimpleTree [110] built cylindrical tree models in C++ by voxel-grid and Euclidean clustering, it also developed crown calculation tool to estimate canopy volume; cylinder fitting was proved robust in shape fitting for tree trunk and branches [111]; AdTree [112] was another skeleton related approach fitting cylinders to point cloud model of a single tree. These solutions to generate QSMs enable physical trees to be converted digitally.

Alternatives for geometric tree branch primitives are seen in the gaming and animation industry. Animation rigging of characters [113] is also adapted to plant models by motion capture [114]. Trunk and branch segments can be defined as rigid bodies connected with constraints. In this way, force (i.e., gravity and wind) and collision can be calculated by physical engines [115].

1.2.5. Digital Tools and Databases of Trees Used in the Building Industry

Following the trend of digitalization driven by rapid IT development, the building industry is also transforming into digital tracks at all its stages, including planning, design, execution, and management [116]. Formulated data and methods also vary for different working scales and purposes (see Table 1).

Table 1. Comparison of digital tools and tree representations in different scales.

Scales	Urban	District	Single Build Project
topics and purposes	Land use and planning	Thermal comfort, ecosystem services	Structural performance, operation, and maintenance, building economy
tools and databases	GIS, CityGML	Environmental design-decision-support platform	Building Information Modeling
Suitable models for trees	Population model, Raster image	Global representation, Spatial decomposition (voxel cells)	Topological skeleton Cylindrical pipes

In the scale of urban areas, Geographic Information System (GIS) offers a platform to overlay both raster images by satellite remote sensing and vector data by field mapping [117]. Vegetation in this scale is represented by the leaf area index (LAI). It indicates leaf area per unit ground surface area, which is estimated by normalized difference vegetation index NDVI [118,119], measured with red and near-infrared regions of the electromagnetic spectrum. This index is used to document and analyze the change in plant populations on a large scale. Recently, LAI is also used to guide urban development in terms of green space and urban forestry [120]. CityGML is an advanced all-in-one database in open standardized XML. It enables potentially describing all city facilities (3D objects) in various LoDs [121]. Gobeawan [122] very briefly proposed four levels of tree representations in CityGML, namely a plane circle, one single cylinder, a convex hull and detailed leaves and branches. Among these levels, tree models in LoD 1–3 are already seen in multiple CityGML databases, while LoD 4 remains unclearly defined.

On the scale of district and community, design-decision-support platforms [123,124] are built to simulate interactions between buildings, plants, environmental conditions, and human activities [125]. Kirnbauer [126] integrated multiple databases into a decision support system for urban tree planting. For these purposes, trees are mainly represented by their canopy volume and position using either a simple geometry [127] or voxels [26]. In this way, environmental engines like ENVI-met [128] and Grasshopper plug-in PANDO [129] can simulate the tree's shading and air flow affected by the tree canopy. These assist decisions in street section or plant arrangements for outdoor comfort.

In the scale of a building and its construction, building information modeling (BIM) has been developed to integrate all necessary information concerning building facilities through their lifecycle [130]. Such data can be as detailed as materials and their manufacturers. Some attempts are made to utilize BIM on the district scale, where global tree representations (similar to Vos et al. [26]) appear alongside a BIM model [131]. However, trees are not yet regarded as core components for buildings that can be integrated into BIM and utilized for living architecture design and engineering. No unified description is given about the data type, utilization, and purpose of tree models in this scale for architecture design. Therefore, this paper provides one standard and solution to fill this gap (see Section 2.1).

2. Tree Information Modeling

2.1. Definition

Similar to the definition of BIM [132], Tree Information Modeling (TIM) is conceived as a digital representation of the physical and functional characteristics of a tree. We define TIM as a data exchange and knowledge-sharing platform about trees, aiming at a solid basis for decision-making in their planning and life-long management. To avoid disaccord understandings that occurred on BIM during its long development [133], in this paper, TIM does not limit to any specific tool or software to convey and calculate the tree data. It is a framework following the same tree description system (TDS) to create the digital twins of trees in real life. A unified updating version of TDS enables the maximum compatibility of all TIM users. The first version of the TDS is stated below (see also Figure 3): A digital tree consists of basic information tags and a geometric representation; basic information tags should include at least tree species, tree age (or years after the first planting), location by longitude and latitude, date of documentation (not necessary for a virtual tree at its planning phase). Information tags must also support additional attributes such as tree images, transplanting history and results of pulling tests in risk assessment. Geometric representations have three compartments: branch (including trunk), leaf canopy, and root; trunk and branch (incl. aerial roots visible above the ground) are represented by the topological skeleton and cylindrical pipelines; leaf canopy is represented by voxel noted with leaf area density; root underground is represented by iso-density layers. Each geometric element can be attached with additional attributes if they are measured, such as decay, sap flow rate, the concentration of phytohormones and electrical resistivity.

It needs to be clarified that due to current limitations in underground detection, even the proposed rough iso-density layers for representing roots cannot be correctly mapped in practice.

It is also aware that even if such a root model is acquired, this may not be sufficient to serve all demands in analyzing tree roots (i.e., root failure described in Section 1.2.1). We propose this root representation in the current version of TDS as a balance between what needs to be studied and which data could be gathered. Depending on the technology development, the root model could be updated to root density voxels in later TDS versions or even cylindrical pipelines (the same as branch representations) in the future.

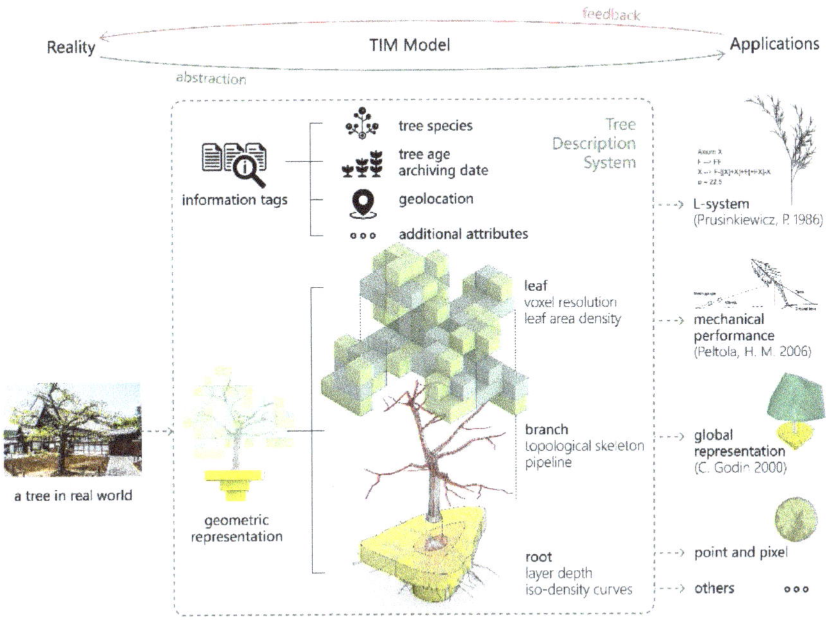

Figure 3. Role of TIM and its data structure. TIM integrates tree information required in multiple professions (see e.g., [48,75,95]). It functions as a media between trees in reality and their digital twins for specific applications.

As for the relationship between BIM and TIM in the building industry, both models function as media between physical objects in the real world and digital applications for planning, evaluating and maintenance. An overview of their geometric representations and key parameters for applications were listed in Table 2. TIM can be applied independently from BIM to digitalize living trees. Data formats in TIM are set with already supported data in BIM software: voxel for leaves, the pipeline for the trunk and branches and closed splines for roots (as shown in Figure 3). When buildings in the future take trees as components and even require their interactions with other building facilities, TIM can be integrated into BIM for more comprehensive uses. This relationship also applies to LIM and CIM on a larger scale, where TIM could benefit the urban green system planning and management as well as evaluate and forecast trees' impact on the urban environment.

Most importantly, although no study about trees yet is built on this novel definition of TIM, TIM models, once initiated or even partly initiated (i.e., missing root data), can be directly applied to related professions introduced in 1.2 because TIM has stored the geometry of trees in a high LoD. Other professions working on more abstracted geometric datasets can interpret TIM into their corresponding forms, such as L-system (see Section 2.2.2), rigid body, a sphere canopy or a pixel. In this way, existing studies, and methods in the fields of forestry science, FSPM and building environment (see Section 1.2) can all be applied to TIM. In this perspective, TIM works as a platform for merging these studies.

Table 2. Comparison between common tree models used in multiple fields and TIM.

	Common Geometric Representation (for Structural Model)			Common Physiological Parameters (for Functional Models)	Common Environmental Factors
	Leaf	Branch (Incl. Stem)	Root		
Forestry science	Crown as an elliptical sphere by its height and diameter	Trunk as a cylinder by DBH and crown height	Not involved	Leaf surface area; sap flux;	Climate; temperature; population density
FSPM	Individual leaf as a rectangle by its length, width, and position	L-system with turtle interpretation	L-system with turtle interpretation	Water transit; carbon assimilation and allocation	Gravity; light rays;
Mechanical calculation	Windward area, leaf density and drag	Trunk as a unilaterally fixed, tapered cantilever beam	A joint with viscoelastic properties	Not involved	Wind velocity, temperature, moisture content
Land resource management	Leaf area index in pixels	Not involved	Not involved	Not involved	Near-infrared spectroscopy; red spectrum;
BIM	Crown as an elliptical sphere	Trunk as a cylinder	Not involved	Not involved	Not involved
TIM	Voxel by leaf area density	Topological skeleton and pipelines	Layer by iso-density curves and depth	Water transit; to be developed	To be developed

2.2. Methods for Data Acquisition and Processing

2.2.1. From Reality to TIM—Data Acquisition through Multiple Tools

As described in Section 2.1, a complete TIM model consists of information tags and geometric representation. In the planning phase, virtual data could be directly generated through designing and simulation programs and then fed into TIM. In this case, it is recommended to initiate a field named "virtual" in the information tag and set its value to true. In the maintenance phase, creating a digital twin of one physical tree must apply different methods in gathering required data: for collecting the information tags, geolocation (longitude and latitude of the tree) can be recorded by portable GPS devices at the stem base; tree species, age and archiving data require manual entries before an automatic tree identification program (i.e., possibly driven by deep learning [134]) is developed; a few photos shot in different distances and angles could also be attached as additional attributes for training the automatic tree identification and for manual cognition. For creating the geometric representation of a tree, the branch (incl. trunk or aerial root) and leaf are visible compartments above the ground, while the root is invisible beneath. So, the surveying methods are different.

Documenting the topological geometry of branches consists of 3 steps (see Figure 4). (1) By LiDAR or photogrammetry scanning [135], a point cloud model of the visible compartment (trunk, branch, aerial root and leaf) can be created. At this step, improvements can be made regarding computing time and tolerance to point clouds in low quality (see Section 3.1). (2) To abstract the topological skeleton of branches out of the point cloud, Cornea [136] compared multiple automatic skeletonization methods; L1-medial skeleton [137] is efficient on point cloud that is not over complex containing too large an amount of points; [138] developed an approach to restore a speculative skeleton without segmenting point clouds into branches and leaves; Wu et al. [139] then achieved an accurate median-axis skeleton abstraction based on the foliage–woody separation by convolutional neural networks [140]; Liu et al. [141] developed a neural network to reconstruct tree geometry out of a point cloud robust to noise, outliers and incompleteness; besides, voxel thinning is able to preserve the precise topological structure of tree branches while estimating approximate diameters of branches during the thinning process [142]. (3) After skeletonization, pipelines can be generated by cylinder fitting or calculating the average distance from the trunk surface to the skeleton on perpendicular planes.

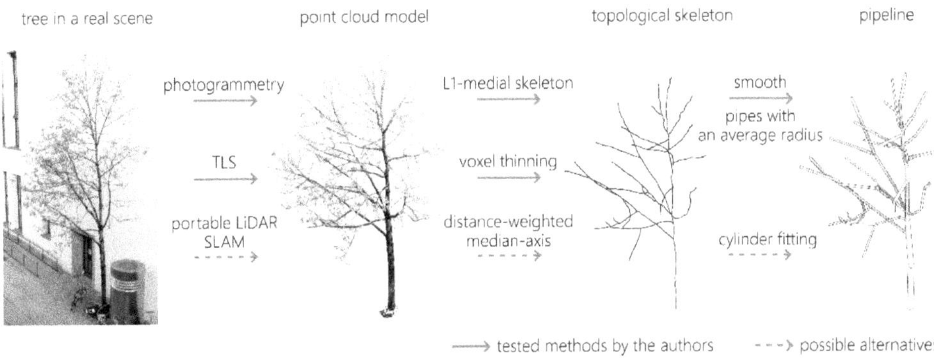

Figure 4. Process to collect topological geometry data of the tree trunk and branches.

Figure 4 illustrates these steps with a tree standing close to a building façade. The arrows with solid lines show several technical roots that are tested by the authors. Arrows with dotted line show alternatives to the same functions. The displayed tree grows an asymmetric canopy against the wall. In this case, sphere geometry as a global representation of a tree canopy is not a precise way to describe it. TIM has the advantage of documenting its main trunk and detailed branches with QSMs.

Voxel-based descriptions of tree canopies were first developed to represent only the volume [143]. To derive LAD at the voxel scale, the Monte-Carlo simulation is a classic approach [144]. Béland, Widlowski [145] proposed the VoxLAD model instead of ray tracing algorithms, enabling the estimation from discrete returning data from any type of TLS; Wu, Phinn [146] used this method on multiple species of fruit trees; Hosoi and Omasa [147] developed voxel-based canopy profiling method to estimate LAD in voxels. The precision according to the voxel size is assessed by Li and Dai [148].

TLS is also applied in scanning tree roots if roots are dug out from the ground [149]. But to detect roots underground, ground-penetrating radar is used. It transmits and receives electromagnetic waves. The returning signals indicate boundaries of overlaying objects [150]. The precision and maximum depth depend on the wave frequency and soil type. Inhomogeneous soil, commonly seen in urban areas, usually produces poor results. New methods like multi-electrode resistivity imaging used for detecting decay inside trunks (see Section 1.2.1) [151] can also show a rough distribution of roots underground, but they are not yet applicable in practice. These data can be processed into 3D root layers in CAD software (i.e., see Gärtner et al. [149] Section 3.2).

Establishing a complete TIM model requires a combination of all methods above. A tree survey is recommended to be conducted in different seasons. For deciduous trees, for example, their LAD can only be documented in summer, while their trunk and branch geometry can only be documented in winter when there is no leaf. Due to such a high standard for completing a TIM model, an incomplete TIM dataset will exist for future applications. Therefore, their access to certain functions should be checked if the required data is missing.

2.2.2. From TIM to Established Applications—Interpreting Pipelines into L-System as an Example

Once trees are documented in TIM, all professions listed in Section 1.2 can extract part of the data from TIM to build their own established model for analysis and simulation. For forestry scientists, for instance, DBH is the diameter of the pipeline at 1.3 m height above ground; tree height is the z-axis coordinate at the upper side of the top canopy voxel; total biomass estimation is the mass sum of leaves, branches, and roots using their volume and average density. Such calculations are similar to measuring real trees, therefore, will not be further explained in this paper. For FSPM studies (see Section 1.2.3), L-system is most used in plant growth simulations. So, it is important to illustrate here how to interpret data, especially branch geometry in TIM, to a tree model written in an L-system.

L-system is a string rewriting mechanism. It recursively replaces certain parts of the strings according to given rules. In this way, it produces patterns with self-similarity, thus being widely used for modeling plants [82]. The method to draw the geometry based on the commands in strings is called turtle interpretation [95]. Typical commands direct a virtual "turtle" moving forward or turning its heading. The trail of the turtle is the geometry to be drawn. However, the original version of such a symbolic L-system has limitations in (1) setting an individual rotating angle and distance for each move, (2) implementing physiological functions for plant growth and (3) operating on the drawn geometry. Therefore, some improvements were developed later (see Section 1.2.4). To illustrate the interpretation from pipelines into one of the L-system models, this study takes language XL as the target format. Language XL is an implementation of relational growth grammars (RGG) [97]. It enables parallel plant description rewriting and geometry generation [152]. Pipelines in TIM can also be interpreted into other variants of the L-system following the same approach but making adaptions to format writing.

Interpreting pipelines into language XL has three steps: (1) translating each pipeline; (2) combining branches in the order of topology; (3) adding defined tree organs to the model.

In the first step, each pipeline in TIM is defined with its geometry and spatial location. The pipeline's geometry consists of its length l and diameter d, which are the same parameters to draw cylinders in XL language using the command $F(l,d)$. The spatial location of a pipeline in TIM is marked with its two ends (i.e., $A(x,\ y,\ z)$ and $B(x\prime,\ y\prime,\ z\prime)$). In XL language, location of an object depends on turtle's state including its position and heading. Turtle's state initiates at the origin point with the default heading shown in Figure 5a. This state will be updated in each step by moving and rotating. Rotation is described by Euler angles along the turtle's local X, Y and Z axis (see also Figure 5a). As cylinder is central symmetric, two degrees of rotation could reach any demanded orientations in a 3D space. In this section, we will use only the turtle's Z and Y axis to perform Euler rotation. As shown in Figure 5b, the command $RH(\alpha)$ and $RU(\beta)$ in XL language rotates the turtle's heading α degrees along the Z axis and β degrees along the Y axis, respectively, in the illustrated direction. Their values can be calculated with equation 1 and 2. A negative number indicates the rotation in a reversed direction. As the rotating sequence affects the results, Z-Y is ruled for all rotating sequences. Starting from the default turtle heading, the rotating angles are calculated with equations 1 and 2. After moving forward along the pipeline, the turtle should make a reversed Y-Z rotation to return to the default turtle heading. This step is crucial for easier connections between individual pipelines because the turtle's headings are identical for all elements. In all, the outcome command for one single pipeline is formed as follows: $RH(\alpha)\ RU(\beta)F(l,d)\ RU(-\beta)RH(-\alpha)$ (see Figure 5c). Such a set of commands for drawing one single pipeline is noted with p_n, where n is the number of the pipeline. When a movement of the turtle is required without drawing a cylinder, the command $F(l,d)$ can be replaced with $M(l)$, where l is still the distance of moving. In this case, the command series is written as $RH(\alpha)\ RU\ (\beta)M(l)\ RU(-\beta)RH(-\alpha)$. This command set is noted with m.

$$\alpha = \begin{cases} \cos^{-1}\frac{x'-x}{\sqrt{(x'-x)^2+(y'-y)^2}} \times 180/\pi\ y'-y \geq 0 \\ -\cos^{-1}\frac{x'-x}{\sqrt{(x'-x)^2+(y'-y)^2}} \times 180/\pi\ y'-y < 0 \end{cases} \alpha \in (-180,\ 180] \quad (1)$$

$$\beta = \cos^{-1}\frac{z'-z}{\sqrt{(x'-x)^2+(y'-y)^2+(z'-z)^2}} \times 180/\pi\ \beta \in [0,\ 180] \quad (2)$$

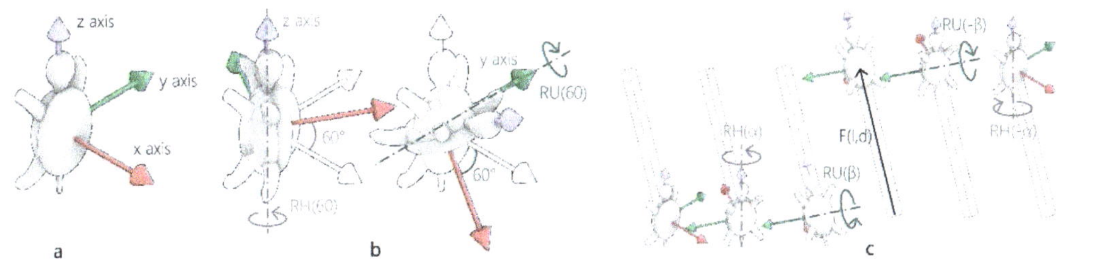

Figure 5. Turtle state and Euler rotation in the interpretation of pipelines. (**a**) the initial heading of the turtle in Language XL; (**b**) rotating the turtle's heading along its z and y axis with command RH (α) and RU (β), respectively; (**c**) visualization of one set of commands for describing one single pipeline.

The second step is to connect these commands together to describe the topology of the tree. Bracket marks "[" and "]" means push and pop the turtle state in XL language. In other words, the turtle state is temporarily stored at each "[" mark. And the turtle will return to this state when it receives the "]" command. By returning to this state, the temporary storing mark is also cleared. So, what is written inside of a bracket does not affect the geometry by following commands. The overall writing strategy for a tree example is illustrated in Figure 6. Tree sections without sub-branch are simply the conduction of pipeline commands in a growth order: older pipelines before the younger (i.e., see $p_{28}p_{29}p_{30}p_{31}$ in Figure 6). At places where multiple sub-branches occur, except one sub-branch writing in the last, the rest branches should be wrapped in the bracket (see colorful and black brackets in Figure 6). Until now, the topological structure of tree branches has been interpreted from TIM data into Language XL.

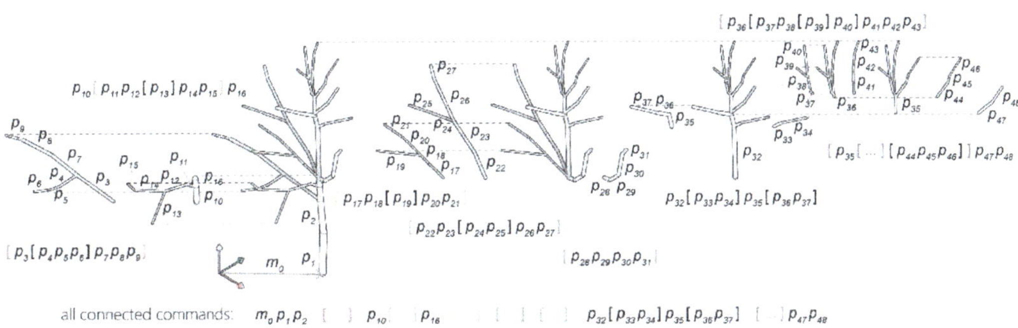

Figure 6. Connecting all turtle commands to represent the topology of a tree.

The third step is to enable functions like tree growth in FSPM by adding defined tree organs and parameters to the string. Different simulation purpose requires different definitions and parameters of these organs. Commonly used ones in FSPM are internode, bud, leaf and so on. TIM is a system for broad applications on trees. So, it will not store all detailed information about every tree organ and their possible parameters. TIM should only save the most common information about a tree to enable compatibility between cross-discipline tree applications. Adding unnecessary parameters to TDS means increasing the cost of establishing the database. In consideration of balancing data needs from different professions, internode (trunk and branch segment) and leaf are contained in the first version of TDS (as described in Section 2.1) because they provide a fundamental description of a tree. Their data acquisition methods are also well developed (see Section 2.2.1). Buds, flowers, and fruits, due to their very targeted use, are not documented in the current TDS version. Therefore, such organs must be manually inserted into the interpreted strings for FSPMs. Even though one tree organ-like internode is stored in TDS, some parameters except its geometry are not compulsory (i.e., the sap flow rate through the internode) in TDS. They can be documented as additional attributes on every geometric element of TIM and then be interpreted into the strings (i.e., "$F(l, d,$ *sapflow rate, other additional attributes...*)"). Some input processes like inserting bud could be, in the future, replaced by programs that make automatic tree organ cognition.

With all three steps, a tree model in TIM is transformed into an FSPM for growth simulation or visualization.

3. A Vision of TIM

To enhance the use and applications of TIM in the future, challenges and opportunities are addressed in this section to guide future works.

3.1. Development in Data Acquisition and Application

In terms of data acquisition, photogrammetry and TLS were tested by the author to acquire detailed branch geometry (see Section 2.2.1). Both methods can obtain shapes of small branches (with radii smaller than 10 mm) in point clouds. However, a photogrammetry survey has certain requirements of stable soft light, calm air and clear background when taking photos. Without proper training, ordinary users cannot acquire a usable point cloud of trees. Although TLS has a lower skill requirement demand on the user, its high costs limit its share and application scenes in the industry. Moreover, point clouds are redundant for getting only the tree geometry. Storage and transmission of such heavily redundant data have little economic value. Not to mention the massive computing resources used for generating and processing the point cloud. Therefore, an in-time skeletonization solution should be considered, where the point cloud is only a temporary media while only cylindrical pipelines for branch geometries are stored. This solution is possibly combined with LiDAR SLAM. Robotic arms carry a portable LiDAR module going around branches to acquire small sections of their geometries and translate them into pipelines immediately.

To allow wider applications on TIM, data acquisition through other approaches about trees should be able to add to the TIM database. For example, an inclinometer and elastometer were used in pull tests for measuring the deformation of tree trunks under a given force [47,48]. These tests ensure tree stability against windstorms. Data in such tests can be added to branch properties and with these data, structural analysis can be implemented on TIM. In this way, tree failures can be warned ahead of meteorological disasters (in reference to Chan and Eng [153]).

For another application example, the leaf area density of trees in voxels is estimated by their branch pipelines in TIM. Because the shape and density of the tree canopy is very closely related to branch geometry. When this branch-to-leaf relationship for different species can be quantitatively described, by scanning the topological geometry of branches in winter, both branch and canopy data can be estimated. This would spare the work of scanning the canopy again in summer. More importantly, such an application associates the manipulation of branch geometries with targeted functions provided by the canopy.

These suggested applications pave the way for a design workflow for treating trees as a core element in the built environment (see Section 3.3).

3.2. Merging the Bottom-Up Simulation with the Global Status of the Tree

Tree growth models in the field of forest science are based on collected data containing a global description of trees under different environmental conditions. So, their models are relatively reliable in forecasting global indicators of trees like biomass and DBH in regard to different ages and species etc. of a tree. This is the advantage of top-down modeling. On the contrary, plant models in the field of FSPM commonly use plant organs (such as internode and bud) as agents to simulate plant behaviors. These are bottom-up models. They have the advantage of reproducing featured patterns in plant growth (see Section 1.2.3). However, their global performances rely on parameters in their physiological process. Some of the settings may not match well with the empirical data. Simulations using TIM in branching scales are also agent-based (bottom-up) models. So, it is a challenge to match the results of this method to empirical data on a global scale. The gap and difference between these two approaches are expected to be better observed and studied if more tree data are collected and shared under the proposed TDS. These findings can lead to (1) modifying equations for improving the quality of bottom-up models, (2) inspiring new theories to explain and model emergent behaviors on trees, and (3) explaining with a deeper insight into patterns in top-down models. One day, tree growth simulation might merge bottom-up and top-down approaches as a unified system.

3.3. Design Workflow Assisted by TIM

A design workflow in a project is as important as the design itself. Achieving multifunctional use of urban trees requires more than datasets and methods. Designers and planners must be able to engage in tree planning by taking advantage of digital tools. As voxels are intuitive in showing spatial distributions of leaves and are used in environmental simulations [26], we hold a strong vision that designers could design tree canopies and their rough density in voxels for urban space (see Figure 7). This design serves as a target parameter for tree status in the long term (i.e., 10–20 years).

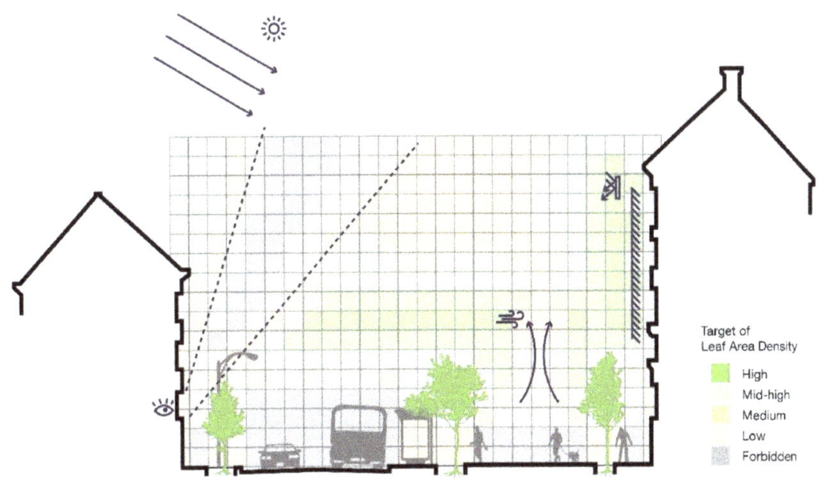

Figure 7. An example of designing street green space with voxels containing leaf area density: the north-facing apartments have access to sunlight and sky view; the motorway and south-facing façade are shaded against high radiation; ventilation at the sidewalk is not blocked. Designers can set target values such as rough leaf area densities to these voxels for various design purposes.

To reach such target parameters, trees must grow under specific manipulation and guidance (especially pruning and bending). Each manipulation of trees impacts the later outcome of leaf area density. So, a feedback workflow is required, consisting of scanning, simulation and decision making for maintenance strategy (see Figure 8): (1) based on the current tree status, boundary conditions are input to simulate the future status of this tree; (2) this simulation will be examined and corrected based on the tree growth in reality; (3) this simulation will also be compared to target parameters in design; (4) different manipulation methods will be virtually tested in TIM to get the best solution for approaching the design target. These four steps are repeated multiple times until getting close to the target parameters. The target parameter, boundary conditions as well as simulation method enable modification during any step in the loop. These changes will not affect previously made manipulations on the tree because tree scans and simulations can be performed at any point again to restart the loop. Every step in this workflow relies on data and methods in TIM. Simulating the tree's reaction to manipulations will be the focus of the next step in this research. A decision-making mechanism to deal with possible conflicts between short-term and long-term outcomes would be a step further. With these steps, the proposed design workflow is developed closer to real application in industry.

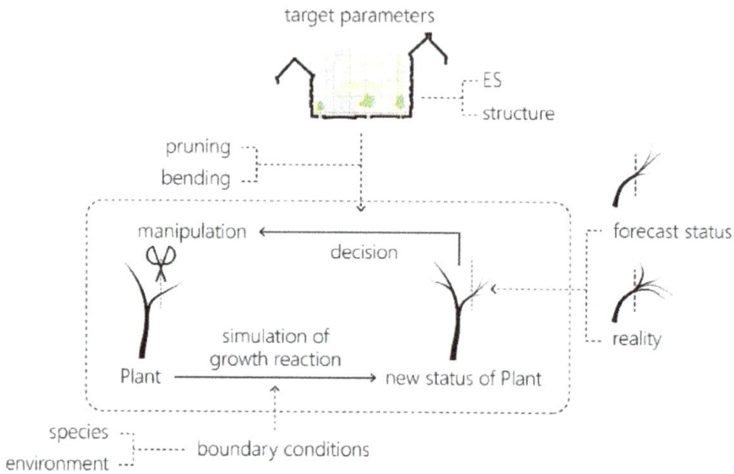

Figure 8. A proposed feedback structure about the iterative survey-simulation-manipulation procedure for designing with trees.

4. Conclusions

In both architecture design competition and practice, integrating trees in building and engineering systems has become the trend worldwide. Trees differ from traditional building materials by their dynamic growth and requirements for constant maintenance. Therefore, efficiently managing trees in a built environment and enabling their multifunctional uses is the key.

In reference to BIM, we proposed TIM (Tree Information Modeling), serving as a bridge between trees in reality and digital applications in multiple professions, including arboriculture, forestry science, biology, animation and the building industry. A complete TIM model consists of information tags and geometric representations for its root, branch (incl. trunk and aerial root), and canopy respectively. It is described in this paper how these data could be collected with various approaches (such as LiDAR scanning, skeletonization and additional measurement). It is also shown how the topological geometry of branches in the TIM model is interpreted to L-system for implementing widely established tree simulations based on that system.

By bringing all related knowledge and data together, TIM can achieve an accurate evaluation of trees grown in various specially trained forms in the city. For gardening companies, nurseries, and urban planners, they can acquire prediction of water and nutrition consumptions through the lifecycle of the trees, estimating their economic benefits and financial expenses; for constructors and civil engineers, TIM can provide them information regarding a minimum space needed for tree canopy and root growth. For arborists, risks of failure in extreme weather or when trees suffer accident damage can be assessed from TIM data. For studying urban forestry, TIM has accurate geometrical data to estimate trees' impact on microclimate through cooling and evaporation. For architects and landscape designers, TIM can assist with species selection, planting layout and branch configuration. With TIM, trees can interact more with other artificial materials and components without causing unpredictable consequences. All these applications then lead to a longer life expectancy of trees in a densely built urban environment and enable the design and management of gray and green infrastructure in harmony.

Despite these benefits of TIM, limitations were also seen when such standards were applied to industry (in reference to limitations of BIM in implementation [154]). Firstly, it enforces a higher learning and training cost for all participants working with TIM. A tree nursery, for example, may not immediately benefit much from TIM data sources but can invest more to adapt their original data storage form and workflow. Secondly, when tree data are commonly structured and packaged as TIM suggests, it raises a higher threshold (especially for the public) to retrieve, interpret and reuse the data. Lastly, it lacks contractual arrangements yet to specify the property right, legal access and liabilities for tree data.

Finally, efforts still need to be made in efficient data acquisition and discovery of more application scenes on TIM. The gap between the results of branch-scale simulation and a global tree status must be matched. With all these efforts, the goal is to achieve an iterative workflow to manage urban trees towards a design proposal quantified with target parameters.

Author Contributions: Conceptualization, Q.S. and F.L.; methodology, Q.S.; software, Q.S.; writing—original draft preparation, Q.S.; writing—review and editing, Q.S., T.R., A.D. and F.L.; visualization, Q.S.; supervision, F.L. and T.R.; project administration, Q.S. and F.L.; funding acquisition, F.L. and T.R. All authors have read and agreed to the published version of the manuscript.

Funding: This study is funded by the German Research Foundation (DFG) under the project numbers DFG-GZ: LU2505/2-1 RO4283/2-1 and PR 292/23-1.

Institutional Review Board Statement: Not applicable.

Informed Consent Statement: Not applicable.

Data Availability Statement: Not applicable.

Acknowledgments: Our previous work cooperated with Wilfried Middleton about topological skeleton abstraction for inosculated structures inspired the idea of TIM. Haoyu Fang provides the primary data of the tree in Figures 4 and 6.

Conflicts of Interest: The authors declare no conflict of interest.

Abbreviations

Abbreviations	Full Term
BIM	Building Information Modeling
CAD	Computer Aided Design
CIM	City Information modeling
DBH	Diameter at Breast Height
ESS/ES	Ecosystem Services
FSPM	Functional Structural Plant Model
GIS	Geographic Information System
IoT	Internet of Things
LAD	Leaf Area Density
LAI	Leaf Area Index
LiDAR	Light Detection And Ranging
LIM	Landscape Information Modelling
LoD	Level of Detail
QSM	Quantitative Structure Models
SfM	Structure from Motion
SLAM	Simultaneous Localization and Mapping
TDS	Tree Description System
TIM	Tree Information Modeling
TLS	Terrestrial Laser Scanning
UHI	Urban Heat Islands

References

1. Schroeder, H.W. Does beauty still matter? Experiential and utilitarian values of urban trees. In *Trees, People and the Built Environment, Proceedings of the Urban Trees Research Conference, Birmingham, UK, 13–14 April 2011*; Institute of Chartered Foresters: Edinburgh, UK, 2011; pp. 159–165.
2. Hong, X.-C.; Wang, G.-Y.; Liu, J.; Dang, E. Perceived Loudness Sensitivity Influenced by Brightness in Urban Forests: A Comparison When Eyes Were Opened and Closed. *Forests* **2020**, *11*, 1242. [CrossRef]
3. Irmak, M.A.; Yilmaz, S.; Mutlu, E.; Yilmaz, H. Assessment of the effects of different tree species on urban microclimate. *Environ. Sci. Pollut. Res.* **2018**, *25*, 15802–15822. [CrossRef] [PubMed]
4. Rahman, M.A.; Franceschi, E.; Pattnaik, N.; Moser-Reischl, A.; Hartmann, C.; Paeth, H.; Pretzsch, H.; Rötzer, T.; Pauleit, S. Spatial and temporal changes of outdoor thermal stress: Influence of urban land cover types. *Sci. Rep.* **2022**, *12*, 1–13. [CrossRef]
5. Lazar, N.; Chithra, K. A comprehensive literature review on development of Building Sustainability Assessment Systems. *J. Build. Eng.* **2020**, *32*, 101450. [CrossRef]
6. Edmunds, A. *Espalier Fruit Trees: Their History and Culture*; Pomona Books: Hebden Bridge, West Yorkshire, UK, 1986.
7. Nicholson, J. Results of coppicing, pollarding and pruning experiments to stimulate Strychnos nux-vomica fruit production. *Indian For.* **1937**, *63*, 588–597.
8. Wilson, P.J. Botanical diversity in the hedges and field margins of lowland Britain. In *The Ecology of Hedgerows and Field Margin*; Routledge: London, UK, 2019; p. 3554.
9. Höpfl, L.; Hensel, D.S.; Hensel, M.; Ludwig, F. Initiating research into adapting rural hedging techniques, hedge types, and hedgerow networks as novel urban green systems. *Land* **2021**, *10*, 529. [CrossRef]
10. Torquati, B.; Giacchè, G.; Venanzi, S. Economic analysis of the traditional cultural vineyard landscapes in Italy. *J. Rural. Stud.* **2015**, *39*, 122–132. [CrossRef]
11. Beckmann, R. *Die Hausschutzhecken im Monschauer Land Unter Besonderer Berücksichtigung Ihrer klimatischen Auswirkungen*; Dümmler Verlag: Bonn, Germany, 1982.
12. Kabir, E.; Guikema, S.; Kane, B. Statistical modeling of tree failures during storms. *Reliab. Eng. Syst. Saf.* **2018**, *177*, 68–79. [CrossRef]
13. Roy, S.; Byrne, J.; Pickering, C. A systematic quantitative review of urban tree benefits, costs, and assessment methods across cities in different climatic zones. *Urban For. Urban Green.* **2012**, *11*, 351–363. [CrossRef]
14. Rötzer, T.; Moser-Reischl, A.; Rahman, M.A.; Grote, R.; Pauleit, S.; Pretzsch, H. Modelling urban tree growth and ecosystem services: Review and perspectives. *Prog. Bot.* **2020**, *82*, 405–464.
15. MVRDV. Salón Verde. 2021. Available online: https://www.mvrdv.nl/projects/453/salon-verde (accessed on 29 June 2022).
16. Sandigliano, T. Street Tree Pods—WeVux. 2019. Available online: https://wevux.com/street-tree-pods0055922/ (accessed on 29 June 2022).
17. Belcher, R.N.; Fornasari, L.; Menz, S.; Schroepfer, T. Birds use of vegetated and non-vegetated high-density buildings—A case study of Milan. *J. Urban Ecol.* **2018**, *4*, juy001. [CrossRef]
18. IngenhovenArchitects. Ingenhoven Architects—Kö-Bogen II Düsseldorf. 2020. Available online: https://www.ingenhovenarchitects.com/projekte/weitere-projekte/koe-bogen-ii-duesseldorf/description (accessed on 29 June 2022).
19. Ludwig, F.; Schwertfreger, H.; Storz, O. Living systems: Designing growth in Baubotanik. *Archit. Des.* **2012**, *82*, 82–87. [CrossRef]
20. Leuzinger, S.; Vogt, R.; Körner, C. Tree surface temperature in an urban environment. *Agric. For. Meteorol.* **2010**, *150*, 56–62. [CrossRef]
21. Moser-Reischl, A.; Rötzer, T.; Pauleit, S.; Pretzsch, H. Urban Tree Growth Characteristics of Four Common Species in South Germany. *Arboric. Urban For.* **2021**, *47*, 150–169. [CrossRef]
22. Franceschi, E.; Moser-Reischl, A.; Rahman, M.A.; Pauleit, S.; Pretzsch, H.; Rötzer, T. Crown Shapes of Urban Trees-Their Dependences on Tree Species, Tree Age and Local Environment, and Effects on Ecosystem Services. *Forests* **2022**, *13*, 748. [CrossRef]
23. Sun, G.; Webster, C.; Zhang, X. Connecting the city: A three-dimensional pedestrian network of Hong Kong. *Environ. Plan. B Urban Anal. City Sci.* **2021**, *48*, 60–75. [CrossRef]
24. Loughran, K. Parks for profit: The high line, growth machines, and the uneven development of urban public spaces. *City Community* **2014**, *13*, 49–68. [CrossRef]
25. Santiago, J.-L.; Rivas, E.; Sanchez, B.; Buccolieri, R.; Esposito, A.; Martilli, A.; Vivanco, M.G.; Martin, F. Impact of Different Combinations of Green Infrastructure Elements on Traffic-Related Pollutant Concentrations in Urban Areas. *Forests* **2022**, *13*, 1195. [CrossRef]
26. Vos, P.E.; Maiheu, B.; Vankerkom, J.; Janssen, S. Improving local air quality in cities: To tree or not to tree? *Environ. Pollut.* **2013**, *183*, 113–122. [CrossRef]
27. Roman, L.A.; Scatena, F.N. Street tree survival rates: Meta-analysis of previous studies and application to a field survey in Philadelphia, PA, USA. *Urban For. Urban Green.* **2011**, *10*, 269–274. [CrossRef]
28. Browell, M.F. Tree Risk Assessment. *Arboric. J.* **1996**, *20*, 3–12. [CrossRef]
29. Aktas, C.B.; Bilec, M.M. Impact of lifetime on US residential building LCA results. *Int. J. Life Cycle Assess.* **2012**, *17*, 337–349. [CrossRef]

30. Marsh, R. Building lifespan: Effect on the environmental impact of building components in a Danish perspective. *Archit. Eng. Des. Manag.* **2017**, *13*, 80–100. [CrossRef]
31. Tucker, D.P.H.; Wheaton, T.A.; Muraro, R. Citrus Tree Pruning Principles and Practices. In *1994: University of Florida Cooperative Extension Service*; Institute of Food and Agricultural Sciences: Gainesville, FL, USA, 1994.
32. Yang, X.; Gan, T.; Zheng, H.; Cai, Q.; Chen, Y. Design of Control System for a new Intelligent Tree-climbing and Pruning Robot. In *Journal of Physics: Conference Series*; IOP Publishing: Bristol, UK, 2022.
33. Nawari, N.O. BIM standard in off-site construction. *J. Archit. Eng.* **2012**, *18*, 107–113. [CrossRef]
34. Bottaccioli, L.; Aliberti, A.; Ugliotti, F.; Patti, E.; Osello, A.; Macii, E.; Acquaviva, A. Building energy modelling and monitoring by integration of IoT devices and building information models. In Proceedings of the 2017 IEEE 41st Annual Computer Software and Applications Conference (COMPSAC), Turin, Italy, 4–8 July 2017.
35. Tang, S.; Shelden, D.R.; Eastman, C.M.; Pishdad-Bozorgi, P.; Gao, X. A review of building information modeling (BIM) and the internet of things (IoT) devices integration: Present status and future trends. *Autom. Constr.* **2019**, *101*, 127–139. [CrossRef]
36. Souza, L.; Bueno, C. City Information Modelling as a support decision tool for planning and management of cities: A systematic literature review and bibliometric analysis. *Build. Environ.* **2022**, *207*, 108403. [CrossRef]
37. München, L. Digitaler Zwilling München. 2019. Available online: https://muenchen.digital/twin/ (accessed on 29 June 2022).
38. Veglianti, E.; Magnaghi, E.; De Marco, M.; Li, Y. Smart city in China: The state of art of Xiong an new area. In *Organizing Smart Buildings and Cities*; Springer: Cham, Switzerland, 2021; pp. 81–97.
39. Ahmad, A.M.; Aliyu, A.A. The need for landscape information modelling (LIM) in landscape architecture. In Proceedings of the 13th Digital Landscape Architecture Conference, Weimar, Germany, 31 May–2 June 2012.
40. Bedker, P.J. *How to Prune Trees*; US Department of Agriculture, Forest Service, Northeastern Area, State & Private Forestry: Radnor, PA, USA, 1995; Volume 1.
41. ATTC. *ATTC-Tree Care: Additional Technical Terms of Contract and Guidelines for Tree Care*; Dujesiefken, D., Ed.; Forschungsgesellschaft Landschaftsentwicklung Landschaftsbau e. V. (FLL): Bonn, Germany, 2017; p. 88.
42. Lilly, S.J.; Gilman, E.F.; Smiley, T.E. *Best Management Practices—Pruning*, 3rd ed.; International Society of Arboriculture: Atlanta, United States, 2019.
43. Clark, J.R.; Matheny, N. The research foundation to tree pruning: A review of the literature. *Arboric. Urban For.* **2010**, *36*, 110–120. [CrossRef]
44. Purcell, L. Tree Pruning: What Do Trees Think? In *Forestry and Natureal Resources*; Purdue Extension: West Lafayette, IN, USA, 2017.
45. Gilman, E.F.; Kempf, B.; Matheny, N.; Clark, J. *Structural Pruning: A Guide for the Green Industry*; Urban Tree Foundation: Visalia, CA, USA, 2013.
46. Pokorny, J.D. *Urban Tree Risk Management: A Community Guide to Program Design and Implementation*; USDA Forest Service, Northeastern Area, State and Private Forestry: Newtown, PA, USA, 2003; Volume 3.
47. Detter, A.; Rust, S. *Aktuelle Untersuchungsergebnisse zu Zugversuchen. Jahrbuch der Baumpflege*; Haymarket Media: Braunschweig, Germany, 2013; pp. 87–100.
48. Peltola, H.M. Mechanical stability of trees under static loads. *Am. J. Bot.* **2006**, *93*, 1501–1511. [CrossRef]
49. Rust, S.; van Wassenaer, P. Tools for tree risk assessment. In *Routledge Handbook of Urban Forestry*; Routledge: London, UK, 2017; pp. 489–499.
50. Abbas, S.; Kwok, C.Y.T.; Hui, K.K.W.; Li, H.; Chin, D.C.; Ju, S.; Heo, J.; Wong, M.S. Tree tilt monitoring in rural and urban landscapes of Hong Kong using smart sensing technology. *Trees For. People* **2020**, *2*, 100030. [CrossRef]
51. Yang, Z.; Hui, K.; Abbas, S.; Zhu, R.; Kwok, C.; Heo, J.; Ju, S.; Wong, M. A review of dynamic tree behaviors: Measurement methods on tree sway, tree tilt, and root–plate movement. *Forests* **2021**, *12*, 379. [CrossRef]
52. Blodgett, G. Frederick law olmsted: Landscape architecture as conservative reform. *J. Am. Hist.* **1976**, *62*, 869–889. [CrossRef]
53. Konijnendijk, C.C. A decade of urban forestry in Europe. *For. Policy Econ.* **2003**, *5*, 173–186. [CrossRef]
54. Konijnendijk, C.C.; Ricard, R.M.; Kenney, A.; Randrup, T.B. Defining urban forestry—A comparative perspective of North America and Europe. *Urban For. Urban Green.* **2006**, *4*, 93–103. [CrossRef]
55. Miller, R.W.; Hauer, R.J.; Werner, L.P. *Urban Forestry: Planning and Managing Urban Greenspaces*; Waveland Press: Long Grove, IL, USA, 2015.
56. Ball, R.; Bussey, S.C.; Patch, D.; Simson, A.; West, S. Country Report on the United Kindom. In *COST Action E12 Research and Development in Urban Forestry in Europe*; Forrest, M., Konijnendijk, C., Randrup, T., Eds.; Office for Official Publications of the European Communities: Luxembourg, 1999; pp. 325–340.
57. Konijnendijk, C.C.; Annerstedt, M.; Nielsen, A.B.; Maruthaveeran, S. *Benefits of Urban Parks. A systematic Review*; A Report for IFPRA; International Federation for Parks and Recreation Administration, IFPRA: Copenhagen, Denmark; International Federation for Parks and Recreation Administration, IFPRA: Alnarp, Sweden, 2013.
58. Monk, C.D.; Child, G.I.; Nicholson, S.A. Biomass, litter and leaf surface area estimates of an oak-hickory forest. *Oikos* **1970**, *21*, 138–141. [CrossRef]
59. Gülçin, D.; van den Bosch, C.C.K. Assessment of above-ground carbon storage by urban trees using LiDAR data: The case of a university campus. *Forests* **2021**, *12*, 62. [CrossRef]
60. McPherson, E.G.; Peper, P.J. Urban tree growth modeling. *J. Arboric. Urban For.* **2012**, *38*, 175–183. [CrossRef]

61. Dahlhausen, J.; Rötzer, T.; Biber, P.; Uhl, E.; Pretzsch, H. Urban climate modifies tree growth in Berlin. *Int. J. Biometeorol.* **2018**, *62*, 795–808. [CrossRef]
62. McPherson, E.G.; van Doorn, N.S.; Peper, P.J. *Urban Tree Database and Allometric Equations*; Gen. Tech. Rep. PSW-GTR-253; US Department of Agriculture, Forest Service, Pacific Southwest Research Station: Albany, CA, USA, 2016; 86p.
63. Escobedo, F.J.; Kroeger, T.; Wagner, J.E. Urban forests and pollution mitigation: Analyzing ecosystem services and disservices. *Environ. Pollut.* **2011**, *159*, 2078–2087. [CrossRef]
64. Nowak, D.J. *Understanding i-Tree: Summary of Programs and Methods*; General Technical Report NRS-200; US Department of Agriculture, Forest Service, Northern Research Station: Madison, WI, USA, 2020; 100p.
65. Rötzer, T.; Rahman, M.; Moser-Reischl, A.; Pauleit, S.; Pretzsch, H. Process based simulation of tree growth and ecosystem services of urban trees under present and future climate conditions. *Sci. Total Environ.* **2019**, *676*, 651–664. [CrossRef]
66. Vogt, J.; Gillner, S.; Hofmann, M.; Tharang, A.; Dettmann, S.; Gerstenberg, T.; Schmidt, C.; Gebauer, H.; Van de Riet, K.; Berger, U.; et al. Citree: A database supporting tree selection for urban areas in temperate climate. *Landsc. Urban Plan.* **2017**, *157*, 14–25. [CrossRef]
67. Thurner, S.; Hanel, R.; Klimek, P. *Introduction to the Theory of Complex Systems*; Oxford University Press: Oxford, UK, 2018.
68. Smith, A.M.; Stitt, M. Coordination of carbon supply and plant growth. *Plant Cell Environ.* **2007**, *30*, 1126–1149. [CrossRef] [PubMed]
69. Lastdrager, J.; Hanson, J.; Smeekens, S. Sugar signals and the control of plant growth and development. *J. Exp. Bot.* **2014**, *65*, 799–807. [CrossRef] [PubMed]
70. Kang, M.; Cournède, P.; de Reffye, P.; Auclair, D.; Hu, B. Analytical study of a stochastic plant growth model: Application to the GreenLab model. *Math. Comput. Simul.* **2008**, *78*, 57–75. [CrossRef]
71. Wilensky, U.; Rand, W. *An Introduction to Agent-Based Modeling: Modeling Natural. Social, and Engineered Complex Systems with NetLogo*; MIT Press: Cambridge, MA, USA, 2015.
72. Kurth, W. Morphological models of plant growth: Possibilities and ecological relevance. *Ecol. Model.* **1994**, *75*, 299–308. [CrossRef]
73. Sievänen, R.; Nikinmaa, E.; Nygren, P.; Ozier-Lafontaine, H.; Perttunen, J.; Hakula, H. Components of functional-structural tree models. *Ann. For. Sci.* **2000**, *57*, 399–412. [CrossRef]
74. Louarn, G.; Song, Y. Two decades of functional–structural plant modelling: Now addressing fundamental questions in systems biology and predictive ecology. *Ann. Bot.* **2020**, *126*, 501–509. [CrossRef]
75. Godin, C. Representing and encoding plant architecture: A review. *Ann. For. Sci.* **2000**, *57*, 413–438. [CrossRef]
76. Norman, J.; Welles, J. Radiative transfer in an array of canopies 1. *Agron. J.* **1983**, *75*, 481–488. [CrossRef]
77. Baker, C. The development of a theoretical model for the windthrow of plants. *J. Theor. Biol.* **1995**, *175*, 355–372. [CrossRef]
78. Deleuze, C.; Houllier, F. A Transport Model for Tree Ring Width. *Silva Fennica.* **1997**, *31*, 239–250. [CrossRef]
79. Sperry, J.S.; Adler, F.R.; Campbell, G.S.; Comstock, J.P. Limitation of plant water use by rhizosphere and xylem conductance: Results from a model. *Plant Cell Environ.* **1998**, *21*, 347–359. [CrossRef]
80. Almeida, D.R.A.D.; Stark, S.C.; Shao, G.; Schietti, J.; Nelson, B.W.; Silva, C.A.; Gorgens, E.B.; Valbuena, R.; Papa, D.D.A.; Brancalion, P.H.S. Optimizing the remote detection of tropical rainforest structure with airborne lidar: Leaf area profile sensitivity to pulse density and spatial sampling. *Remote Sens.* **2019**, *11*, 92. [CrossRef]
81. Knutzen, J.; Saito, S.; Nakajima, M. Generating climbing plants using l-systems. In *Proceedings of the Korean Society of Broadcast Engineers Conference*; The Korean Institute of Broadcast and Media Engineers: Daejeon, Republic of Korea, 2009; pp. 784–789.
82. Prusinkiewicz, P.; Lindenmayer, A. *The Algorithmic Beauty of Plants*; Springer Science & Business Media: New York, NY, USA, 2012.
83. Salminen, H.; Saarenmaa, H.; Perttunen, J.; Sievänen, R.; Väkevä, J.; Nikinmaa, E. Modelling trees using an object-oriented scheme. *Math. Comput. Model.* **1994**, *20*, 49–64. [CrossRef]
84. Armstrong, M.A. *Basic Topology*; Springer Science & Business Media: New York, NY, USA, 2013.
85. Jain, R.; Kasturi, R.; Schunck, B.G. *Machine Vision*; McGraw-Hill: New York, NY, USA, 1995; Volume 5.
86. Zhou, Y.; Kaufman, A.; Toga, A.W. Three-dimensional skeleton and centerline generation based on an approximate minimum distance field. *Vis. Comput.* **1998**, *7*, 303–314. [CrossRef]
87. Tănase, M.; Veltkamp, R.C. A straight skeleton approximating the medial axis. In *European Symposium on Algorithms*; Springer: Berlin/Heidelberg, Germany, 2004.
88. Wang, K.; Li, X.; Trinder, J.C. Mathematical morphological analysis of typical cyclone eyes on ocean SAR. In Proceedings of the 2014 IEEE Geoscience and Remote Sensing Symposium, Quebec City, QC, Canada, 13–18 July 2014.
89. Bender, E.A.; Williamson, S.G. *Lists, Decisions and Graphs*; S. University of California: San Diego, CA, USA, 2010.
90. Prathik, A.; Uma, K.; Anuradha, J. An Overview of application of Graph theory. *Int. J. ChemTech Res.* **2016**, *9*, 242–248.
91. Pradal, C.; Godin, C. MTG as a Standard Representation of Plants in FSPMs. In Proceedings of the 9th International Conference on Functional-Structural Plant Models: FSPM2020, Online, 5–9 October 2020; Institute of Horticultural Production Systems: Hannover, Germany, 2020; pp. 86–87.
92. Ahuja, R.K.; Magnanti, T.L.; Orlin, J.B. Chapter iv network flows. *Handb. Oper. Res. Manag. Sci.* **1989**, *1*, 211–369.
93. Fourcaud, T.; Lac, P. *Mechanical Analysis of the Form and Internal Stresses of a Growing Tree by the Finite Element Method*; ASME: New York, NY, USA, 1996.
94. Waldrop, M. The chips are down for Moore's law. *Nat. News Feature* **2016**, *530*, 144–147. [CrossRef]

95. Prusinkiewicz, P. Graphical applications of L-systems. In Proceedings of the Graphics Interface and Vision Interface '86, Vancouver, BC, Canada, 26–30 May 1986; pp. 247–253.
96. Karwowski, R.; Prusinkiewicz, P. The L-system-based plant-modeling environment L-studio 4.0. In *Proceedings of the 4th International Workshop on Functional-Structural Plant Models*; UMR AMAP: Montpellier, France, 2004.
97. Hemmerling, R.; Kniemeyer, O.; Lanwert, D.; Kurth, W.; Buck-Sorlin, G. The rule-based language XL and the modelling environment GroIMP illustrated with simulated tree competition. *Funct. Plant Biol.* **2008**, *35*, 739–750. [CrossRef]
98. Fournier, C.; Pradal, C.; Louarn, G.; Combes, D.; Soulié, J.C.; Luquet, D.; Boudon, F.; Chelle, M. Building modular FSPM under OpenAlea: Concepts and applications. In *6th International Workshop on Functional-Structural Plant Models*; University of California: Berkeley, CA, USA, 2010.
99. Boudon, F.; Pradal, C.; Cokelaer, T.; Prusinkiewicz, P.; Godin, C. L-Py: An L-system simulation framework for modeling plant architecture development based on a dynamic language. *Front. Plant Sci.* **2012**, *3*, 76. [CrossRef]
100. Cieslak, M.; Seleznyova, A.N.; Hanan, J. A functional-structural kiwifruit vine model. In Proceedings of the 2009 Third International Symposium on Plant Growth Modeling, Simulation, Visualization and Applications, Beijing, China, 9–13 November 2009. [CrossRef]
101. Allen, M.; Prusinkiewicz, P.; DeJong, T. Using L-systems for modeling source–sink interactions, architecture and physiology of growing trees: The L-PEACH model. *New Phytol.* **2005**, *166*, 869–880. [CrossRef]
102. Fumey, D.; Lauri, P.; Guedon, Y.; Godin, C.; Costes, E. Effects of pruning on the apple tree: From tree architecture to modeling. In Proceedings of the IX International Symposium on Integrating Canopy, Rootstock and Environmental Physiology in Orchard Systems, Geneva, NY, USA, 4–8 August 2008.
103. Elfiky, N.M.; Akbar, S.A.; Sun, J.; Park, J.; Kak, A. Automation of dormant pruning in specialty crop production: An adaptive framework for automatic reconstruction and modeling of apple trees. In Proceedings of the IEEE Conference on Computer Vision and Pattern Recognition Workshops, Boston, MA, USA, 7–12 June 2015.
104. Palubicki, W.; Horel, K.; Longay, S.; Runions, A.; Lane, B.; Měch, R.; Prusinkiewicz, P. Self-organizing tree models for image synthesis. *ACM Trans. Graph. (TOG)* **2009**, *28*, 1–10. [CrossRef]
105. Ludwig, F. *Baubotanik: Designing with Living Material, in Materiality and Architecture*; Routledge: London, UK, 2016; pp. 206–216.
106. Alhasan, A.; Younkin, K.; White, D.J. *Comparison of Roadway Roughness Derived from LIDAR and SFM 3D Point Clouds*; Iowa State University, Institute for Transportation: Ames, IA, USA, 2015.
107. Lau, A.; Bentley, L.P.; Martius, C.; Shenkin, A.; Bartholomeus, H.; Raumonen, P.; Malhi, Y.; Jackson, T.; Herold, M. Quantifying branch architecture of tropical trees using terrestrial LiDAR and 3D modelling. *Trees* **2018**, *32*, 1219–1231. [CrossRef]
108. Raumonen, P.; Kaasalainen, M.; Åkerblom, M.; Kaasalainen, S.; Kaartinen, H.; Vastaranta, M.; Holopainen, M.; Disney, M.; Lewis, P. Fast automatic precision tree models from terrestrial laser scanner data. *Remote Sens.* **2013**, *5*, 491–520. [CrossRef]
109. Delagrange, S.; Jauvin, C.; Rochon, P. PypeTree: A tool for reconstructing tree perennial tissues from point clouds. *Sensors* **2014**, *14*, 4271–4289. [CrossRef] [PubMed]
110. Hackenberg, J.; Spiecker, H.; Calders, K.; Disney, M.; Raumonen, P. SimpleTree—An efficient open source tool to build tree models from TLS clouds. *Forests* **2015**, *6*, 4245–4294. [CrossRef]
111. Åkerblom, M.; Raumonen, P.; Kaasalainen, M.; Casella, E. Analysis of geometric primitives in quantitative structure models of tree stems. *Remote Sens.* **2015**, *7*, 4581–4603.
112. Du, S.; Lindenbergh, R.; Ledoux, H.; Stoter, J.; Nan, L. AdTree: Accurate, detailed, and automatic modelling of laser-scanned trees. *Remote Sens.* **2019**, *11*, 2074. [CrossRef]
113. Baran, I.; Popović, J. Automatic rigging and animation of 3d characters. *ACM Trans. Graph. (TOG)* **2007**, *26*, 72-es. [CrossRef]
114. Hu, S.; He, P.; He, D. Motion Capture and Estimation of Dynamic Properties for Realistic Tree Animation. In *International Workshop on Next Generation Computer Animation Techniques*; Springer: Cham, Switzerland, 2017.
115. Quigley, E.; Yu, Y.; Huang, J.; Lin, W.; Fedkiw, R. Real-time interactive tree animation. *IEEE Trans. Vis. Comput. Graph.* **2017**, *24*, 1717–1727. [CrossRef]
116. Talamo, C.; Bonanomi, M.M. The impact of digitalization on processes and organizational structures of architecture and engineering firms. In *Digital Transformation of the Design, Construction and Management Processes of the Built Environment*; Daniotti, B., Gianinetto, M., Della Torre, S., Eds.; Springer: Cham, Switzerland, 2020; pp. 175–185.
117. Komzák, J.; Slavík, P. Scaleable GIS data transmission and visualisation. In Proceedings of the Seventh International Conference on Information Visualization, London, UK, 18 July 2003.
118. Yan, G.; Hu, R.; Wang, Y.; Ren, H.; Song, W.; Qi, J.; Chen, L. Scale effect in indirect measurement of leaf area index. *IEEE Trans. Geosci. Remote Sens.* **2016**, *54*, 3475–3484. [CrossRef]
119. Xie, Q.; Dash, J.; Huang, W.; Peng, D.; Qin, Q.; Mortimer, H.; Casa, R.; Pignatti, S.; Laneve, G.; Pascucci, S.; et al. Vegetation indices combining the red and red-edge spectral information for leaf area index retrieval. *IEEE J. Sel. Top. Appl. Earth Obs. Remote Sens.* **2018**, *11*, 1482–1493. [CrossRef]
120. Ren, Z.; Du, Y.; He, X.; Pu, R.; Zheng, H.; Hu, H. Spatiotemporal pattern of urban forest leaf area index in response to rapid urbanization and urban greening. *J. For. Res.* **2018**, *29*, 785–796. [CrossRef]
121. Kutzner, T.; Chaturvedi, K.; Kolbe, T.H. CityGML 3.0: New functions open up new applications. *PFG–J. Photogramm. Remote Sens. Geoinf. Sci.* **2020**, *88*, 43–61. [CrossRef]

122. Gobeawan, L.; Lin, E.S.; Tandon, A.; Yee, A.T.K.; Khoo, V.H.S.; Teo, S.N.; Yi, S.; Lim, C.W.; Wong, S.T.; Wise, D.J.; et al. Modeling Trees for Virtual Singapore: From Data Acquisition to CITYGML Models. *Int. Arch. Photogramm. Remote Sens. Spat. Inf. Sci.* **2018**, *XLII-4/W10*, 55–62. [CrossRef]
123. Lee, S.-K.; Koenig, R.; Petzold, F. Computational Support for Interactive Exploration of Urban Design Variants. In Proceedings of the 18th International Conference, CAAD Futures 2019, Daejeon, Republic of Korea, 26–29 June 2019; pp. 259–273.
124. Seifert, N.; Mühlhaus, M.; Petzold, F. Urban strategy playground: Rethinking the urban planner's toolbox. *Int. J. Archit. Comput.* **2020**, *18*, 20–40. [CrossRef]
125. Reinhard, K. Urban design synthesis for building layouts based on evolutionary many-criteria optimization. *Int. J. Archit. Comput.* **2015**, *13*, 257–269. [CrossRef]
126. Kirnbauer, M.; Kenney, W.; Churchill, C.; Baetz, B. A prototype decision support system for sustainable urban tree planting programs. *Urban For. Urban Green.* **2009**, *8*, 3–19. [CrossRef]
127. Gabriel, M.; Fellner, J.; Reitberger, R.; Lang, W.; Petzold, F. Voxel based method for real-time calculation of urban shading studies. In Proceedings of the 12th Symposium on Simulation for Architecture and Urban Design (SimAUD), Virtual Event, USC, Los Angeles, CA, USA, 15–17 April 2021.
128. Huttner, S.; Bruse, M. Numerical modeling of the urban climate—A preview on ENVI-met 4.0. In Proceedings of the 7th International Conference on Urban Climate ICUC-7, Yokohama, Japan, 29 June–3 July 2009.
129. Chokhachian, A.; Hiller, M. PANDO: Parametric tool for simulating soil-plant-atmosphere of tree canopies in grasshopper. In Proceedings of the 11th Annual Symposium on Simulation for Architecture and Urban Design, Virtual Event, USC, Los Angeles, CA, USA, 25–27 May 2020.
130. Motamedi, A.; Hammad, A.; Asen, Y. Knowledge-assisted BIM-based visual analytics for failure root cause detection in facilities management. *Autom. Constr.* **2014**, *43*, 73–83. [CrossRef]
131. Rafiee, A.; Dias, E.; Fruijtier, S.; Scholten, H. From BIM to geo-analysis: View coverage and shadow analysis by BIM/GIS integration. *Procedia Environ. Sci.* **2014**, *22*, 397–402. [CrossRef]
132. Messner, J.; Anumba, C.; Dubler, C.; Goodman, S.; Kasprzak, C.; Kreider, R.; Leicht, R.; Saluja, C.; Zikic, N. *BIM Project Execution Planning Guide (v. 2.2)*; Computer Integrated Construction Research Program, Pennsylvania State University: University Park, PA, USA, 2019.
133. Latiffi, A.A.; Brahim, J.; Fathi, M.S. The development of building information modeling (BIM) definition. In *Applied Mechanics and Materials*; TRANS TECH PUBL: Zurich, Switzerland, 2014.
134. Lasseck, M. Image-based Plant Species Identification with Deep Convolutional Neural Networks. In *CLEF (Working Notes), CEUR Workshop Proceedings Vol-1866*; Conference and Labs of the Evaluation Forum: Dublin, Ireland, 2017.
135. St-Onge, B.; Vega, C.; Fournier, R.A.; Hu, Y. Mapping canopy height using a combination of digital stereo-photogrammetry and lidar. *Int. J. Remote Sens.* **2008**, *29*, 3343–3364. [CrossRef]
136. Cornea, N.D.; Silver, D.; Min, P. Curve-skeleton properties, applications, and algorithms. *IEEE Trans. Vis. Comput. Graph.* **2007**, *13*, 530. [CrossRef] [PubMed]
137. Huang, H.; Wu, S.; Cohen-Or, D.; Gong, M.; Zhang, H.; Li, G.; Chen, B. L1-medial skeleton of point cloud. *ACM Trans. Graph.* **2013**, *32*, 65:1–65:8. [CrossRef]
138. Guo, J.; Xu, S.; Yan, D.-M.; Cheng, Z.; Jaeger, M.; Zhang, X. Realistic procedural plant modeling from multiple view images. *IEEE Trans. Vis. Comput. Graph.* **2018**, *26*, 1372–1384. [CrossRef] [PubMed]
139. Wu, B.; Zheng, G.; Chen, Y.; Yu, D. TreeSke: A Structural-Lossless Skeleton Extraction Method for Point Cloud Data of Canopy Woody Materials. Preprint. Available online: https://www.researchsquare.com/article/rs-78239/v1 (accessed on 11 November 2022).
140. Jin, S.; Su, Y.; Gao, S.; Wu, F.; Ma, Q.; Xu, K.; Hu, T.; Liu, J.; Pang, S.; Guan, H.; et al. Separating the structural components of maize for field phenotyping using terrestrial LiDAR data and deep convolutional neural networks. *IEEE Trans. Geosci. Remote Sens.* **2019**, *58*, 2644–2658. [CrossRef]
141. Liu, Y.; Guo, J.; Benes, B.; Deussen, O.; Zhang, X.; Huang, H. TreePartNet: Neural decomposition of point clouds for 3D tree reconstruction. *ACM Trans. Graph.* **2021**, *40*, 1–16. [CrossRef]
142. Middleton, W.; Shu, Q.; Ludwig, F. Representing living architecture through skeleton reconstruction from point clouds. *Sci. Rep.* **2022**, *12*, 1–13. [CrossRef]
143. Vonderach, C.; Voegtle, T.; Adler, P. Voxel-based approach for estimating urban tree volume from terrestrial laser scanning data. *Int. Arch. Photogramm. Remote Sens. Spat. Inf. Sci.* **2012**, *39*, 451–456. [CrossRef]
144. Van der Zande, D.; Stuckens, J.; Verstraeten, W.W.; Muys, B.; Coppin, P. Assessment of light environment variability in broadleaved forest canopies using terrestrial laser scanning. *Remote Sens.* **2010**, *2*, 1564–1574. [CrossRef]
145. Béland, M.; Widlowski, J.-L.; Fournier, R.A. A model for deriving voxel-level tree leaf area density estimates from ground-based LiDAR. *Environ. Model. Softw.* **2014**, *51*, 184–189. [CrossRef]
146. Wu, D.; Phinn, S.; Johansen, K.; Robson, A.; Muir, J.; Searle, C. Estimating changes in leaf area, leaf area density, and vertical leaf area profile for mango, avocado, and macadamia tree crowns using terrestrial laser scanning. *Remote Sens.* **2018**, *10*, 1750. [CrossRef]
147. Hosoi, F.; Omasa, K. Voxel-based 3-D modeling of individual trees for estimating leaf area density using high-resolution portable scanning lidar. *IEEE Trans. Geosci. Remote Sens.* **2006**, *44*, 3610–3618. [CrossRef]

148. Li, S.; Dai, L.; Wang, H.; Wang, Y.; He, Z.; Lin, S. Estimating leaf area density of individual trees using the point cloud segmentation of terrestrial LiDAR data and a voxel-based model. *Remote Sens.* **2017**, *9*, 1202. [CrossRef]
149. Gärtner, H.; Wagner, B.; Heinrich, I.; Denier, C. 3D-laser scanning: A new method to analyze coarse tree root systems. *For. Snow Landsc. Res.* **2009**, *82*, 95–106.
150. Hruska, J.; Čermák, J.; Šustek, S. Mapping tree root systems with ground-penetrating radar. *Tree Physiol.* **1999**, *19*, 125–130. [CrossRef] [PubMed]
151. Amato, M.; Basso, B.; Celano, G.; Bitella, G.; Morelli, G.; Rossi, R. In situ detection of tree root distribution and biomass by multi-electrode resistivity imaging. *Tree Physiol.* **2008**, *28*, 1441–1448. [CrossRef]
152. Kniemeyer, O.; Buck-Sorlin, G.; Kurth, W. GroIMP as a platform for functional-structural modelling of plants. *Frontis* **2007**, *22*, 43–52.
153. Chan, W.-L.; Eng, Y.; Ge, Z.; Lim, C.W.C.; Gobeawan, L.; Poh, H.J.; Wise, D.J.; Burcham, D.C.; Lee, D.; Cui, Y.; et al. Wind loading on scaled down fractal tree models of major urban tree species in Singapore. *Forests* **2020**, *11*, 803. [CrossRef]
154. Enshassi, M.A.; al Hallaq, K.A.; Tayeh, B.A. Limitation factors of building information modeling (BIM) implementation. *Open Constr. Build. Technol. J.* **2019**, *13*, 189–196. [CrossRef]

forests

MDPI

Article

The Influence of Plant Community Characteristics in Urban Parks on the Microclimate

Yu Bao [1], Ming Gao [2], Dan Luo [3,*] and Xudan Zhou [1]

1 College of Forestry and Grassland Science, Jilin Agricultural University; Jilin Provincial Key Laboratory of Tree and Grass Genetics and Breeding, Changchun 130118, China
2 School of Architecture, Harbin Institute of Technology; Key Laboratory of Cold Region Urban and Rural Human Settlement Environment Science and Technology, Ministry of Industry and Information Technology, Harbin 150006, China
3 School of Architecture and Urban Planning, Chongqing University; Key Laboratory of New Technology for Construction of Cities in Mountain Areas, Chongqing 400044, China
* Correspondence: luodan@cqu.edu.cn; Tel.: +86-023-6512-0702

Abstract: The hot and humid feeling of the urban environment enhances residents' discomfort indices. Although the cooling and humidifying effects of plant communities in various urban parks are significant, there is still insufficient evidence for the effects of plant community characteristics on temperature and humidity. In this study, 36 typical plant communities in the Changchun Water Culture and Ecological Park in China were selected in the summer (21–23 August 2020) from 8:00 to 18:00 for three days when it was sunny and windless. We obtained plant community characteristics through field measurements and drone recordings to explore the relationship between plant community characteristics and the mechanism of temperature and humidity. The study observed that (1) the canopy density and three-dimensional green amount were significantly related to the benefits of cooling and humidification. When the canopy density is between 0.7 and 0.8 and the three-dimensional green volume is above 4 m^3/m^2, the greatest benefit is achieved; (2) the discomfort index is between 0.6 and 0.8, and the three-dimensional green volume is 4 m^3/m^2–6 m^3/m^2 minimum; and (3) the changes in temperature and humidity are different for different types of plant communities, which lead to differences in people's perceptions of environmental comfort. The tree–grassland and tree–shrub–grass types had the most apparent improvement effects on comfort. The results show that in the design process of urban park plants, emphasis is placed on plant community configuration with apparent cooling and humidification effects, which can improve the comfort of tourists in hot and humid environments. The research results provide theoretical support for sustainable urban green space development.

Keywords: urban green space; plant community; outdoor thermal comfort; microclimate; canopy density; tridimensional green biomass

Citation: Bao, Y.; Gao, M.; Luo, D.; Zhou, X. The Influence of Plant Community Characteristics in Urban Parks on the Microclimate. *Forests* **2022**, *13*, 1342. https://doi.org/10.3390/f13091342

Academic Editors: Thomas Rötzer, Stephan Pauleit, Mohammad A Rahman and Astrid Reischl

Received: 25 July 2022
Accepted: 21 August 2022
Published: 23 August 2022

1. Introduction

The deterioration of the urban environment and the unique nature of the underlying city surface have changed the thermal environment, forming the urban heat island (UHI) effect [1,2], reducing the comfort of urban residents, aggravating the negative impact of the urban environment, and causing more significant difficulties to the daily work and lives of the residents [3]. In recent years, solving the urban heat problem has become an urgent issue for worldwide urban development planning. The relevant studies observed that urban green spaces can effectively alleviate the UHI effect, reduce the temperature in urban spaces, and act as urban cold islands [4,5]. In addition, urban green spaces can meet citizens' spiritual, cultural, leisure, and entertainment needs, and provide various ecosystem services, such as the ecological adjustment and maintenance of biodiversity [6,7]. Moreover, it plays a vital role in biodiversity [8,9].

In recent years, the analysis of the improvement of the temperature and humidity in green spaces has begun to be refined from green park spaces to small-scale areas. The current study examines small-scale plant communities and different plant species. The relevant studies show that different plant types have different effects on temperature improvement [10,11]. The research conducted on hawthorn, *Robinia pseudoacacia* L., *Sorbifolia*, and other plants shows that the difference in their temperature-improvement effect was nearly four times greater [12,13]. It was observed that different plant communities, such as grasslands, woodlands, and ornamental shrubs, have different effects on improving temperature values in different environments [14]. Further research shows that the leaf characteristics of different types of plants affect the cool temperature of shaded air space [15].

With the progress of research methods, the research process tends to be quantitative. Canopy density, three-dimensional green yield, leaf area index, and other indicators began to be applied to studies on the environmental temperature and humidity changes [16]. For example, canopy coverage was adopted as an indicator to evaluate forest ecosystem services in an urban environment. A study conducted on tree cover and vertical leaves [17] indicated that trees with larger leaf area indices (LAI) were conducive to better property values. In other words, the more trees present with a larger leaf area index, the higher the attributed value of the property. The biomass and tree–shrub cover have a neutral effect, whereas replacing trees with grass cover results in a low value. In the current study, we determine the leaf area index, crown height, and plants, among which, for example, the height and crown width impact the plants' cooling effects [18].

The cooling effect of plants on the environment can improve the comfort of urban residents, resulting in a sense of belonging and identity [4,19]. There are many indexes used for evaluating aspects of comfort, such as standard effective temperature (SET) [20], effective physiological equivalent temperature [5] (PET), and universal thermal climate index (UTCI) [21]. It has been shown that SET can be used to accurately evaluate the thermal comfort of people in a stable, indoor environment. We investigate the thermal comfort experienced on university campuses and observe that the average temperature is 20.6 °C in the current study. The comfortable temperature range is from 19.5 °C to 21.8 °C [22]. An in-depth study observed that green space had a positive effect on human comfort, not only because the transpiration of plants can reduce the air temperature, but also because plants can block part of the direct radiation of the sun [23–26]. By adding green spaces to the simulation, it was observed that the areas with higher greenery rates had higher comfort levels, and the correlation analysis concludes that if the park is entirely covered by green space, it is in a complete thermal-comfort state [27].

To summarize, there is a strong recognition that urban green spaces can alleviate urban thermal environment problems, and the cooling effect of urban parks is remarkable. It was observed that the scale, distance, and other factors of green space affect the cooling levels of green space. However, in urban forests and green space planning and designing, the connection between the configuration of plant community structures and ecological service functions is still insufficient. It is necessary to thoroughly study the relationship and mechanism of the spatial layout of vegetation, the structure of plants, and temperature and humidity in relation to comfort for human beings. In turn, the ecological function of urban green spaces will be enhanced, and the quality of living in that environment will be improved through the optimal allocation of plant communities.

In the current study, we propose the following research questions:

Research question 1 (RQ1): Are there significant differences in the effects of plant characteristics on the temperature and humidity in urban parks?

Research question 2 (RQ2): If the answer to research question 1 is positive, are there significant differences between plant community characteristics and comfort? What characteristics of plants reduce feelings of discomfort? Which ones improved?

Research question 3 (RQ3): If the answer to research question 2 is positive, does the vegetation's spatial arrangement affect the determined conclusions?

2. Methods

2.1. Study Area

Changchun City Water Culture and Ecological Park, 43°51′ N, 123°21′ E, is located at the intersection of Jingshui Road and Yatai Street in Changchun City, China (Figure 1, location and scope of the study). The park covers an area of 30.2 ha. It is an urban brownfield reconstruction project that won the ASLA 2019 comprehensive design award. The original site was the first water-purification plant in Changchun, built during the period of Puppet Manchuria. Changchun's water supply culture has experienced 80 years of historical evolution, and 300,000 square meters of scarce ecological green space have appeared in the city's hinterland. There are 17 families and 36 genera of woody plants at the site. The environment is pleasant and adjacent to a residential area and is thus highly recognized by the public. A representative mix of artificial and natural plant communities was selected for the study. Moreover, plant communities are widely used in cities and parks. The plant community is rich in layers, and the tree branches are higher than 3M, which is suitable for people to move freely in the forest space.

Figure 1. Location and scope of the study.

2.2. Data Acquisition

In the current study, four routes were selected in the Water Culture Ecological Park in Changchun City, as presented in Figure 2. Each route had 9 sample points out of a total of 36. The sample points were divided into five plant community types: arbor, arbor–grass, arbor–shrub–grass, shrub–grass, and grassland. The plot range for arbor, arbor–grass, and arbor–shrub–grass-type plant communities was set as 20 m × 20 m, while the shrub–grass and grassland types were set as 10 m × 10 m. Two groups of non-vegetation coverage areas were designated as reference groups 100 m outside of the park, as presented in Figure 2 and the reference group location map. To avoid interference, the distance between each sample plot was more than 10 m, and there was no water source within 50 m. The high-density population areas were avoided as much as possible. The measurement date was from 21 to 23 August 2020, from 7:50 to 18:10 every day—there were three consecutive days with clear sky conditions and there was no rainfall or irrigation during the sampling times. The details are presented in Table 1. The instruments used in the acquisition process are presented in Table 2.

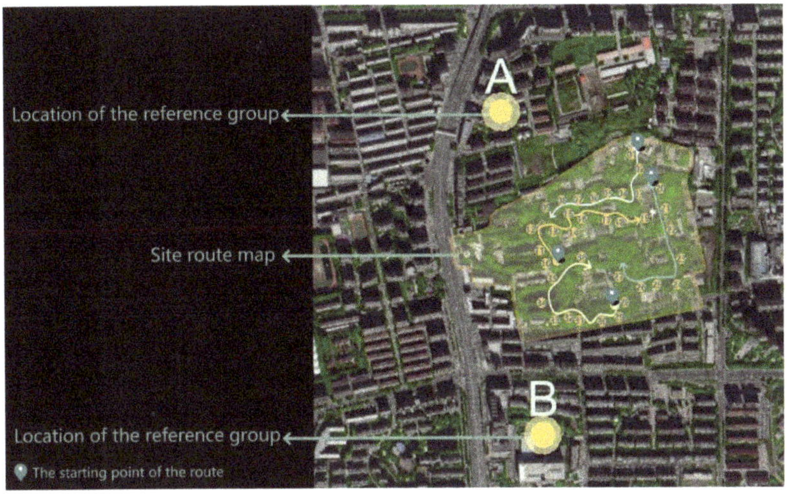

Figure 2. Site route and reference group location, A and B are both reference groups.

Table 1. Weather information for Changchun from 21 to 23 August 2020.

Measurement Date	Temperature Condition	Weather Condition	Wind Direction
21 August 2020	26 °C/14 °C	Sunny	Northeasterly wind, levels 1–2
22 August 2020	27 °C/ 17 °C	Sunny/cloudy	Southwesterly wind, levels 1–2
23 August 2020	27 °C/17 °C	Sunny/cloudy	Southwesterly wind, level 3

Table 2. List of experimental instruments.

Instrument Name	Model and Origin	Experimental Use	Parameter Range
Aerial unmanned aerial vehicle (UAV)	Phantom 4 Pro/China	Takes photos of the sample to obtain the real picture	Maximum altitude: 6000 m Fov84° 20 megapixels Photo resolution: 5472 × 3648/4864 × 3648/5472 × 3078
Temperature and humidity recorder	Tes-1361c/Taiwan	Measures the temperature and humidity of the sample	Measurement range: Humidity: 10%–95% R.H Temperature: −2–20 °C −60 °C/−4 ° F − + 140 ° f Measurement accuracy: Humidity: ±3% R.H − ±5% R.H Temperature: ±0.8 °C, ±1.5° f

2.2.1. Temperature and Humidity

A temperature and humidity recorder was used at a vertical height of 1.5 m above the ground. The flow measurement was conducted for three consecutive days from 7:50 to 18:10 and at every other hour for each sample plot. It took approximately 20 min for each line to complete the measurement. Five groups of values at four corners and middle points were obtained for each sample point, and the average value was the temperature and humidity value of the sample.

2.2.2. Canopy Density

Canopy density refers to the ratio of the total projected area (crown width) of the arbor (shrub) crown under direct sunlight to the entire scope of the forestland (stand) [28]. The unmanned aerial vehicle (UAV) collected vertical projection pictures of the sample points. The sample areas of the arbor, arbor–grass, and arbor–shrub–grass-type plant communities were 20 m × 20 m, and the flight height was 40 m. The shrub–grass and grassland types were 10 m × 10 m and the flight height was 20 m.

2.2.3. Tridimensional Green Biomass

Green biomass refers to the three-dimensional area of green plants in a certain area, and tridimensional green biomass density refers to the proportion of plant stems and leaves in a unit space. Infrared rangefinders were used to measure the plot's tree- and shrub-height characteristics. The characteristics, such as diameter at breast height, base diameter, crown width of trees, shrub diameter, height, and other characteristics of shrubs and herbs, were measured individually. The four corners and the focal point of the diagonal line were set at 2 m × 2 m in the five positions of the grassland community.

2.3. *Data Processing*

The calculation method of the discomfort index (DI) with the highest adaptability in the outdoor environment was adopted for the thermal comfort degree [29], and the calculation formula was as follows:

$$DI = Tair - 0.55(1 - 0.01RH)(Tair - 14.5) \tag{1}$$

where DI is the discomfort index, TAIR is the air temperature (°C), and RH is the relative humidity (%). According to this procedure, the comfort level distribution for each square point can be obtained, and the comfort level can be classified according to the interval division of the discomfort index according to the criteria presented in Table 3 [30,31]. The greater the discomfort, the lower the comfort level of the human body.

Table 3. Division of discomfort index and human comfort.

Grade	Temperature Humidity Effect on Discomfort Index (DI)	Sensory Level
1	<21.0	No discomfort.
2	21.0–23.9	A small number of people felt uncomfortable. Discomfort expressed by <50% of the population.
3	24.0–26.9	Most people did not feel comfortable. Discomfort expressed by >50% of the population.
4	27.0–28.9	Most people did not feel comfortable. Discomfort expressed by the majority of the population.
5	29.0–31.9	Almost everyone felt uncomfortable. Discomfort expressed by all.
6	>32.0	Risk of heatstroke. Stages of medical alarm.

The cooling and humidifying effects of the plant community are expressed by the average temperature percentage (T_p) and average humidity percentage (H_p), respectively. The calculation formula is as follows:

$$T_p = \frac{\sum_{i=1}^{n} \frac{T_{ci} - T_i}{T_{ci}} \times 100\%}{n} \tag{2}$$

$$H_p = \frac{\sum_{i=1}^{n} \frac{H_i - H_{ci}}{H_i} \times 100\%}{n} \tag{3}$$

where T_{ci} is the temperature value of the control plot at the i-th time in °C; T_i is the temperature value of the community plot at the i-th time in °C; H_{ci} is the relative humidity value of the control plot at the i-th time (%); H_i is the community plot; and n is the recording-time period.

3. Results

3.1. Influence of Canopy Density on Temperature and Humidity Effect

3.1.1. Relationship between Canopy Density and Cooling Effect

To explore the relationship between a plant community's canopy density and temperature, the quadratic curve fitting of the two variables is presented in Figure 3. The results show that with the increase in canopy density, the cooling capacity of the plant community increases within one day, with a correlation coefficient of 0.868, significant at $p < 0.01$. When the canopy density is greater than 0.91, the fitting curve tends to be stable, and the cooling level of the plant community reaches the maximum level.

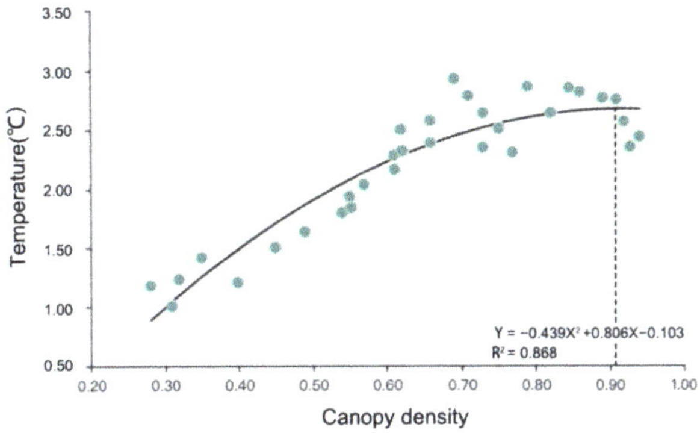

Figure 3. Relationship between temperature and canopy density.

3.1.2. Relationship between Canopy Density and Humidification Effect

By fitting a quadratic curve between the plant community's canopy density and humidity, their relationship is presented in Figure 4. As the plant community's canopy density increases, so does the humidification benefit it creates. The correlation coefficient was 0.413, with a significant $p < 0.01$. In comparison to the abilities of plant canopy density and the humidification effect, the canopy density produces a better cooling effect.

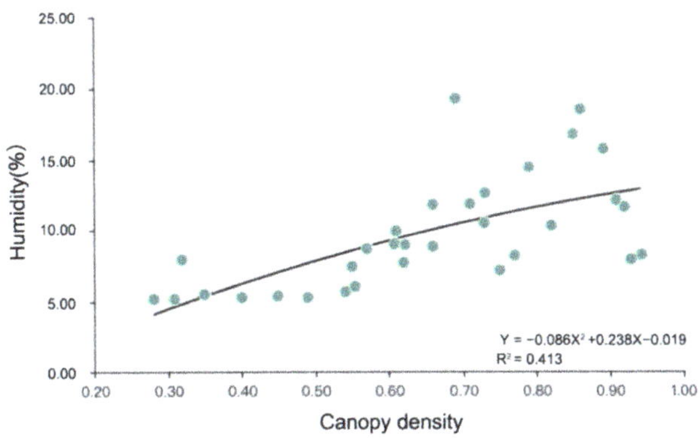

Figure 4. Relationship between humidification and canopy density.

3.2. Influence of Tridimensional Green Biomass on Temperature and Humidity Effect

3.2.1. Relationship between Tridimensional Green Biomass and Cooling Effect

To answer research question 1 (RQ1), the tridimensional green biomass was fitted with a delicate logarithmic temperature change curve, as presented in Figure 5. The correlation coefficient was 0.761, with a significant $p < 0.01$. Especially when the three-dimensional green biomass was more remarkable than 4 m^3/m^2, for every increase of 1 m^3/m^2 for the three-dimensional green biomass, the significant cooling benefit was less than the increase in the three-dimensional green biomass less than 4 m^3/m^2.

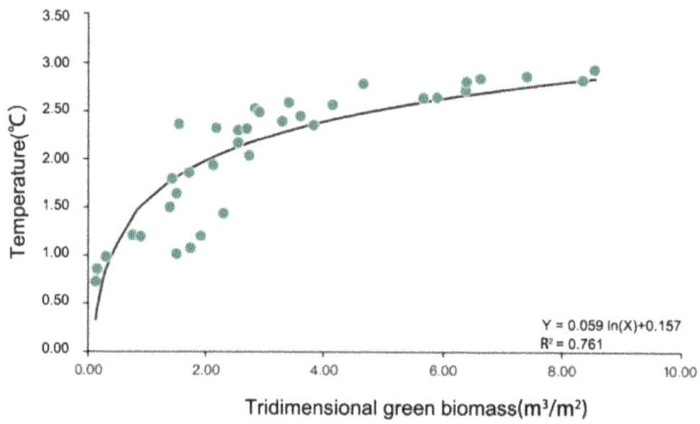

Figure 5. Relationship between temperature and tridimensional green biomass.

3.2.2. Relationship between Tridimensional Green Biomass and Humidification Effect

A quadratic curve fitted the tridimensional green biomass and humidity change. The results are presented in Figure 6. The correlation coefficient between the tridimensional green biomass and humidity was 0.840, with a significant $p < 0.01$. The humidification benefit of the plant community increased with the increase in the tridimensional green biomass density. The effect was better when the density of the tridimensional green biomass was greater than 2 m^3/m^2.

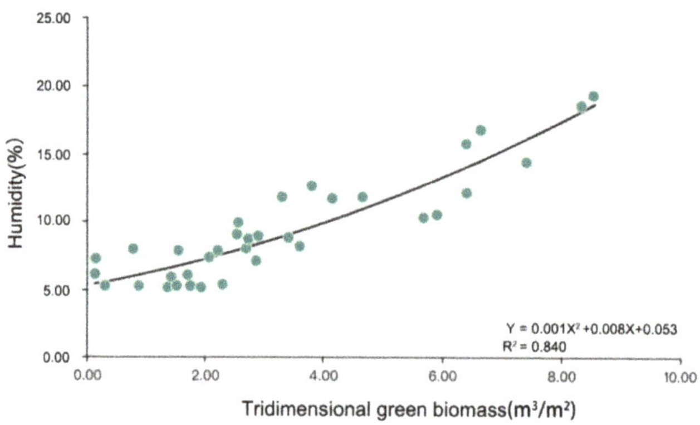

Figure 6. Relationship between humidification and tridimensional green biomass.

3.3. The Relationship between Discomfort and Plant Community Characteristics

3.3.1. The Relationship between Discomfort and Canopy Density

As presented in Figure 7, the plants' discomfort indices and canopy densities are used as scatter points, and they are fitted by a quadratic curve. The effect of the plant community's canopy closure on the discomfort index was analyzed. The results show that as the canopy density of the plant community increased, the discomfort index decreased. When the canopy density reached 0.7, the discomfort index was at its lowest level. The canopy density of the plant community was between 0.6 and 0.8, and the range of the discomfort index was relatively suitable, which could effectively create a comfortable environment.

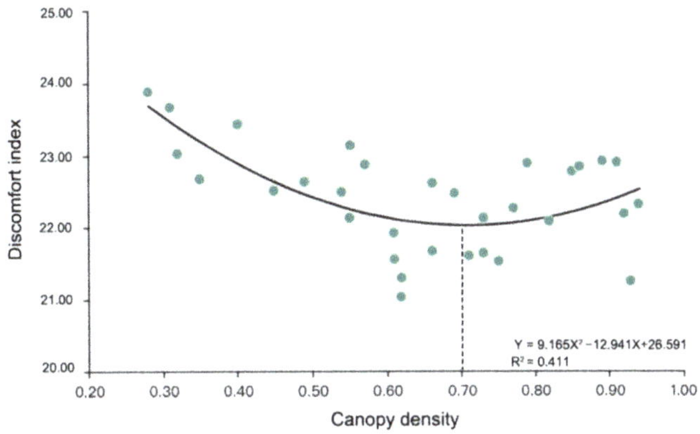

Figure 7. Relationship between discomfort index and canopy density.

3.3.2. The Relationship between Discomfort and Tridimensional Green Biomass

Using a scatter plot, a quadratic curve fitting was performed between the discomfort index and tridimensional green biomass to explore their relationship. As shown in Figure 8, the discomfort index first decreased and then increased with the increase in plant tridimensional green biomass. The discomfort index was relatively low when the density of the tridimensional green biomass was 4–6 m^3/m^2, which provides an answer for research question 2 (RQ2).

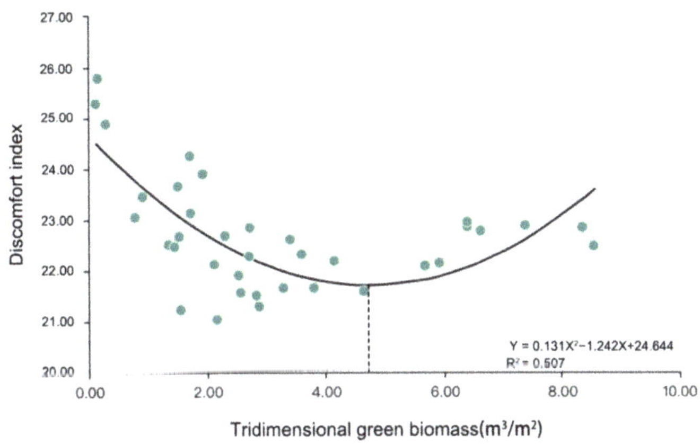

Figure 8. Relationship between discomfort index and tridimensional green biomass.

3.4. The Relationship between Discomfort and Plant Community Structure

3.4.1. Vertical Structure

In order to answer research question 3 (RQ3), the analysis of variance (one-way ANOVA) was used to study the differences in community types for discomfort factors. Community type was significant at a 0.01 level for comfort (F = 26.323, p = 0.000), as presented in Table 4. The compared results of the groups' average scores with noticeable differences were grassland type > shrub–grass type > arbor type > arbor–grassland type > arbor–shrub–grass type.

Table 4. Discomfort and the results for the variance analysis of plant community structures.

	Community Type (Mean ± SD)					F	p
	Arbor Type (n = 6)	**Arbor–Shrub–Grass Type (n = 20)**	**Arbor–Grassland Type (n = 3)**	**Shrub–Grass Type (n = 3)**	**Grassland Type (n = 4)**		
Discomfort index	22.89 ± 0.35	22.07 ± 0.61	22.28 ± 0.31	23.52 ± 0.45	25.04 ± 0.65	26.323	0.000 **

** $p < 0.01$.

3.4.2. Plane Layout

We explored the relationship between the layout of the plants and the discomfort index. According to the plant community's quadratic structure, the plant plane layout was divided into five forms. The data comparison and analysis of the discomfort index were conducted, as presented in Table 5. We integrated the plant floor plans according to the typical quadrats. The order of discomfort index in the plane layout was grass type > adaptive type > determinant type > encircling type > encircling type, which contributes further to answering research question 3.

Table 5. Plant community's plane characteristics.

Item	Schematic Diagram of Plane Layout	Real-Life Example	Average Discomfort Index
Adaptive type			22.49
Encircling type			21.93
Grass type			24.70
Encircling type			21.91
Determinant type			22.47

4. Discussion

4.1. Temperature and Humidity Effects under the Influence of Plant Characteristics

Compared to the urban environment, urban vegetation has been proven to play an essential role in mitigating the heat island effect. The intense transpiration of green plants can play a specific role in cooling and humidifying effects. Additionally, the leaves of plants can partially block and absorb solar radiation, so the internal environment of urban parks has trends of low temperature and high humidity [32,33]. In summer, they feel more comfortable and present distinct differences during different periods.

During the day, the temperature change and plant canopy density presented a positive, upward trend. They reached a critical point when the canopy density was 0.80. The cooling capacity reached a maximum value of 2.61 °C, similar to the previous results [5]. In addition, the high plant canopy density had no significant effect on the cooling and humidification of the local microclimate, which could lead to the formation of excessively high numbers of green closures that would result in the spread of the local, brutal, and hot climate. On the other hand, when the tridimensional green biomass of the plants was approximately 4 m^3/m^2, the humidification effect was the best. The green biomass density of the plants was closely related to the shaded areas they produced, which absorbed and blocked solar heat radiation [34]. However, the effect of excessive green biomass on the microclimate was not apparent.

4.2. Discomfort under the Influence of the Plant Community

The current study obtained the discomfort index under the influence of different plant communities, grassland type > shrub–grass type > arbor type > arbor–grassland type > arbor–shrub–grass type, which is roughly the same as the results obtained by previous studies [28,29,35]. During the summertime, the canopy density of small-scale plants is an essential indicator for reducing temperature intensity. Since the influence of lawns on the microclimate is weak, plants can be added to areas with low thermal comfort levels [30].

It is worth noting that dense vegetation communities surrounded subsidence depression, and the wind speed was low. It is easy to create a microclimate with low temperatures and high humidity levels [4]. On the other hand, the factors affecting the local microclimate may have been related to the surrounding large-scale fields [36]. The relevant studies show that hard paving materials are affected by thermal conductivity, specific heat capacity, and surface reflectivity, and have a certain warming effect on the surrounding environment [37]. Although the quadratic point is a certain distance away, it is still affected by its thermal radiation.

4.3. Impact on the Design of the Park's Climate and Environment

The sustainability of urban green spaces has been proven time and time again. Increasing urban green spaces will improve the quality of urban life [38] and, by improving the microclimate and reducing urban pollution levels [39], create a space that is beneficial to the health of human beings, allowing them to perform exercises for fitness purposes, thereby reducing the risk of some chronic diseases.

In previous studies, increasingly more people began to pay attention to the critical role of green park spaces in cooling and humidifying the environment. According to our research experiments and analysis of the results, we concluded that different types of plant communities have apparent differences in the regulation of temperature and humidity levels. We observed that multi-layered plant communities were most effective in terms of their cooling and humidification effects. This plant community is diverse in structure and rich in species [40]. The point-like tree layout had a more pronounced cooling effect, which may be attributed to the synergistic effect of the trees and the underground cover, so we encourage the mixing of trees and grasses and the expansion of the urban forest belt [41]. This can be used as part of the basis for the design of parks and green space plants. In the spatial layout of vegetation, a higher level of vegetation canopy density is also essential for regulating temperature and humidity levels. It seems that the perceived comfort levels in different areas should also be an essential part of the design of a park environment.

4.4. Limitations and Future Research

This study highlighted the importance of plant community types and spatial layouts in parks for assessing the benefits and discomfort levels created by temperature and humidity effects. However, it is still necessary to study the relevant influencing factors, such as wind and building environments, in the broader range of the physical environment of a park's green space to strengthen the accuracy of the results produced by this study. A park's green space is an open-space environment, not a relatively stable environment inside a laboratory, which is restricted and affected by multiple factors. In addition, in order to describe the discomfort factors, the most apparent temperature and humidity levels sensed by the human body were selected. Therefore, the influence of the external, physical environment was considered in the green space, and the environment simulation was conducted in an indoor laboratory to couple and superimpose more accurate quantitative research as the goal of future studies.

5. Conclusions

This study investigated five different plant community structure types and the relationship between temperature and humidity benefits in urban parks. Aiming to address the internal temperature and humidity changes in urban parks can ensure an improved perception of the environment by people in green park spaces. The study further revealed the importance of urban green spaces for ecosystem services. The degree of effect of urban green spaces on environmental cooling and humidification was determined. The average cooling effect in the green space was significantly related to plant canopy closure and three-dimensional green volume. In order to create a more ecologically beneficial urban park and green space, more attention should be paid to the planting designs in the parks.

Secondly, environmental temperature and humidity changes affected peoples' perceptions of environmental comfort in different ways. Through the cooling and humidification effects of different plant communities and their impact on human comfort, it was observed that the arbor and grassland types, as well as the types of trees, shrubs, and grasses, impacted individuals' comfort levels. The effect of improvement was the most apparent outcome. The construction of the same type of urban park in the same area could increase the allocation ratio of these two types of plant communities. The cooling and humidifying capacity of the park could be enhanced to reduce the discomfort of tourists during outdoor leisure activities. Therefore, scientifically planned urban green spaces can provide multifunctional habitats for ecosystem services and increase human wellbeing in a more effective manner.

Author Contributions: Conceptualization, Y.B., M.G. and D.L; methodology, M.G.; software, Y.B.; validation, M.G. and D.L.; formal analysis, M.G. and X.Z.; investigation, Y.B. and X.Z.; resources, Y.B., X.Z. and D.L.; data curation, Y.B. and X.Z.; writing—original draft preparation, Y.B.; writing—review and editing, Y.B., M.G. and X.Z.; visualization, M.G.; supervision, Y.B., X.Z. and D.L.; funding acquisition, D.L. All authors have read and agreed to the published version of the manuscript.

Funding: This research was funded by the China Postdoctoral Science Foundation (grant number 2017M622964), Jilin Province Science and Technology Development Plan Project (grant number 20210203013SF).

Data Availability Statement: Not applicable.

Acknowledgments: We would especially like to thank the graduate students who participated in our research.

Conflicts of Interest: The authors declare no conflict of interest.

References

1. Oke, T.R. The energetic basis of the urban heat island. *Q. J. R. Meteorol. Soc.* **1982**, *108*, 1–24. [CrossRef]
2. Priyadarsini, R. Urban Heat Island and its Impact on Building Energy Consumption. *Adv. Build. Energy Res.* **2009**, *3*, 261–270. [CrossRef]
3. Soga, M.; Gaston, K.J. Extinction of experience: The loss of human–nature interactions. *Front. Ecol. Environ.* **2016**, *14*, 94–101. [CrossRef]
4. Lai, D.; Liu, W.; Gan, T.; Liu, K.; Chen, Q. A review of mitigating strategies to improve the thermal environment and thermal comfort in urban outdoor spaces. *Sci. Total Environ.* **2019**, *661*, 337–353. [CrossRef] [PubMed]
5. Yan, H.; Wu, F.; Dong, L. Influence of a large urban park on the local urban thermal environment. *Sci. Total Environ.* **2018**, *622–623*, 882–891. [CrossRef]
6. Shuhao, L.; Chang, S.; Ruochen, Y.; Jianye, Z.; Kun, L.; Kwangmin, H.; Shiro, T.; Junhua, Z. Using Crowdsourced Big Data to Unravel Urban Green Space Utilization during COVID-19 in Guangzhou, China. *Land* **2022**, *11*, 990.
7. Escobedo, F.J.; Kroeger, T.; Wagner, J.E. Urban forests and pollution mitigation: Analyzing ecosystem services and disservices. *Environ. Pollut.* **2011**, *159*, 2078–2087. [CrossRef]
8. García-Martínez, M.; Vanoye-Eligio, V.; Leyva-Ovalle, O.R.; Zetina-Córdoba, P.; Mejía, M. Diversity of Ants (Hymenoptera: Formicidae) in a Sub-Montane and Sub-Tropical Cityscape of Northeastern Mexico. *Sociobiology* **2019**, *66*, 44–47. [CrossRef]
9. Rosas-Mejía, M.; Llarena-Hernández, C.; Núez-Pastrana, R.; Vanoye-Eligio, V.; García-Martínez, M. Value of a Heterogeneous Urban Green Space for Ant1 Diversity in a Highland City in Central Eastern Mexico. *Southwest. Entomol.* **2020**, *45*, 461–474. [CrossRef]
10. Peters, E.B.; McFadden, J.P.; Montgomery, R.A. Biological and environmental controls on tree transpiration in a suburban landscape. *J. Geophys. Res. Biogeosci.* **2010**, *115*, G04006. [CrossRef]
11. Rötzer, T.; Rahman, M.A.; Moser-Reischl, A.; Pauleit, S.; Pretzsch, H. Process based simulation of tree growth and ecosystem services of urban trees under present and future climate conditions. *Sci. Total Environ.* **2019**, *676*, 651–664. [CrossRef] [PubMed]
12. Moser-Reischl, A.; Rahman, M.A.; Pauleit, S.; Pretzsch, H.; Rötzer, T. Growth patterns and effects of urban micro-climate on two physiologically contrasting urban tree species. *Landsc. Urban Plan.* **2019**, *183*, 88–99. [CrossRef]
13. Rahman, M.A.; Armson, D.; Ennos, A.R. A comparison of the growth and cooling effectiveness of five commonly planted urban tree species. *Urban Ecosyst.* **2015**, *18*, 371–389. [CrossRef]
14. Fung, C.K.W.; Jim, C.Y. Microclimatic resilience of subtropical woodlands and urban-forest benefits. *Urban For. Urban Green.* **2019**, *42*, 100–112. [CrossRef]
15. Lin, B.-S.; Lin, Y.-J. Cooling Effect of Shade Trees with Different Characteristics in a Subtropical Urban Park. *HortScience* **2010**, *45*, 83–86. [CrossRef]
16. Kabisch, N.; Qureshi, S.; Haase, D. Human–environment interactions in urban green spaces—A systematic review of contemporary issues and prospects for future research. *Environ. Impact Assess. Rev.* **2015**, *50*, 25–34. [CrossRef]
17. Escobedo, F.J.; Adams, D.C.; Timilsina, N. Urban forest structure effects on property value. *Ecosyst. Serv.* **2015**, *12*, 209–217. [CrossRef]
18. Morakinyo, T.E.; Kong, L.; Lau, K.L.; Yuan, C.; Ng, E. A study on the impact of shadow-cast and tree species on in-canyon and neighborhood's thermal comfort. *Build. Environ.* **2000**, *115*, 1–17. [CrossRef]
19. Lai, D.; Guo, D.; Hou, Y.; Lin, C.; Chen, Q. Studies of outdoor thermal comfort in northern China. *Build. Environ.* **2014**, *77*, 110–118. [CrossRef]
20. Gagge, A.P.; Fobelets, A.P.; Berglund, L.G. A standard predictive index of human response to the thermal environment. *ASHRAE Trans.* **1986**, *92*, 709–731.
21. Hadianpour, M.; Mahdavinejad, M.; Bemanian, M.; Nasrollahi, F. Seasonal differences of subjective thermal sensation and neutral temperature in an outdoor shaded space in Tehran, Iran. *Sustain. Cities Soc.* **2018**, *39*, 751–764. [CrossRef]
22. Liu, J.; Yang, X.; Jiang, Q.; Qiu, J.; Liu, Y. Occupants' thermal comfort and perceived air quality in natural ventilated classrooms during cold days. *Build. Environ.* **2019**, *158*, 73–82. [CrossRef]
23. Shashua-Bar, L.; Pearlmutter, D.; Erell, E. The influence of trees and grass on outdoor thermal comfort in a hot-arid environment. *Int. J. Climatol.* **2011**, *31*, 1498–1506. [CrossRef]
24. Ali-Toudert, F.; Mayer, H. Effects of asymmetry, galleries, overhanging facades and vegetation on thermal comfort in urban street canyons. *Sol. Energy* **2007**, *81*, 742–754. [CrossRef]
25. Streiling, S.; Matzarakis, A. Influence of single and small clusters of trees on the bioclimate of a city: A case study. *J. Arboric.* **2003**, *29*, 309–316. [CrossRef]
26. Matzarakis, A.; Streiling, S. Stadtklimatische Eigenschaften von Bumen. *Gefahrst. Reinhalt. Luft* **2004**, *64*, 307–310.
27. Yang, A.S.; Juan, Y.H.; Wen, C.Y.; Chang, C.J. Numerical simulation of cooling effect of vegetation enhancement in a subtropical urban park. *Appl. Energy* **2017**, *192*, 178–200. [CrossRef]
28. Bowler, D.E.; Buyung-Ali, L.; Knight, T.M.; Pullin, A.S. Urban greening to cool towns and cities: A systematic review of the empirical evidence. *Landsc. Urban Plan.* **2010**, *97*, 147–155. [CrossRef]
29. Cohen, P.; Potchter, O.; Matzarakis, A. Daily and seasonal climatic conditions of green urban open spaces in the Mediterranean climate and their impact on human comfort. *Build. Environ.* **2012**, *51*, 285–296. [CrossRef]

30. Yang, Y.; Zhou, D.; Wang, Y.; Ma, D.; Chen, W.; Xu, D.; Zhu, Z. Economical and outdoor thermal comfort analysis of greening in multistory residential areas in Xi'an. *Sustain. Cities Soc.* **2019**, *51*, 101730. [CrossRef]
31. Georgi, N.J.; Zafiriadis, K. The impact of park trees on microclimate in urban areas. *Urban Ecosyst.* **2006**, *9*, 195–209. [CrossRef]
32. Taleghani, M. Outdoor thermal comfort by different heat mitigation strategies—A review. *Energy Rev.* **2018**, *81*, 2011–2018. [CrossRef]
33. Smithers, R.J.; Doick, K.J.; Burton, A.; Sibille, R.; Steinbach, D.; Harris, R.; Groves, L.; Blicharska, M. Comparing the relative abilities of tree species to cool the urban environment. *Urban Ecosyst.* **2018**, *21*, 851–862. [CrossRef]
34. Wang, Y.; Bakker, F.; de Groot, R.; Wörtche, H.; Leemans, R. Effects of urban green infrastructure (UGI) on local outdoor microclimate during the growing season. *Environ. Monit. Assess.* **2015**, *187*, 732. [CrossRef] [PubMed]
35. Zhang, B.; Xie, G.-D.; Gao, J.-X.; Yang, Y. The cooling effect of urban green spaces as a contribution to energy-saving and emission-reduction: A case study in Beijing, China. *Build. Environ.* **2014**, *76*, 37–43. [CrossRef]
36. Oke, T.R.; Mills, G.; Christen, A.; Voogt, J.A. *Urban Climates*; Cambridge University Press: Cambridge, UK, 2017.
37. Taha, H.; Akbari, H.; Rosenfeld, A.; Huang, J. Residential cooling loads and the urban heat island—The effects of albedo. *Build. Environ.* **1988**, *23*, 271–283. [CrossRef]
38. Ss, A.; Ja, B. Role of geospatial technology in understanding urban green space of Kalaburagi city for sustainable planning. *Urban For. Urban Green.* **2019**, *46*, 126450.
39. Makhelouf, A. The effect of green spaces on urban climate and pollution. *J. Environ. Health Sci. Eng.* **2009**, *6*, 35–40.
40. Zhang, Z.; Lv, Y.; Pan, H. Cooling and humidifying effect of plant communities in subtropical urban parks. *Urban For. Urban Green.* **2013**, *12*, 323–329. [CrossRef]
41. Amani-Beni, M.; Zhang, B.; Xie, G.-D.; Xu, J. Impact of urban park's tree, grass and waterbody on microclimate in hot summer days: A case study of Olympic Park in Beijing, China. *Urban For. Urban Green.* **2018**, *32*, 1–6. [CrossRef]

forests

Article

Research on Thermal Comfort of Underside of Street Tree Based on LiDAR Point Cloud Model

Xuguang Zhang [1,†], Yakai Lei [1,†], Rui Li [1], Aidan Ackerman [2], Nan Guo [1], Yonghua Li [1], Qiusheng Yang [1,*] and Yang Liu [1,*]

[1] College of Landscape Architecture and Art, Henan Agricultural University, Zhengzhou 450002, China; zyh1067662256@163.com (X.Z.); lykfjyl@henan.edu.cn (Y.L.); lirui0423@163.com (R.L.); nyaunu@163.com (N.G.); liyhhany@163.com (Y.L.)

[2] College of Environmental Science and Forestry, State University of New York, Syracuse, NY 14250, USA; acackerm@esf.edu

* Correspondence: qsyang@henan.edu.cn (Q.Y.); liuyang1991@henau.edu.cn (Y.L.); Tel.: +86-139-0383-8327 (Q.Y.); +86-185-3991-0220 (Y.L.)

† These authors contributed equally to this work.

Abstract: As a major part of the urban green space system, street trees play a corresponding role in adjusting the thermal comfort of the environment and alleviating heat island effects. The correlation between the morphological structure and microclimate factors in the lower canopy of street trees was studied, using data that were captured with vehicle-borne LiDAR to model the morphological structure and geometric canopy features of six key street tree species in the built-up area of Zhumadian City, Henan Province. The regulating ability and differences of canopy geometry on cooling, humidification, shading, and Physiologically Equivalent Temperature (PET) were studied. Research shows that: (1) Canopy Volume (CV), Canopy Area (CA), Canopy Diameter (CD), and Tree Height (TH) have a linear negative correlation with air temperature, relative humidity, and luminosity. TH had significant effects on the air temperature and relative humidity (R^2 = 0.90, 0.96), and CV and CD had significant effects on luminosity (R^2 = 0.70, 0.63). (2) The oval-shaped plant (*Platanus acerifolia (Aiton) Willdenow*) had a strong cooling and shading ability, with an average daily cooling of 2.3 °C and shading of 318 cd/m^2. The spire-shaped plant (*Cedrus deodara (Roxb.) G. Don*) had a strong ability to humidify, with an average daily humidification of 4.5%. (3) The oval-shaped and spire-shaped plants had a strong regulation ability on PET, and the daily average regulation values were 40.5 °C and 40.9 °C, respectively. (4) The CD of the oval-shaped plant had a significant effect on PET (R^2 = 0.49), and the TH of the spire-shaped plant had a significant effect on PET (R^2 = 0.80), as well as a significantly higher CV and Leaf Area Index (LAI) than other street tree species. Therefore, selecting oval and spire canopy-shaped plants with a thick canopy, dense leaves, and high CD and TH values as street trees can provide significant advantages in cooling, humidifying, and shading, and can effectively adjust human comfort in the lower canopy understory. This study is the first to apply LiDAR technology to the regulation of urban microclimate. The research results provide a theoretical basis and quantitative reference for street tree design from the perspective of outdoor thermal comfort evaluation and play a guiding role in the application of LiDAR to urban forestry research.

Keywords: street tree; lidar point cloud model; morphological structure; canopy geometry; thermal comfort; microclimate

Citation: Zhang, X.; Lei, Y.; Li, R.; Ackerman, A.; Guo, N.; Li, Y.; Yang, Q.; Liu, Y. Research on Thermal Comfort of Underside of Street Tree Based on LiDAR Point Cloud Model. *Forests* **2022**, *13*, 1086. https://doi.org/10.3390/f13071086

Academic Editors: Thomas Rötzer, Stephan Pauleit, Mohammad A Rahman and Astrid Reischl

Received: 9 May 2022
Accepted: 7 July 2022
Published: 11 July 2022

1. Introduction

Street trees are a significant part of urban green space and can improve the microclimate of road spaces, providing benefits such as reducing air temperature, increasing relative humidity, attenuating solar and ground radiation, and improving ventilation. During the summer months, the average air temperature in the space under the street trees can be reduced by about 1.7–3.3 °C, compared with the ambient temperature in the open pavement space [1]. In the summer months in subtropical and tropical areas, an increase in tree planting area of 25% may effectively reduce the temperature by about 3.3–5.6 °C [2]. Additionally, the variation of spatial microclimate under street trees is also impacted by the tree species that are present [3]. The morphological structural characteristics of street trees may affect their cooling, humidification, and shading functions. CV, CA, and leaf density can promote canopy transpiration and reduce air temperature [4–6]. Canopy structure, leaf shape, and leaf color may affect visible and solar infrared light penetration levels [7]. Lastly, the cooling effect of small-leaved species is generally stronger than that of large-leaved species [8]. In contrast, canopy geometry is a significant factor in regulating microclimate and human comfort. Canopy shadows may not only reduce glare and prevent environmental diffuse reflection, but also may increase the heat exchange buffer layer between the vertical space of the building and the transverse space of the street, thereby reducing the temperature of the underlying surface, as well as reducing the wind speed [9,10]. At present, most research focuses on exploring the microclimate regulation ability of urban street trees in tropical and arid regions; research on the microclimate regulation of urban street trees in temperate regions is still insufficient [11,12].

Vehicle-borne LiDAR as an active remote sensing technology has been widely used in urban construction management. Compared with airborne LiDAR, it can perform a complete acquisition of surface feature information, which is conducive to the collection of morphological and structural information below the vegetation canopy. Vehicle-borne LiDAR point clouds can be used to obtain more accurate morphological and structural information of vegetation than airborne LiDAR, allowing the extraction of phenotypic parameter information such as LAI, in contrast with traditional field surveys of plant resources [13]. Popescu used local filtering techniques to separate trees from deciduous, coniferous, and mixed forests to extract individual tree structural parameters [14]. Hyyppa calculated the CV value of individual trees based on an extracted point cloud [15,16]. Li developed a top-down regional growth Point Cloud Segmentation (PCS) to segment complex mixed forests [17]. Tao used Comparative Shortest-Path (CSP) to verify its accuracy, and the segmentation accuracy was as high as 94% [18].

This study innovatively combines LiDAR technology to explore the impact and intensity differences of the morphological structure and canopy geometry of urban street trees in typical temperate regions on microclimate regulation. Based on LiDAR point cloud models of main roads in the high-density urban area of Zhumadian, the CV, CA, CD, TH, and Diameter at Breast Height (DBH) eigenvalues of six key street trees were extracted and their canopy geometric characteristics summarized. The microclimate characteristic values of air temperature, relative humidity, and luminosity in the lower space of street trees were obtained through field measurement, and the regulation effect of different morphological structure characteristics on microclimate factors was explored. Following this, the PET index measuring human comfort was used as the standard to quantify the intensity of microclimate regulation, and the regulation benefit of canopy geometric characteristics on PET index was discussed, providing a quantitative reference for selecting street tree forms with better human comfort in temperate regions.

2. Materials and Methods

2.1. Study Area

The study area is located in Zhumadian City, Henan Province (32°18′~33°35′ N, 113°10′~115°12′ E). Zhumadian City is a typical continental monsoon type semi-humid climate with four distinct seasons. Its annual average temperature is 14.8 °C~15.4 °C; precipitation is 850 mm~980 mm; wind speed is 1.8 m/s~2.4 m/s; and cumulative total solar radiation is 112~120 KWh/cm^2 (Figure 1) [19]. Six typical main roads in the main urban area were selected, including: Tongda Road, Zhengyang Road, Leshan Avenue, Cedar Avenue, Yulan Road, and Landmark Avenue. Additionally, a control check (CK) was selected in an open space of Tianzhong Academy located on Landmark Avenue. Zhengyang Road and Leshan Avenue run north-south (N-S), while Tongda Road, Cedar

Avenue, Yulan Road, and Landmark Avenue run east-west (E-W) (Figure 2). The street trees within the study area can all be classified as mature (≥10 years) and are planted in a linear arrangement. The selected road boundary conditions (aspect ratio, length, building density) are generally consistent throughout the study area.

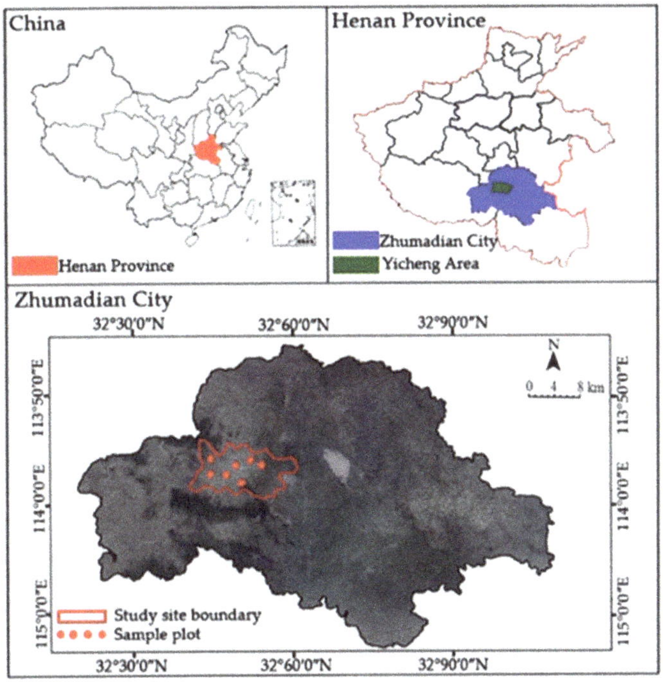

Figure 1. Location analysis.

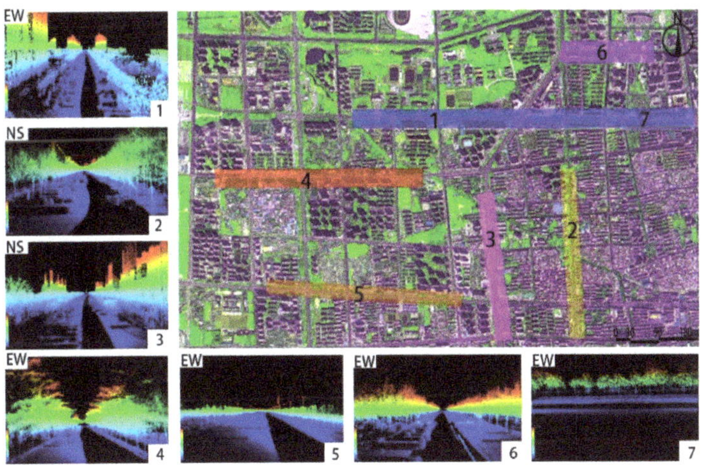

(1) Tongda Road; (2) Zhengyang Road; (3) Leshan Road; (4) Xuesong Road; (5) Yulan Road; (6) Zhidi Road; (7) CK

Figure 2. Test road point cloud quadrat.

2.2. Data Acquisition

2.2.1. LiDAR Point Cloud Data Acquisition

A vehicle-borne LiDAR mobile scanning measurement was carried out on a test road using the AS-900HL mobile scanning system with the RIEGL VUX-1 UAV laser scanner, and equipped with an Inertial Measurement Unit (IMU) and Global Navigation Satellite System (GNSS). The measurement enabled capture of the high-precision color point clouds of streets and surrounding buildings, the extraction of key plant features, and 3D modeling of the surrounding environment. The horizontal field of view of the laser scanner is $0°$~$360°$, the 112 scanning frequency is 10–200 Hz, the accuracy is better than 10 mm, and the weight is 3.75 g. The scanner rotates around the direction of traffic to scan the street environment. The captured HDR panorama has a resolution of 6000×4000, reaching 24.3 megapixels. The operation of the vehicle-borne LiDAR is carried out automatically according to the fixed point, and the operator controls the scanning to ensure the accuracy of the data. 1. The LiDAR Delta2A completes the map construction and determines the running track. 2. The operator selects the starting point and the end point. Cross driving was carried out on the test area, and the route was set as a closed loop to prevent deviation from the scanning route. This is critically important, as deviation from the scanning route has been shown to cause high rates of missing point cloud information that can lead to data deformation [20]. 3. The multi-scan method is used to complete the scan within 4 h.

2.2.2. Measurement of Microclimatic Factors

According to the weather records of the local meteorological bureau in August 2021, the climatic factors of air temperature, relative humidity, and luminosity have rich diurnal changes, and local microclimate characteristics are obvious. The 1st, 5th, and 8th are the most typical, which can represent the climatic characteristics of the entire month of August. On 1, 5 and 8 August 2021, a period of clear weather and peak crowd activity (10:00–16:00) was selected for data collection. According to the existing forestry and garden plant survey and site review in Zhumadian city, 6 key street tree species were selected as the test objects (Figure 3), and the microclimate factors were measured according to the road corresponding to the planting of 6 tree species. An LM-8000 environmental measuring instrument was used to collect the air temperature, relative humidity, and luminosity data at the vertical height of 1.5 m from the ground at 7 measuring points, and black ball temperature data were collected with the JTR04 thermometer. The test height is set at 1.5 m where the human body is most sensitive to temperature and humidity, and air temperature and relative humidity are important factors that affect human comfort in environmental factors. The measurement location is in the space under the tree canopy in the middle of the road. In addition, the control point (CK) was set in the open space of Tian Zhong Academy of Landmark Avenue to facilitate the comparison of the differences in climatic factors of different tree species. During the sampling period, the actual measurement was performed once an hour. For each measurement, four groups of data were captured in four directions at the same measurement point. The measurement time of each group of data lasted 2–3 min, and the average value of each group of data was recorded [21]. Methodology for conducting microclimate research: The microclimate test is carried out according to the previous research standards. In this paper, the method of combining on-the-spot measurement, LiDAR scanning, and model simulation is used to study the road climate environment in Zhumadian city. It is mainly divided into four parts: 1. Field measurement of climate factors. 2. LiDAR scanning to extract tree species canopy information. 3. RayMan model to simulate human comfort value. 4. Data analysis. The basic parameters of each sampling equipment are shown in Table 1.

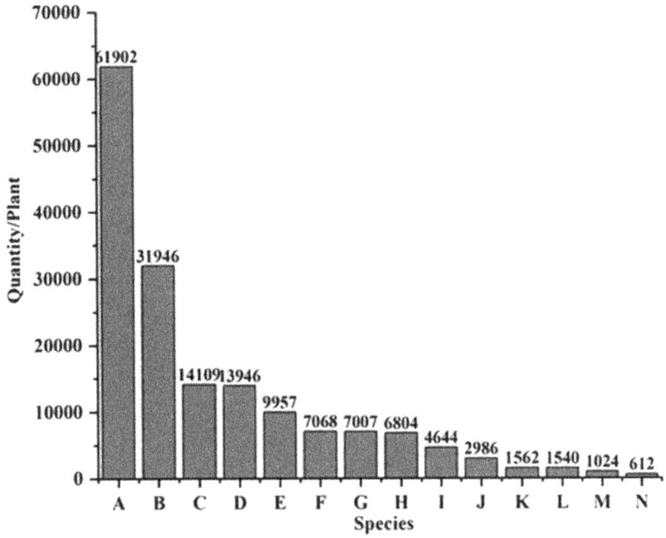

A: *Cinnamomum camphora (Linn) Presl*; B: *Platanus acerifolia (Aiton) Willdenow*;
C: *Sophora japonica Linn.*; D: *Koelreuteria paniculata*; E: *Cedrus deodara (Roxb.)*;
F: *Metasequoia glyptostroboides*; G: *Ligustrum lucidum*; H: *Ginkgo biloba* L.;
I: *Fraxinus chinensis*; J: *Albizia julibrissin Durazz.*; K: *Salix matsudana Koidz*;
L: *Magnolia grandiflora* L.; M: *Liquidambar formosana Hance*;
N: *Liriodendron chinense (Hemsl.) Sargent*

Figure 3. Statistics of key tree species in Zhumadian City.

Table 1. Test equipment.

Equipment Name	Measurement Parameters	Measuring Range	Measurement Accuracy
Luchang LM-8000 temperature, relative humidity, wind speed, and illumination four-in-one environmental measuring instrument	Air temperature Relative humidity Wind speed Luminosity	−100~1300 °C 10~95%RH 0.4~30.0 m/s	±1% ±4%RH 0.1 m/s 0~2000 FC
JTR04 Black ball thermometer	Black bulb temperature	−20~125 °C	±0.5 °C
AS-900HL Mobile Scanning System	Canopy structure	120 m	±1 cm
CI-110 Canopy Analyzer	LAI	Adjustable viewing angle 150°, 180°	3,000,000 pixels

2.3. Data Processing

2.3.1. Morphological and Structural Characteristics of Street Trees

After the vehicle-borne LiDAR scan was completed, the original point cloud was processed by LiDAR360 software. The processing included several steps: 1. Denoising to remove redundant noise from the point cloud; the number of neighborhood points was set to 10, and the standard deviation multiple was set to 5, in order to improve the quality of the point cloud. 2. Ground point filtering to separate ground points from the point cloud; the grid size was set to 0.5, and the thickness of the ground point was set to 0.3. This provided a basis for the accuracy of the next step (the CSP algorithm). 3. Regional ground point classification to reclassify buildings within, ground, and plants point clouds. 4. Ground point normalization to remove the influence of topographic relief on point cloud elevation. 5. Use of the CSP algorithm to separate single trees from the overall street tree point cloud;

the clustering threshold was set to 0.2, which controlled single tree detection and single tree canopy point cloud growth. The larger the value, the higher the segmentation efficiency, but too high affects the segmentation effect. The minimum number of clustering points was set to 500. This value mainly affects the growth of the point cloud of a single tree canopy. The smaller the value, the better the segmentation effect. By adjusting the two values, the optimal segmentation effect can be achieved. 6. After the single wood is divided, the information of the single wood is generated in a CSV table; followed by the extraction of CSV tables to include the morphological and structural parameters of single trees, including CV, CA, CD, TH, and DBH values [18]; and the manual correction of the separation results using seed point editing. 7. Obtaining the LAI value of tree canopy at each measuring point using the CI-110 canopy analyzer to supplement the CSV table.

2.3.2. Classification of Street Canopy Geometric Features

The street tree canopy in the study area exhibited varying geometric characteristics in the front, side, and top perspectives, creating a specific impact on the shade of the canopy [22,23]. According to the existing forestry and garden plant survey and site review in Zhumadian city, six tree species (Figure 3) of key street trees were selected as test objects [24]. Due to the morphological similarity among the major tree species, six different morphological tree species were selected based on both the trees' garden ornamental characteristics, as well as the overall the number of each tree species within the study area. Following the selection of six key tree species, point cloud contour fitting was carried out for each species. The 100 points cloud files of the six street trees were extracted from the original point cloud model. LiDAR360 software was then used to raster and intercept three views of point cloud of each street tree, with each view then imported into Adobe Photoshop 2017 software using the planar pixel grid. The grid size was set to 30 × 30 pixels, with a geometric feature fitting carried out for each view of the street tree. Following this, the geometric feature contour each tree was drawn (Figure 4). The fitting results fit into a range of geometric categories: cylindrical (*Cinnamomum camphora (Linn) Presl*); oval (*Platanus acerifolia (Aiton) Willdenow*); semi-circular (*Koelreuteria paniculata*); spire (*Cedrus deodara (Roxb.) G. Don*); spherical (*Magnolia grandiflora* L.); and triangular (*Liquidambar formosana Hance*).

2.3.3. Quantization of Physiological Equivalent Temperature (PET)

PET refers to an indoor or outdoor temperature corresponding to the skin temperature and internal temperature of the human body reaching the same thermal state as that of a typical indoor environment. In simple terms, the higher the PET value, the hotter the weather. In this paper, PET was used as the evaluation index of spatial thermal comfort in the lower part of the plant. After the measured microclimate factors were input into RayMan1.2, time data, geographic data, personal data, clothing activity data, and Tmrt value were added to simulate the measured microclimate data (Table 2), and the PET index was obtained.

Table 2. RayMan model input data.

Data Name	Data Content	Data Parameter
Time data	Simulation date, simulation time	1 August 2021–3 August 2021
Geographic data	Location, latitude, and longitude	Location: Zhumadian City, Henan Province, China; latitude and longitude 32°18′~33°35′ N,113°10′~115°12′ E
Climate data	Air temperature, relative humidity, wind speed, cloud cover, luminosity, average radiant temperature	The air temperature, relative humidity, and luminosity are measured data, the wind speed is 0.1/s, the cloud cover is 1, and the average radiation temperature can be calculated by it
Personal data	Height, weight, age, gender	Height: 175 cm; weight 75 kg; age 35 years old; gender: male
Clothing activity data	Clothing thermal resistance, activity	Clothing thermal resistance 0.6, activity 120 W

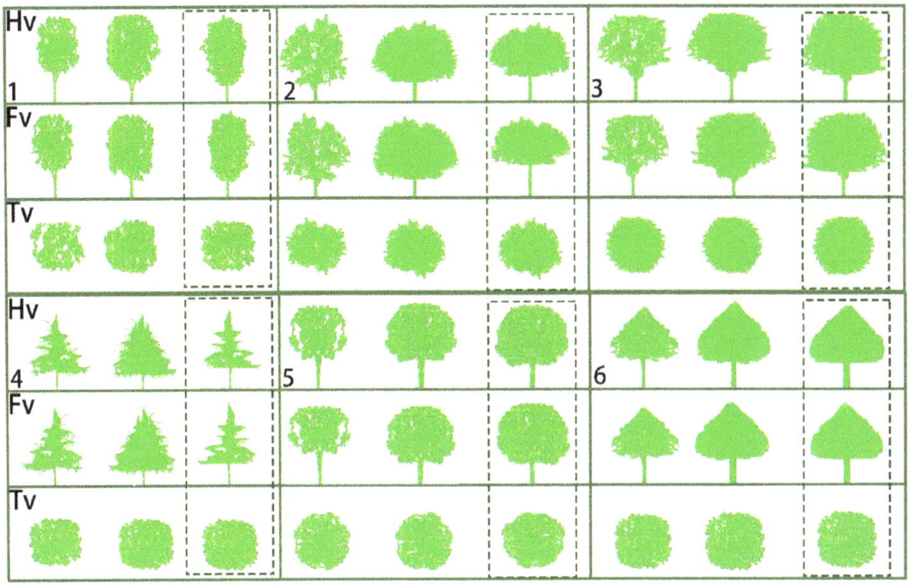

Hv: horizontal view; Fv: front view; Tv: top view. ⌊⌋: Results of morphological fitting:
1: *Cinnamomum camphora (Linn) Presl* (cylindrical); 2: *Platanus acerifolia (Aiton) Willdenow* (ellipse); 3: *Koelreuteria paniculata* (semicircle); 4: *Cedrus deodara* (Roxb.) G. Don (spired); 5: *Magnolia grandiflora* L. (spheroidal); 6: *Liquidambar formosana Hance* (triangle)

Figure 4. Three-view geometrical fitting of street tree.

A RayMan Model

In this study, the RayMan model was used to simulate and calculate the PET value of each measuring point. This model is a micro-scale model developed by Professor Matzarakis Andreas and his team at the University of Freiburg, Germany, for calculating thermal comfort parameters [25]. It is used to calculate thermal radiation flux and to evaluate and simulate the thermal environment. By inputting meteorological parameters such as cloud cover, wind speed, air temperature, relative humidity, and human body information (such as the age, gender, height, and weight of the tested person), the PET value of the human body under different space–time conditions is calculated. (https://www.urbanclimate.net/rayman/introraymanpro.htm, accessed on 3 April 2021)

B Mean Radiant Temperature

Mean Radiant Temperature (Tmrt) is an important parameter for evaluating the thermal comfort of the human body, and it is also a necessary parameter for calculating the input of different thermal indices, including PMV, SET, and PET. Tmrt converts the physiological effects of short-wave and long-wave radiation fluxes in the environment on human beings into units of degrees Celsius, representing the uniform temperature of the environment around a hypothetical black body radiation. The calculation formula of Tmrt according to Lintp and Hienn [26] is:

$$T_{mrt} = \left[\left(T_g + 273 \right)^4 + \frac{1.10 \times 10^8 {V_a}^{0.6}}{\varepsilon D^{0.4}} \left(T_g - T_a \right) \right]^{\frac{1}{4}} - 273 \quad (1)$$

Among them, T_g is the black ball temperature (°C); T_a is the air temperature; V_a is the wind speed (m/s); and D is the black ball diameter (m) (D = 0.05 m in this study). According to previous studies, the emissivity of this paper is chosen to be 0.95 [27].

3. Results

3.1. Correlation Analysis between Morphological Structure Characteristics of Street Trees and Microclimate Factors

SPSS 22.0 was used to analyze the main microclimate factors of six street trees. The results showed that the influence of street trees on air temperature, relative humidity, and luminosity showed extremely significant differences ($p < 0.01$). The correlation analysis between morphological structure characteristics and microclimate factors (Table 3) showed that: CV, CA, CD, and TH were negatively correlated with air temperature, relative humidity, and luminosity ($p < 0.05$). There was no obvious correlation between DBH and air temperature, relative humidity, and luminosity. The linear regression analysis of morphological structure characteristics and microclimate factors showed that TH had a significant effect on air temperature and relative humidity, with an R^2 of 0.90 and 0.96, respectively. CV and CD had significant effects on luminosity, with an R^2 of 0.70 and 0.63, respectively (Figure 5).

Table 3. Correlation analysis between morphological structure characteristics of street trees and microclimate factors.

		CV	CA	CD	TH	DBH
Air temperature	Pearson correlation	−0.182 *	−0.238 **	−0.278 **	−0.228 *	0.071
Relative humidity	Pearson correlation	−0.156 *	−0.194 *	−0.236 **	−0.250 **	−0.005
Luminosity	Pearson correlation	−0.281 **	−0.213 *	−0.142 *	−0.177 *	−0.158

**: Correlation is significant at 0.01 level. *: Correlation is significant at 0.05 level.

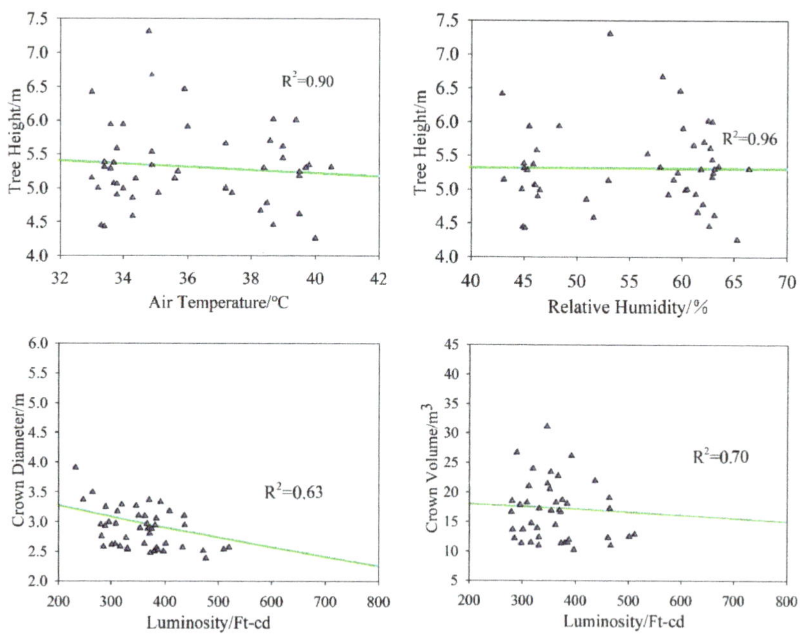

Figure 5. Linear relationship between morphological structure characteristics.

3.2. Thermal Environment Analysis of Street Canopy Geometry

According to the analysis of the climatic factors of different morphological trees, different morphological tree species have obvious differences in air temperature, relative humidity, and luminosity. The air temperature generally showed a trend of first rising and then falling. Relative humidity showed a trend of decline after an initial rise, with the distribution of values as a whole tending to be "U" shaped. Luminosity showed a single peak distribution trend and then a decreasing trend. These three factors all reached the peak value or valley during 12:00–13:00 (Figures 6–8). The oval plants showed the best performance in cooling and shading, with an average daily cooling of 2.3 °C, a humidification of 4.1%, and shading of 318 cd/m^2; the plants with the best performance in increasing humidity were spire-shaped, with an average daily humidification of 4.5%, cooling by 2 °C, and shading of 307 cd/m^2. The results show that the different morphological structure parameters of the plants themselves may lead to obvious differences in the cooling, humidification, and shading of the plants [28], which is consistent with the results of the correlation analysis in Section 3.1 above.

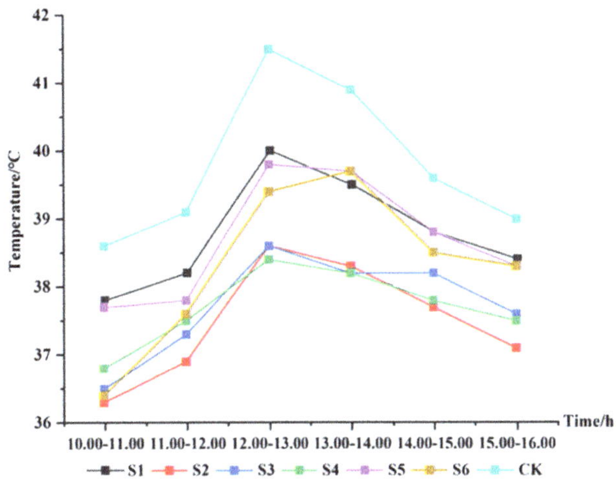

S1: *Cinnamomum camphora (Linn) Presl*; S2: *Platanus acerifolia (Aiton) Willdenow*; S3: *Koelreuteria paniculata*; S4: *Cedrus deodara Roxb.*; S5: *Magnolia grandiflora* L.; S6: *Liquidambar formosana Hance*

Figure 6. Air temperature diurnal variation of the tree species (S1–S6).

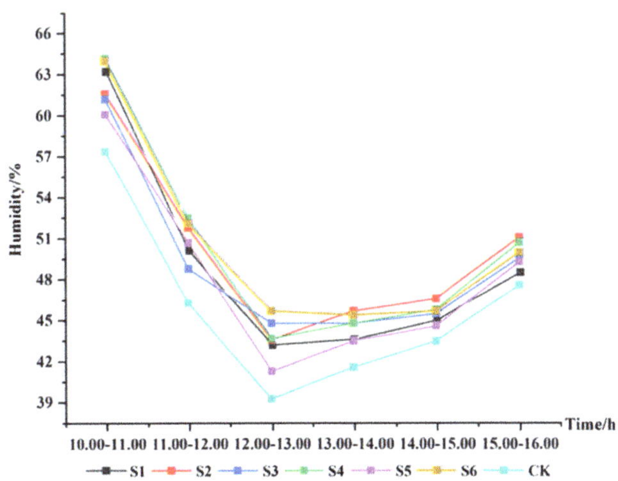

S1: *Cinnamomum camphora (Linn) Presl*; S2: *Platanus acerifolia (Aiton) Willdenow*; S3: *Koelreuteria paniculata*; S4: *Cedrus deodara Roxb.*; S5: *Magnolia grandiflora* L.; S6: *Liquidambar formosana Hance*

Figure 7. Relative humidity diurnal variation of the tree species (S1–S6).

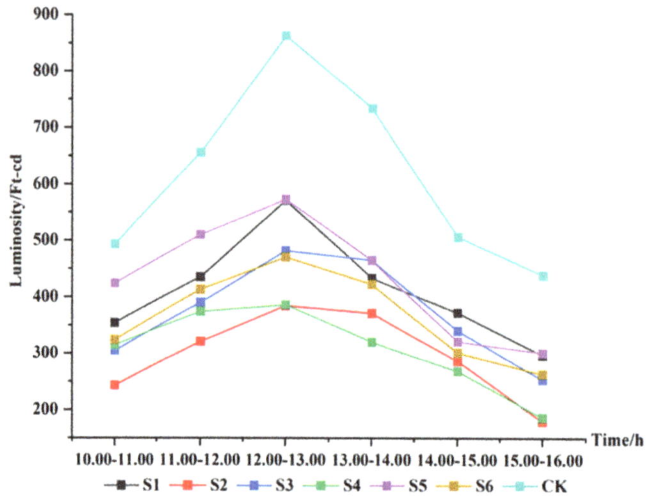

S1: *Cinnamomum camphora (Linn) Presl*; S2: *Platanus acerifolia (Aiton) Willdenow*; S3: *Koelreuteria paniculata*; S4: *Cedrus deodara Roxb.*; S5: *Magnolia grandiflora* L.; S6: *Liquidambar formosana Hance*

Figure 8. Luminosity diurnal variation of the tree species (S1–S6).

3.3. Analysis of Street Canopy Geometry Features and PET Index

The fitting relationship between plants with different canopy geometric characteristics and the PET index in the lower canopy space showed that (Table 4): the oval and spire-shaped plants had a strong ability to regulate PET, with daily average PET values of 40.5 °C and 40.9 °C, which were 1.2 °C and 0.8 °C lower than the average regulated PET values, respectively. The triangular plants showed a weak regulation ability to PET. The daily mean value of PET was 42.7 °C, 1 °C higher than the average adjusted PET value. From calculations of the average value of the morphological and structural characteristics of street trees (Table 5), it can be seen that the oval and spire shapes with a strong PET adjustment ability had high CV and LAI values. The CV and LAI values were 110 m^3 and 1.90, respectively. Among them, the CV values of the oval and spire shapes were 2.80 and 1.33 times the average CV value, respectively, and the LAI value was 1.07 and 1.11 times the average LAI value, respectively. In addition, the morphological structure characteristics of the oval and spire-shaped plants that had a significant effect on PET were CD (R^2 = 0.49) and TH (R^2 = 0.80), respectively (Figure 9).

Table 4. Diurnal variation characteristics of PET in plants with different canopy geometries.

Shape	10:00–11:00	11:00–12:00	12:00–13:00	13:00–14:00	14:00–15:00	15:00–16:00	Average PET
Cylindrical	41.5	42.1	43.5	42.3	41.9	40.5	41.9
Oval	38.1	39.8	42.9	42.8	41.1	38.7	40.5
Semi-circular	39.3	41.2	44.4	44.2	42.6	40.7	42
Spire	39.8	41.3	42.7	41.9	40.9	39.3	40.9
Spherical	41.5	42.6	43.3	43.1	43	40.4	42.3
Triangle	39.4	42	45.3	45.4	42.3	41.8	42.7

Table 5. Morphological structure characteristics of different tree shape.

Shape	Average DBH/m	Average TH/m	Average CD/m	Average CA/m^2	Average CV/m^3	Average LAI
Cylindrical	0.1	2.8	6.7	17	11	1.75
Oval	0.30	11	15.6	50	231	1.82
Semicircle	0.15	5.1	10.3	15	17	1.65
Spire	0.18	9.7	13.4	30	110	1.90
Spherical	0.15	2.9	6.9	18	18	1.87
Triangle	0.20	6.1	10.1	33	108	1.25

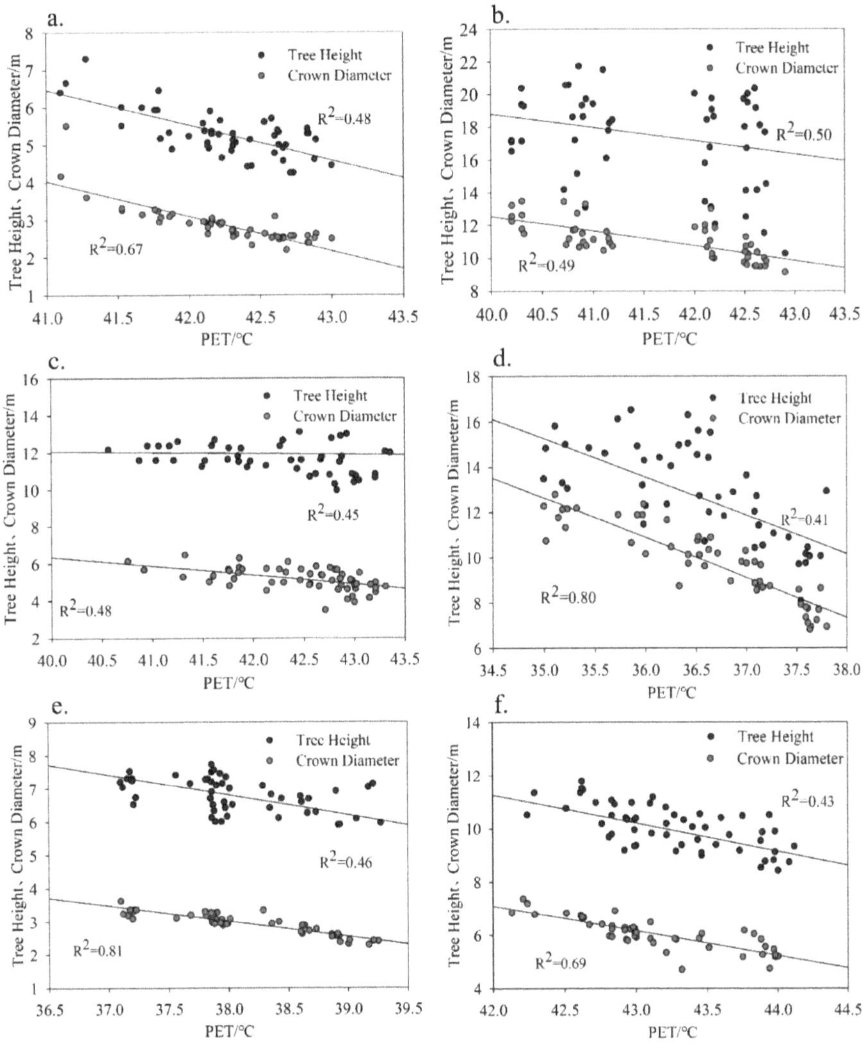

(**a**). *Cinnamomum camphora (Linn) Presl* (cylindrical); (**b**). *Platanus acerifolia (Aiton) Willdenow* (ellipse); (**c**). *Koelreuteria paniculata* (semicircle); (**d**). *Cedrus deodara (Roxb.G.Don)* (spired); (**e**). *Magnolia grandiflora* L. (spheroidal); (**f**). *Liquidambar formosana Hance* (triangle).

Figure 9. Linear relationship between canopy geometric features and PET.

4. Discussions

4.1. Influence of Street Tree Morphological Structure Characteristics and Canopy Geometry on Thermal Environment

4.1.1. Influence of Street Tree Morphological Structure Characteristics on Thermal Environment

The TH of the air temperature, relative humidity, CV, and CD to the photometric were characterized by a significant negative correlation, and further explained the influence of the shape and structure of street trees on the microclimate factors that are displayed in ventilation cooling, shading, and humidification [29]. The TH of street trees had a significant impact on the wind environment in the lower space of the plant. Under the same CD, the increase in TH was conducive to improving the ventilation conditions in the lower space of the canopy, improving the air heat circulation and diffusion, and thus reducing the air temperature in the lower space [30]. CV and CD can be quantified as the three-dimensional volume of the canopy, which comprehensively reflects the ability of the canopy to absorb and refract solar radiation [31]. The canopy can effectively block and refract visible light, and a high CV value can reduce the light transmittance of the lower space of the canopy and reduce the luminosity value. At the same time, high CV and CD values can effectively expand the spatial projection area under the canopy when the solar radiation angle is constant [32]. The TH of the street trees affects the ventilation conditions of the lower space of the canopy, and the CV and CD affect the shadowing area of the canopy. This indicates that selecting tall street trees with thick canopies and dense branches and leaves can improve the ventilation conditions of the lower space of the canopy and improve the microclimate environment.

4.1.2. Influence of Street Canopy Geometry on Thermal Environment

This research demonstrated that oval and spire-shaped plants perform best in regulating a thermal environment. Oval-shaped plants affect cooling and shading effects, and spire-shaped plants affect shading effects. The main reason was that the cooling, humidifying, and shading effect of the plant was significantly positively correlated with the LAI value [33], where the higher LAI value of plants corresponded to significantly increased cooling effects. During the day, the plant leaves absorb solar radiation and convert the solar radiation into latent heat by transpiration, allowing plant leaves to effectively reduce the temperature within the lower space of the plant. The oval plants had a high LAI value, which also more effectively reduced the temperature of the lower space of the plant. In addition, the shading effect of plants is mainly achieved through the geometric characteristics of the canopy. Compared with other plants, the oval-shaped "big on the top and small on the bottom" type plants have higher CV and CD values, which can intercept 80–90% of solar radiation and more effectively improve the shading effect of the lower space of the plant [34]. Compared with other plants, the LAI value of the spire-shaped plant was relatively high, leading to increased transpiration as the ambient temperature increased. The unique geometric characteristics and the TH and LAI value of the spire shapes can limit the water loss through transpiration and keep the humidity in the lower part of the plant at a high level [35]. Therefore, the LAI, CV, and CD values of the plant itself and the geometric characteristics of the canopy play a unique role in cooling, humidifying, and shading, which is consistent with the results obtained in this paper. In selecting street tree species, it is helpful to note that oval and spire trees provide good cooling, humidification, and shading in summer.

4.2. Influence of Street Canopy Geometry on PET

Variation in plant morphology can affect the PET index by changing the air temperature, relative humidity, luminosity, and wind speed of the microclimate factors in the lower space of the tree canopy. The oval and spire-shaped trees had a strong ability to adjust to PET. The CD of the oval-shaped plants ($R^2 = 0.49$) and the TH of the spire-shaped plants ($R^2 = 0.80$) all had significant effects on PET (Figure 9). Compared with other street tree species, the oval shaped trees had higher LAI values. The oval and spire-shaped trees had obvious effects on the diurnal variation of temperature, humidity, and luminosity (Figures 6–8). The difference in the geometry of the canopy layer on the street had obvious differences in the adjustment ability of PET: oval plants have strong lateral extension. When the TH is the same, a higher CD value leads to a larger shadowing area of the lower space of the plant, and in turn, a lower air temperature [34,36]. This is consistent with the significant effect of oval plants' CD on PET that was observed in this paper. Most of the spire-shaped plants were coniferous trees, and the space below the canopy was generally more limited than other trees (which was not conducive to airflow diffusion) and had a weak ventilation capacity [30]. The spire-shaped plant can adjust the ventilation environment of the lower space of the plant by changing the TH value and increasing the humidity. This is consistent with the results that were obtained in this paper [37]. Peters E B found that the LAI of plants had a significant effect on the temperature in the lower space of the canopy, and trees with the same morphological structure and canopy geometric characteristics could produce a temperature difference of up to 6 °C due to differences in LAI [38]. For the oval canopy, the expansion of LAI could enhance the transpiration capacity of plants, increase the relative humidity of the air in the lower space of the canopy, and further improve thermal comfort [39,40]. Due to the characteristics of needle leaves, LAI amplification did not significantly improve leaf evapotranspiration, although it could significantly increase leaf density, enhance shading ability, weaken visible light penetration, and improve thermal comfort [22].

5. Conclusions

In summer, the morphological and structural characteristics and geometric canopy characteristics of plants had an impact on the thermal comfort of the lower space, and the six plants had significant differences in cooling, humidification, shading, and thermal comfort adjustment. The correlation analysis between plant morphological structure characteristics and microclimate factors showed that TH was significantly negatively correlated with air temperature and relative humidity (R^2 = 0.90, 0.96), and CV and CD were significantly negatively correlated with luminosity (R^2 = 0.70, 0.63). The oval-shaped plants had the ability to regulate microclimates, with an average daily cooling of 2.3 °C and shading of 318 cd/m^2; the spire-shaped plants had a strong humidification ability, with an average daily humidification of 4.5%. The linear regression analysis on the geometric characteristics of plant canopy and PET showed that the geometric characteristics of oval-shaped and spire-shaped plant canopies had a strong regulation ability on PET, and the daily average regulation values of PET were 40.5 °C and 40.9 °C, respectively. The oval plants' CD had a significant effect on PET with an R^2 of 0.49, the spire-shaped TH had a significant effect on PET with an R^2 of 0.80, and both showed high CV and LAI values.

In this paper, we focused on the correlation between plant morphological structure and microclimate factors in the main urban area of Zhumadian city in summer, with an analysis of the linear relationship between plant canopy geometry and PET. Based on the microclimate benefits of different plants, this research suggests that oval and spire-shaped plants with a thick canopy, dense leaves in summer, and high CD and TH values should be chosen in the selection of street tree species. On the one hand, it can serve as a reference for future research on different regions and different types of street trees to improve the thermal comfort of road space, and guide the selection of road tree species scientifically. On the other hand, the research proved that the application of LiDAR technology can quickly obtain high-density point clouds, extract plant information, provide data support for urban forestry to cope with extreme weather, and play a guiding role in the application of LiDAR technology to different scientific fields.

Author Contributions: Conceptualization, X.Z., Y.L. (Yang Liu) and Q.Y.; data curation, X.Z. and R.L.; funding acquisition, Y.L. (Yang Liu); investigation, X.Z.; project administration, Y.L. (Yang Liu) and Y.L. (Yonghua Li); resources, Y.L. (Yang Liu) and Y.L. (Yakai Lei); supervision, Y.L. (Yang Liu), Y.L. (Yakai Lei), N.G. and A.A.; validation, Y.L. (Yonghua Li) and Q.Y.; writing—review and editing, X.Z., Y.L. (Yang Liu) and A.A. All authors have read and agreed to the published version of the manuscript.

Funding: This research was funded by: The Key Projects of the Henan Provincial Department of Education, grant number (21A220002); The Henan Province Science and Technology Research Project, grant number (212102310581). The Henan Province Science and Technology Research Project, grant number (222102520031).

Institutional Review Board Statement: Not applicable.

Informed Consent Statement: Not applicable.

Data Availability Statement: Not applicable.

Acknowledgments: This research is supported by Henan Provincial Joint International Research Laboratory of Landscape Architecture.

Conflicts of Interest: The authors declare no conflict of interest.

References

1. Taha, H.; Akbari, H.; Rosenfeld, A. Vegetation Canopy Micro-Climate: A Field-Project in Davis, California. *J. Appl. Meteorol. Climatol.* **1989**, *1*, 13–15.
2. Akbari, H.; Taha, H. The impact of trees and white surfaces on residential heating and cooling energy use in four Canadian cities. *Energy* **1992**, *2*, 141–149. [CrossRef]
3. Souch, C.A. The effect of trees on summertime below canopy urban climates: A case study Bloomington, Indiana. *J. Arboric.* **1993**, *19*, 303–312. [CrossRef]
4. Hsieh, C.M.; Jan, F.C.; Zhang, L. A simplified assessment of how tree allocation, wind environment, and shading affect human comfort. *Urban For. Urban Green.* **2016**, *18*, 126–137. [CrossRef]
5. Kong, F.; Yan, W.; Zheng, G. Retrieval of three-dimensional tree canopy and shade using terrestrial laser scanning (TLS) data to analyze the cooling effect of vegetation. *Agric. For. Meteorol.* **2016**, *217*, 22–34. [CrossRef]
6. Zhang, Z.; Lv, Y.; Pan, H. Cooling and humidifying effect of plant communities in subtropical urban parks. *Urban For. Urban Green* **2013**, *12*, 323–329. [CrossRef]
7. de Abreu-Harbicha, L.V.; Labaki, L.C.; Matzarakis, A. Effect of tree planting design and tree species on human thermal comfort in the tropics. *Landsc. Urban Plan.* **2015**, *138*, 99–109. [CrossRef]
8. Doick, K.; Hutchings, T. *Air Temperature Regulation by Urban Trees and Green Infrastructure*; Forestry Commission: Bristol, UK, 2013; pp. 1–11.
9. Shahidan, M.F.; Jones, P.J.; Gwilliam, J.; Salleh, E. An evaluation of outdoor and building environment cooling achieved through combination modification of trees with ground materials. *Build. Environ.* **2012**, *58*, 245–257. [CrossRef]
10. Bo, H.; Lin, B. Numerical studies of the outdoor wind environment and thermal comfort at pedestrian level in housing blocks with different building layout patterns and trees arrangement. *Renew. Energy* **2015**, *1*, 18–27.
11. Kotzenn, B. An investigation of shade under six different tree species of the Negev desert towards their potential use for enhancing micro-climatic conditions in landscape architectural development. *J. Arid. Environ.* **2003**, *55*, 231–274. [CrossRef]
12. Rosdi, K.; Ainuddin, A.N. Microclimatic modification of three timber species stands on ex-tin mining land. *Malays. For.* **2004**, *67*, 44–49.
13. Chen, Q.; Gong, P.; Baldocchi, D. Estimating Basal Area and Stem Volume for Individual Trees from Lidar Data. *Photogramm. Eng. Remote Sens.* **2007**, *73*, 1355–1365. [CrossRef]
14. Popescu, S.C.; Wynne, R.H.; Nelson, R.F. Estimating plot-level tree heights with lidar: Local filtering with a canopy-height based variable window size. *Comput. Electron. Agric.* **2003**, *37*, 71–95. [CrossRef]
15. Hyyppa, J.; Kelle, O.; Lehikoinen, M. A segmentation-based method to retrieve stem volume estimates from 3-D tree height models produced by laser scanners. *IEEE Trans. Geosci. Remote Sens.* **2001**, *39*, 969–975. [CrossRef]
16. Kato, A.; Moskal, L.M.; Schiess, P. Capturing tree crown formation through implicit surface reconstruction using airborne lidar data. *Remote Sens. Environ.* **2016**, *113*, 1148–1162. [CrossRef]
17. Li, W.; Guo, Q.; Jakubowski, M.K. A New Method for Segmenting Individual Trees from the Lidar Point Cloud. *Photogramm. Eng. Remote Sens.* **2012**, *78*, 75–84. [CrossRef]
18. Tao, S.; Wu, F.; Guo, Q.; Wang, Y.; Li, W.; Xue, B.; Hu, X.; Li, P.; Tian, D.; Li, C.; et al. Segmenting tree crowns from terrestrial and mobile LiDAR data by exploring ecological theories. *ISPRS J. Photogramm. Remote Sens.* **2015**, *110*, 66–76. [CrossRef]
19. Zhao, X. *Zhumadian Yearbook*; Zhongyuan Publishing Media Group Zhongyuan Media Co., Ltd. Zhongzhou Ancient Books Publishing House: Zhengzhou, China, 2020; pp. 15–20.
20. Zhao, J. Street-level road change detection using vehicle lidar scanning. *Surv. Mapp. Spat. Geogr. Inf.* **2021**, *44*, 1–3.

21. Chen, K.; Liang, T.; Gan, Y. Influence of the building and green space layout for microclimate in Zhengzhou residential. *J. Henan Agric. Univ.* **2016**, *50*, 674–682.
22. Wei, X.; Hao, R.; Zhang, M.; Shen, H.; Qiu, Y.; Geng, H. Simulation of the impact of tree canopy spatial structure on microclimate. *J. Zhejiang A&F Univ.* **2019**, *36*, 783–792.
23. Chen, Y. *Garden Arborology*; China Forestry Press: Beijing, China, 2014; pp. 109–115.
24. Wu, Z. Investigation and application of tree species on the street in Zhumadian City. *Mod. Hortic.* **2016**, *2*, 158.
25. Badescu, V. Verification of some very simple clear and cloudy sky models to evaluate global solar irradiance. *Sol. Energy* **1997**, *61*, 251–264. [CrossRef]
26. Liao, C.; Tsai, K.T.; Huang, Y.; Lin, T. Effects of thermal comfort and adaptation on park attendance regarding different shading levels and activity types. *Build. Environ.* **2013**, *59*, 599–611.
27. Yang, L.; Liu, J.; Ren, J.; Zhu, X.; An, F. Research on Outdoor Thermal Comfort of Campus under Special Weather in Transition Season. *J. Shandong Jianzhu Univ.* **2021**, *36*, 75–96.
28. Xiao, X.; Chen, G.; Dong, L.; Yan, H. Study on the influence of four plant community types on the cooling effect in hot and humid regions of East China. *Landsc. Archit.* **2019**, *26*, 94–99.
29. Lin, Y.; Tsai, K. Screening of Tree Species for Improving Outdoor Human Thermal Comfort in a Taiwanese City. *Sustainability* **2017**, *9*, 340. [CrossRef]
30. Xu, M.; Hong, B.; Jiang, R. Research on the influence of campus street trees on the thermal comfort of outdoor pedestrians in summer. *Chin. Landsc. Archit.* **2020**, *36*, 139–144.
31. Bao, N.; Zhang, S.; Mo, X. Review of the research on tree canopy structure of arbor species. *Eucalyptus Sci. Technol.* **2021**, *38*, 68–74.
32. Wu, R.; Yan, H.; Shu, Y.; Shi, Y.; Bao, Z. Microclimate characteristics of bamboo plants in summer and their effects on human comfort. *Chin. Gard.* **2019**, *35*, 112–117.
33. Wu, X.; Lin, Y.; Yan, H.; Hao, X. Research on the correlation between cooling and humidification effect of urban green space and its structural characteristics. *Chin. J. Ecol. Agric.* **2008**, *6*, 1469–1473.
34. Wu, Y. Tree-shading and avenue-tree planting. *J. Hortic.* **1963**, *2*, 295–308, 335–336.
35. Qin, Z.; Li, Z.; Cheng, F.; Sha, H. The regulating effect of canopy structure of Luan tree community on its environmental temperature and humidity in summer. *Chin. J. Appl. Ecol.* **2015**, *26*, 1634–1640.
36. Wu, L.; Wang, Z. Research on the dynamic of canopy shade. *J. Nanjing For. Univ. (Natural Science Edition)* **1991**, *2*, 61–66.
37. Zhao, X.; Li, G.; Gao, T. Summer thermal comfort effect and morphological characteristics adjustment mechanism of typical street trees in Harbin. *Landsc. Archit.* **2016**, *12*, 74–80.
38. Peters, E.B.; McFadden, J.P.; Montgomery, R.A. Biological and environmental controls on tree transpiration in a suburban landscape. *J. Geophys. Res.* **2010**, *115*, G04006. [CrossRef]
39. Zheng, Y.; Zhu, S.; Fang, M.; Wu, H.; Yan, H.; Shao, F. Relationship between different plant types and temperature and humidity effects in urban parks. *J. Northwest For. Univ.* **2020**, *35*, 243–249.
40. Shahidan, M.F.; Shariff, M.; Jones, P.; Salleh, E.; Abdullah, A.M. A comparison of *Mesua ferrea* L. and *Hura crepitans* L. for shade creation and radiation modification in improving thermal comfort. *Landsc. Urban Plan.* **2010**, *97*, 168–181. [CrossRef]

forests

Article

Growth Response of Nine Tree Species to Water Supply in Planting Soils Representative for Urban Street Tree Sites

Alexander Schütt [1,*], Joscha Nico Becker [1], Christoph Reisdorff [2] and Annette Eschenbach [1]

[1] Institute of Soil Science, CEN Center for Earth System Research and Sustainability, Universität Hamburg, 20146 Hamburg, Germany; joscha.becker@uni-hamburg.de (J.N.B.); annette.eschenbach@uni-hamburg.de (A.E.)
[2] Institute of Plant Science and Microbiology, Universität Hamburg, 22609 Hamburg, Germany; christoph.reisdorff@uni-hamburg.de
* Correspondence: alexander.schuett@uni-hamburg.de

Abstract: In urban environments, newly planted street trees suffer from poor site conditions and limited water availability. It is challenging to provide site conditions that allow the trees to thrive in the long term, particularly under climate change. Knowledge about the hydrological properties of artificial urban planting soils related to the response of tree species-specific growth is crucial, but still lacking. Therefore, we established a three-year experimental field setup to investigate the response of nine tree species (135 individuals) to two common urban planting soils and a loamy silt reference. We determined and measured soil hydrological parameters and monitored tree growth. Our results revealed low plant available water capacities (6% and 10% v/v) and hydraulic conductivity restrictions with the drying of the sandy-textured urban planting soils. Therefore, tree species that are investing in fine root growth to extract water from dry soils might be more successful than trees that are lowering their water potential. Tree growth was overall evidently lower in the urban planting soils compared with the reference and differed between and within the species. We showed that using unfavorable planting soils causes severe, species-specific growth deficits reflecting limited above-ground carbon uptake as a consequence of low water availability.

Keywords: tree growth; soil water monitoring; relative extractable water; soil water tension; soil water content; urban water management; soil texture; structural soil

Citation: Schütt, A.; Becker, J.N.; Reisdorff, C.; Eschenbach, A. Growth Response of Nine Tree Species to Water Supply in Planting Soils Representative for Urban Street Tree Sites. *Forests* **2022**, *13*, 936. https://doi.org/10.3390/f13060936

Academic Editors: Thomas Rötzer, Stephan Pauleit, Mohammad A Rahman and Astrid Reischl

Received: 20 April 2022
Accepted: 13 June 2022
Published: 15 June 2022

1. Introduction

Establishing newly planted street trees is a critical step to achieve green infrastructure goals and to maintain the tree population in cities. This is particularly challenging, since growing conditions for urban street trees are highly unfavorable [1–3]. Due to anthropogenic transformations and translocations, urban soils exhibit a high degree of spatial heterogeneity. Artificial sandy and stony substrates [4,5], high degrees of soil compaction [1], surface sealing [6,7], and vertical and horizontal limitations of root distribution [8] alter tree sites and, in particular, street tree sites in comparison with natural environments. However, there has been little research on the impact of artificial urban soils on tree growth and the long-term success of new plantings [9]. Climate change and the urban heat island are prolonging the tree growing season and increasing the inner-city atmospheric water deficit, thus raising the transpirational water demand of the trees [10]. As observed in the summers of 2018 and 2019 [11,12], increased meteorological droughts (i.e., phases of deficiency and uneven distribution of precipitation) and storm events with high run-off shares are likely to occur more frequently in Mid-Europe, causing soil water scarcity in cities [13]. Since soil water replenishment by single and few precipitation events at freshly planted tree sites is less effective [14] and successive and frequent precipitation during summer is predicted to decrease in the future climate [15], climate change is likely to exacerbate water limitation and thus the growing conditions of street trees. As a consequence, urban street trees are likely to grow slower than trees in rural environments and forests [16]. However, direct growth comparisons are sparse, and findings are not consistent [17,18] and seem to be not appropriate to assess the effect level that soil constraints have on the growth of urban street trees. Therefore, on-site comparisons of various tree species and planting soils under identical climate conditions are necessary.

Freshly planted street trees, suffering from transplanting shock, have to adapt rapidly to urban site conditions and reestablish a root:shoot ratio adequate to meet the above-ground demands [19]. This results in low life expectancies of less than 30 years [20] and particularly large premature mortality rates for recently transplanted tree populations from nurseries into urban street sites [21,22]. To achieve fully developed ecosystem services in the long term, these initial growth conditions have to be further improved based on detailed investigations into the interaction of tree growth and urban planting soils with a focus on soil water characteristics. Because street tree sites usually act as multifunctional public spaces, however, the tree's demand itself often takes a secondary role in planning processes. For example, original local or artificial soils are often replaced by technical soil–gravel mixtures (structural planting soils) prior to planting to ensure compaction stability for parking lots and pedestrian infrastructure [23–27]. These structural planting soils contain high percentages of gravel (grain size > 2 mm), amended with soil–compost mixtures (grain size < 2 mm) that are dominated by a sandy texture [5,26]. In addition to often small planting pits [24,28,29] in relation to standards [30], it has to be questioned whether sandy-textured structural soil substrates can ensure sufficient water supply for young street trees [5,14,31], particularly under climate change. However, the hydrology of structural soils and surrounding urban roadside soils and its effect on tree growth potential has not been systematically studied so far [25,29,32].

Tree growth habit is the result of intrinsically determinate reaction patterns to extrinsic factors [16]. Therefore, sandy and coarsely textured planting soils [3,5,33] with comparatively low organic matter contents [34] are expected to significantly affect tree growth rates [1,23,32], but not in a uniform pattern [9]. Depending on the particular hydrological characteristics of their habitat, trees have evolved multiple physiological strategies to counteract substantial vitality loss in situations of limited water availability. Thus, stress reactions can differ quantitatively and/or qualitatively between species. It is still the subject of scientific debates, which strategies of stress reactions are most promising for surviving increasing stress incidences and which main intrinsic factors foster mortality risk [35–38]. Since the drought strategies of particular tree species are yet not fully understood, the suitability of tree species for urban street site planting has been so far mainly determined by meta-studies, assessing the drought resistance of tree species based on the aridity of their natural habitat [39]. A classification according to additional site-specific criteria, such as individual soil requirements for these tree species, is missing [39]. To determine the suitability of tree species and cultivars as urban street trees, long-term observational studies under in situ conditions in different cities have been conducted (e.g., GALK-Straßenbaumtest: http://strassenbaumliste.galk.de/sblistepdf.php (accessed on 2 April 2021); [40–42]). However, a conclusive statement regarding the species-specific growth responses to urban planting soils is still missing.

The aim of the present study was to determine the constraints of artificial urban soils on the growth of different tree species with a focus on soil hydrological properties. In particular, we aim to answer the questions: (a) Which hydrological properties and dynamics are characteristic of the selected representatives of urban planting soils? (b) Did periods of critical soil water availability occur in urban planting soils vs. natural soils under the climatic conditions during 2019–2021? (c) Which growth patterns were measurable overall and in the responses of tree species to specific soil conditions? Our approach for a suchlike assessment is to simulate experimentally urban below-ground conditions at one single site to provide identical climatic conditions. Therefore, we established a large-scale experimental field in a complete block design. A total of 135 standardized eight-year-old trees of nine tree species (DBH: 4.5 cm–5.1 cm) were each planted in pits of 7.5 m^3 filled alternately with urban-typical artificial planting soils and one reference (loamy silt). The tree species selection was conducted based on three criteria: (a) general suitability as an urban tree, (b) underrepresentation in Hamburg's (Germany) street tree population, and (c) availability in the tree nursery. The soil hydrological parameters were measured in the laboratory and in the field. Tree growth was monitored morphometrically and dendrometrically for three years.

2. Materials and Methods

2.1. Study Site

The experiment was carried out at an open field site (6000 m^2) in a tree nursery located 15 km south of Hamburg, Germany (Baumschule Lorenz von Ehren GmbH & Co. KG; 53°24′ N; 9°57′ E). The climate shows a marine influence (Koeppen & Geiger: Cfb), with the highest precipitation in summer, a slightly dryer autumn and winter, and low precipitation in the spring season. The long-term annual means (1981–2010) of precipitation and air temperature are 743 mm and 9.6 °C, respectively [43]. The topography is slightly sloping and the groundwater table is below 20 m from the soil's surface. The predominant soil types at the experimental research site are agricultural-affected Cambisols with fine-textured loamy silts above silty sands to pure sands. The water-holding capacity at field capacity (FC; 60 hPa) ranged between 33% and 38% v/v, and that at wilting point (WP; 15,000 hPa) ranged between 10% and 15% v/v in the upper soil horizons (analysis method in Section 2.4).

2.2. Substrate and Tree Selection

The experimental site was set up in a randomized complete block design of nine tree species planted in three planting soils. Every species–substrate combination was replicated five times (five blocks), resulting in a total number of 135 trees. Two treatment soils represented urban artificial soil conditions along streets: (1) a pure sand ('Sand'), representing artificial urban soils around street tree plantings [5,14], and (2) a one-layer gravel–soil medium (structural planting soil) mixed according to German standardized guidelines [30], similar to 'Amsterdam Tree Soil' [44] and commonly used as backfill material at newly planted tree sites ('FLL'). The reference soil was supposed to represent optimal growing conditions. Therefore, we used a regional harvested natural topsoil similar to the Cambisols at the experimental site, which were considered very suitable for the production of trees in the nursery and for years have provided good yields and quality ('Loamy Silt'). Planting pits with dimensions of 2.5 m × 2.5 m × 1.2 m (7.5 m^3) were excavated for each tree studied. The planting pits were arranged in a 4 × 35 grid with spacing from center to center of 3.5 m × 8 m (Figures S1 and S2).

The tree species were selected on the basis (1) of long-term observational and literature-based studies focusing on the suitability of diverse native and non-native tree species as urban trees (GALK (Table 1) and KLAM [39,45] (Table 1)). Nine tree species were selected from a wide range of tree species and cultivars rated in GALK (n = 178) and KLAM (n = 235). To achieve greater variability in the growth response patterns, we selected not only the highest-scoring tree species, but also the overall tree species that differed in their scores for drought tolerance (Table 1). Additional criteria for tree selection were (2) an underrepresentation within Hamburg's street tree population as the trees were mostly not yet commonly planted and (3) the availability of the same-sized and aged trees cultivated at the same nursery to provide comparability for the former growing conditions. This resulted in the selection of the following nine tree species: *Tilia cordata 'Greenspire'* (*Tc*), *Quercus cerris* (*Qc*), *Quercus palustris* (*Qp*), *Carpinus betulus 'Lucas'* (*Cb*), *Ostrya carpinifolia* (*Oc*), *Gleditsia triacanthos 'Skyline'* (*Gt*), *Liquidambar styraciflua* (*Ls*), *Amelanchier lamarkii* (*Al*), and *Koelreuteria paniculata* (*Kp*) (Table 1).

Table 1. Tree species characteristics. Planting diameter at breast height (DBH in May 2019) and suitability classifications of the investigated tree species for use as street trees. STP HH is the actual percentage of the investigated tree species from Hamburg's total street tree population (state: 31 December 2020). Stock of tree species is listed according to age class: The tree species were determined via the Hamburg tree cadaster and show the street tree population according to age classes classified by planting year after 2010, between 1990 and 2010, and before 1990.

Tree Species 'Cultivar'	Code	DBH	Suitability		STP HH [3]	Stock of Tree Species [3]		
		(cm)	KLAM [1]	GALK [2]	(%)	>2010	1990–2009	<1990
Tilia cordata 'Greenspire'	*Tc*	5.1 ± 0.1	2.1	Well suited	1.3	737	1561	596
Quercus cerris	*Qc*	4.6 ± 0.2	1.2	Well suited	0.4	588	113	263
Quercus palustris	*Qp*	5.3 ± 0.1	2.2	Partly suited	0.9	400	720	924
Carpinus betulues 'Lucas'	*Cb*	4.6 ± 0.1	2.1	In test	<0.1	61	0	0
Ostrya carpinifolia	*Oc*	4.5 ± 0.2	1.1	Suited	0.1	187	90	0
Gleditsia triacanthos 'Skyline'	*Gt*	5.1 ± 0.1	1.2	Well suited	0.1	153	98	1
Liquidambar styraciflua	*Ls*	4.9 ± 0.4	2.2	Suited	0.3	352	155	54
Amelanchier lamarckii	*Al*	5.0 ± 0.1	3.1	n.d.	0.1	52	53	78
Koelreuteria paniculata	*Kp*	5.1 ± 0.2	1.2	Partly suited	<0.1	28	2	0

[1] KLAM = Climate Tree Species Matrix. Source: [39,45]. [2] GALK = Deutsche Gartenamtsleiterkonferenz (GALK e.V.; German Garden Agency Directors Conference). Source: http://strassenbaumliste.galk.de/sblistepdf.php (accessed on 2 April 2021). [3] Source: Free and Hanseatic City of Hamburg, Department for the Environment, Climate, Energy and Agriculture.

In order to create representative conditions for urban tree sites to be planted, the dimensions of the selected trees at the planting date were in accordance with standards. The trees had, therefore, similar initial stem diameters between 4.5 and 5.3 cm at breast height (Table 1). The selected study trees were harvested with a tree digger according to the B&B method (root ball excavated and burlap-wrapped [46]). Therefore, the volume of each root ball was almost similar, with maximum root ball dimensions of 0.5 m × 0.6m × 0.5 m (0.1 m^3). Single trees were planted in the excavated planting pits after the backfilling of the pits with the planting soils (Figures S1 and S2). The planting soils were moderately compacted by consistently trampling throughout the backfilling process. After the first winter, small depressions caused by natural settling at the corners of the planting pits were refilled with the respective soil material. To ensure tree establishment in the early post-transplantation phase, we applied four irrigations using 'tree gator' watering bags (approx. 60 L) during the first year. Fertilization (NPK, Mg, and micronutrients (Mn and Zn)) was applied in spring 2020 and 2021 to minimize differences in the initial nutrient concentrations ('Sand' lowest and 'Loamy Silt' highest) and thus to minimize nutrient-related effects on tree growth.

2.3. Soil Water Monitoring and Meteorological Data

We installed a monitoring setup consisting of soil water potential sensors (SWP; WATERMARK 200SS, Irrometer Inc., Riverside, CA, USA)), volumetric water content probes (VWC; CS-650, Campbell Scientific Ltd., Bremen, Germany; Theta Probe ML2x; Delta-T Devices Ltd., Cambridge, UK), and soil temperature sensors (Irrometer Inc., Riverside, CA, USA). The data were logged at hourly intervals by WATERMARK M900 Monitors for SWP and soil temperature sensors (Irrometer Inc., Riverside, CA, USA), with CR-1000 data loggers for the CS-650 VWC sensors (Campbell Scientific Ltd., Bremen, Germany) at plots with *Qc* and with DL18 data loggers for the ML2x VWC sensors (Ecomatik, Dachau, Germany) at plots with *Al* and *Oc*. The monitoring setup on the experimental site was structured as follows: intensive monitoring at one plot of three tree species per planting soil (9 plots; tree species: *Qc*, *Al*, and *Oc*) and extensive monitoring at one plot of all tree species per planting soil (27 plots). The intensive monitoring covered eight SWP sensors with two replicates per depth (10 cm, 35 cm, root-ball, and 100 cm) and four VWC probes without replication at the same depths. The extensive monitoring covered four SWP sensors without replication, also at the same depths. Temperature sensors were installed next to the SWP and VWC sensors within four randomly chosen plots per planting soil. Soil temperature sensors were installed to allow correction for temperature's effect on SWP measurements. We installed the sensors simultaneous to the backfilling of the planting soils at the predefined depths. For installing the root ball sensors, the root balls were placed at approximately 40 cm depth and covered with the planting soils up to half, where the sensors were placed. Except for the root ball sensors, all sensors were placed with a distance of 75 cm from the horizontal center of the planting pit. The 36 monitoring plots were randomly distributed over the experimental site.

Meteorological data were recorded within one replication block in the middle of the experimental site by a meteorological station (Campbell Scientific, Logan, UT, USA). The air temperature and relative humidity were measured at 2 m height (HMP155A, Vaisala, Vantaa, Finland). For precipitation measurements, we used a tipping bucket rain gauge (52203, R.M. Young Co., Traverse City, MI, USA). All data were recorded within a 15 min interval using a CR-1000 data logger. The atmospheric vapor pressure deficit (VPD) was calculated as the difference between the saturated and actual vapor pressure using the daily means of air temperature and relative humidity.

2.4. Soil Physical Characteristics

We took disturbed and undisturbed soil samples from one plot of each planting soil in November 2019 after the first growing season. The undisturbed soil samples were analyzed in the lab for the fine soil texture composition, coarse soil content (gravel content), total and organic C contents, and soil pH ($CaCl_2$) (Table 1). The water retention characteristics were assessed using undisturbed soil samples (soil cores of 100 cm^3 and 250 cm^3) for lab measurements. To derive the water retention curves, the water contents at characteristic pressure levels of pF 0.5, 1.3, 1.8, 2.1, 2.48, 3.48, and 4.2 were measured as a percentage of weight using pressure plate apparatus. Additionally, the water retention characteristics and unsaturated conductivity were determined with 250 cm^3 undisturbed soil samples using the fully automated measuring and evaluation system HYPROP [47,48]. The bulk density was determined by drying and weighing these undisturbed volumetric soil samples. Based on the calculated daily means of the measured SWP and VWC data, we also plotted the field retention curves for the three planting soils. The van Genuchten parameters for the functions of the three approaches were determined according to [49] to describe the relationship between VWC and SWP. The data pairs of water contents at levels of pF 0.5, 1.3, 3.48, and 4.2 of the pressure plate method were added to supplement the curves of the HYPROP measurements and the field retention curves, as the measuring devices for these methods had limited measurement ranges. In spring 2020, the infiltration capacities of the planting soils were measured using a double-ring infiltrometer. In order to quantitatively evaluate and compare the soil hydrological characteristics of the three planting soils, we considered the variables of the plant-available water capacity (PAWC), field capacity (FC), infiltration capacity (IC), and unsaturated hydraulic conductivity (K_s).

2.5. Vitality Assessment and Tree Growth Measurements

We assessed the tree vitality by visual inspections of each individual tree regularly at the beginning of September 2020 and 2021. Criteria leading to the vitality score (1, very vital–5, strongly impaired) were (1) the overall leaf conditions and (2) the crown density in comparison with a vital tree of the same species. The assessment followed a regular inspection scheme developed by the GALK (Deutsche Gartenamstleiterkonferenz e.V.).

The stem growth and current-year shoot length were used as indicators for tree performance in the three planting soils. We measured the stem circumference of all trees at heights of 100, 130, and 150 cm in May 2019, January 2020, January 2021, and October 2021 to a decimal place using a measuring tape. Shoot growth was measured in September 2020 and September 2021. Five main external, sun-exposed branches per tree were randomly chosen and the shoot increment was measured in length. The annual average stem increment and shoot extension for each tree species per substrate were calculated as the average of all sampled replicates.

To study the tree species' reaction to short-term changes in the environmental conditions by measuring continuously the intra-annual stem diameter variation, we installed 23 electronic point dendrometers (DD-L2, Ecomatik, Dachau, Germany; accuracy of ± 1.97 µm and temperature coefficient < 0.2 µm K^{-1}) in spring 2021. For dendrometer installation, three individual trees per species (with the exception of *Qc* (only $n = 2$) and *Kp* (no measurements)) growing in the planting soil 'Sand' were selected. At each tree, the dendrometers were installed with the dendrometer head on the outer bark at approximately 110 cm stem height. The sensors measured with a frequency of 15 min and were connected to both CR-1000 and DL-18 dataloggers.

2.6. Data Processing

The monitored SWP data were manually corrected for the actual soil temperatures using the in situ soil temperature measurements from the three planting soils [50,51]. The corrected data extended the lower measurement range of −2500 hPa, as 10 °C instead of a default value of 24 °C was initially set as the constant temperature for logger internal temperature correction. To detect water stress conditions and to quantify the annual duration of critical soil water availability, we used the monitored SWP data below −63 hPa and transformed them into the volumetric water content data using the van Genuchten water retention functions [49] determined for each planting soil (VWC_{vG}). Thereafter, the relative extractable soil water content (REW%) was calculated from the VWC_{vG} depth weighted for a soil compartment from 0–35 cm and depth weighted for a soil compartment from 0–100 cm according to [52]:

$$REW\ (\%) = 100 \cdot ((VWC_{actual} - VWC_{WP}) \cdot (VWC_{FC} - VWC_{WP})^{-1}) \quad (1)$$

where VWC_{PWP} is the VWC_{vG} at the wilting point and VWC_{FC} is the VWC_{vG} at field capacity, both measured in the lab using the pressure plate method. In the literature, water stress conditions are assumed to occur when the REW drops below the threshold of 40%, as transpiration is gradually reduced due to increasing stomatal diffusion resistance [53]. Therefore, based on the calculated daily means, we calculated the number of days per growing season when REW < 40% to seasonally characterize water stress according to [52].

The stem circumference measurements were height-related-transformed to the stem diameter increment at breast height (DBH), assuming a circular stem area. Data processing of the intra-annual stem diameter variations (SDV) was performed using R v. 4.2.1 [54] with the package *treenetproc* [55]. The workflow of *treenetproc* converts raw data (L0) into time-aligned time-series data (L1 data, in this study with hourly resolution), cleans the L1 data from outliers and data shifts, and linearly interpolates data gaps < 30 min (L2 data). From the L2 data, we extracted three relevant seasonal proxies [55]. (1) The reversible, water-induced shrinkage and expansion of the stem diameter (maximum daily shrinkage (mds in µm d^{-1})). (2) The irreversible diametric expansion due to woody cell formation with the zero-growth concept, transformed into relative values for better comparison (stem diameter increment (SDI in % of seasonal maximum stem diameter)) [56]. Thus, contrary to [57], no negative growth-due to a smaller diameter maximum on the following day–was considered in the analysis. (3) The tree water deficit (TWD), derived on a daily basis from the maximum precedent stem diameter and actual (daily) maximum stem diameter, represented the missing water in µm d^{-1} [56]. Analyzing the response of intra-annual tree growth on soil hydrological and climate variables is only reliable within the main period of stem growth [58]. Since *treenetproc* returns values for onset and cessation starting from the second year only [55], we determined the length of the growing season of the instrumented trees ($n = 23$) as follows: onset of growth was reached when the actual daily growth rate at least equaled the seasonal average daily growth rate for at least seven consecutive days (period of 1 May 2021 to 30 September 2021). Growth cessation was, therefore, reached when the seasonal average daily growth rate was exceeded for the last time by the actual daily growth rate on at least seven consecutive days. As we expect mds in contrast to SDI to be a parameter reflecting the trees' water status in the short term, we used daily values for mds to analyze relationships with environmental parameters and weekly means for SDI [57].

2.7. Statistical Analysis

The annual duration of temporal water stress conditions (number of days with REW < 40%) was compared between the three planting soils and the years 2020 and 2021 using a two-way ANOVA. The stem diameter increment and shoot growth as mean values for the years 2020 and 2021 were compared separately (a) within a species between the three planting soils and (b) between all species within each planting soil using a one-way ANOVA. All group means were compared using Tukey's post hoc comparison at p-level ≤ 0.05.

For analyzing the relationships of the weekly SDI and mds with the environmental variables REW% in 0–35 cm and 0–100 cm depth, precipitation, and VPD, single-factor correlations were calculated and the Spearman's correlation coefficient were reported. The environmental variables for the correlation analysis with SDI were also calculated as calendar week means.

Statistical analyses were conducted using OriginPro v2021b (OriginLab Corporation, Northampton, MA, USA) and R v4.1.2 [54].

3. Results

3.1. Substrate Characteristics

The fine texture (<2 mm) of the urban planting soils 'Sand' and 'FLL' composed predominantly of sandy grains (0.063–2 mm), whereas the dominant fine texture grain size of the 'Loamy Silt' was silt (2–63 μm) (Table 2). The structural planting soil ('FLL') was composed of 20% v/v gravel. The organic matter content of all investigated soils was below 1% w/w, whereas the 'Sand' had almost no organic matter (<0.1% w/w) and the organic matter content in the 'FLL' was 0.6% w/w. All soils were backfilled into the planting pits loosely compacted to bulk densities between 1.4 and 1.44 g cm^{-1}. The plant available water capacity (PAWC) of the 'Loamy Silt' was highest, with 23% v/v, corresponding to 1725 L per planting pit (Figure 1a). Compared to the 'Loamy Silt', the plant available water capacity of 'FLL' was <50% and <25% for 'Sand' (773 L per planting pit and 450 L per planting pit, respectively). The total pore volumes of the planting soils 'Sand' and 'Loamy Silt' were the same, at approximately 40% v/v, whereas the total pore volume of the planting soil 'FLL' was 32% v/v. The slope of the 'Loamy Silt' soil-water-retention curve was almost linear in the FC range, while in both artificial soils, the retention curve approached an exponential slope. The soil hydraulic conductivity of 'Sand' declined sharply at REW 75% (Figure 1b). The hydraulic conductivity in 'FLL' was constantly lower and fell below the hydraulic conductivity of 'Loamy Silt' at WP (100% REW) already before 70% REW. Although the hydraulic conductivity in 'Loamy Silt' compared with 'Sand' and 'FLL' had constantly lower values in the wet soil (>10% VWC) (Figure 1a), unsaturated conductivity of >10^{-5} cm d^{-1} was provided within the whole range of PAWC until WP.

Table 2. Soil properties of the three planting soils with OM as the organic matter content, BD as the bulk density, IC as the infiltration capacity, FC as the field capacity, and PAWC as the plant available water capacity.

Planting Soil	Fine Tex. < 2 mm			Coarse Tex. > 2 mm		OM	pH	BD	Hydrological Properties		
	Sand (v/v)	Silt (v/v)	Clay (v/v)	(w/v)	(v/v)	(v/v)	$CaCl_2$	(g cm^{-1})	IC (cm min^{-1})	FC (v/v)	PAWC (v/v)
'Sand'	95	4	1	2	n.d.	0.1	6.8	1.4	1.5	9.1	6.0
'FLL'	93	5	2	35	20	0.6	6.5	1.4	1.9	14.6	10.3
'Loamy Silt'	29	61	10	1	n.d.	0.9	5.7	1.4	0.3	33.0	23.0

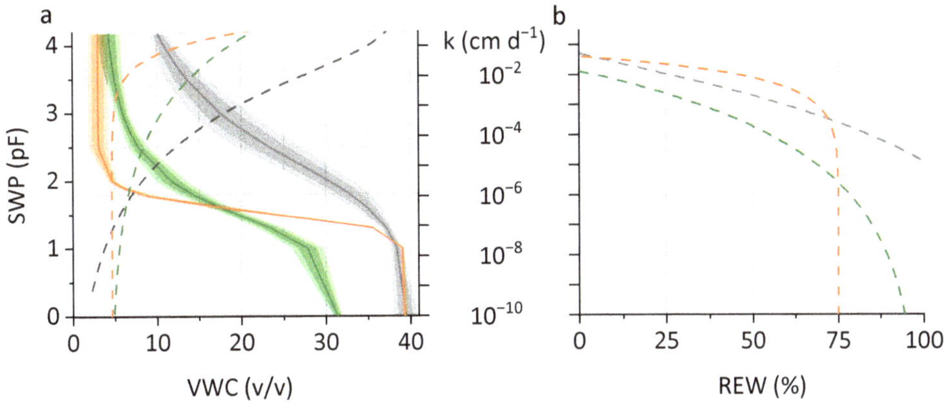

Figure 1. (**a**) Mean planting soil water retention curves (left y-axis) determined by pressure chamber, HYPROP, and field measurements in the three planting soils 'Sand' (orange), 'FLL' (green), and 'Loamy Silt' (grey). SWP is the soil water potential and VWC is the volumetric water content. Shaded areas highlight the standard deviation (light) and the standard error (dark). Dashed lines show the log-transformed fitted hydraulic conductivity (k) (right y-axis) as a function of VWC determined by the HYPROP apparatus; (**b**) log-transformed fitted hydraulic conductivity (k) as a function of the REW (relative extractable water) determined by the HYPROP apparatus.

3.2. Environmental Conditions and Soil Water Availability

Compared to the mean 30-year growing season precipitation (May–September 1981–2010; DWD Station Hamburg-Neuwiedenthal) of 343 mm, the years of 2019 to 2021 were slightly dry years (231 mm, 231 mm, and 299 mm, respectively). With the beginning of the growing season in 2019, the root balls in all planting soils strongly dried below the REW 40% threshold value (Figure 2). Irrigation replenished the root balls for short phases. Until the end of the growing season of 2019, the REW in the planting pits remained high in all three planting soils above the threshold value of 40%. During the growing season of 2020, three distinct dry out phases (each > 16 days) occurred, interrupted by replenishing precipitation events.

The first phase was characterized by strong soil water depletion below the water stress threshold only in the root balls. Subsequently, two dry periods led the average REW to drop below the threshold even within the planting soils, prioritized at 10 and 35 cm depths, indicating that tree roots then tapped water from the planting soils in the top soil layer. In 2021, two dry phases occurred, with the REW within the root balls being almost completely depleted and the soil water storage within all planting pits being minimized. However, in 'FLL' and 'Sand', the soil water storage was replenished by drought-breaking precipitation events, while soil drying in 'Loamy Silt' continued. Despite the higher precipitation in 2021, the soil water storage at 0–100 cm depth was depleted below the threshold of 40% REW on evidently more days ($p \leq 0.001$) compared with 2020, indicating progressing root density. Differences in drought incidence between the planting soils were not significant, but we observed a clear tendency of a reduced amount of water stress days in 'Sand' during 2021 (Figure 3).

3.3. Tree Growth Analysis

Across all investigated species, annual growth was highest in 'Loamy Silt' and lowest in 'Sand' (Figure S3). Overall, the annual stem diameter increment was highest during 2021, whereas the annual shoot growth was greatest during 2020 (Figure S3). The lowest values and no significant differences between species and planting soils were measured in 2019. The growth responses of the investigated tree species in the three planting soils varied strongly (Figure 4a). In 'Loamy Silt', *Qc* and *Tc* showed the highest growth, whereas

Ls and *Kp* showed the lowest growth. For the DBH growth response toward the planting soils, we identified four general patterns:

1. Significant difference in DBH growth between all three planting soils (*Tc* and *Qp*);
2. Significantly less DBH growth only in 'Sand' (No significant difference between 'FLL' and 'Loamy Silt' (*Oc*, *Ls*, and *Al*);
3. Significantly less DBH growth in 'Sand' and 'FLL' than in 'Loamy Silt' (*Qc*, *Cb*, and *Gt*);
4. No difference in DBH growth between all planting soils (*Kp*).

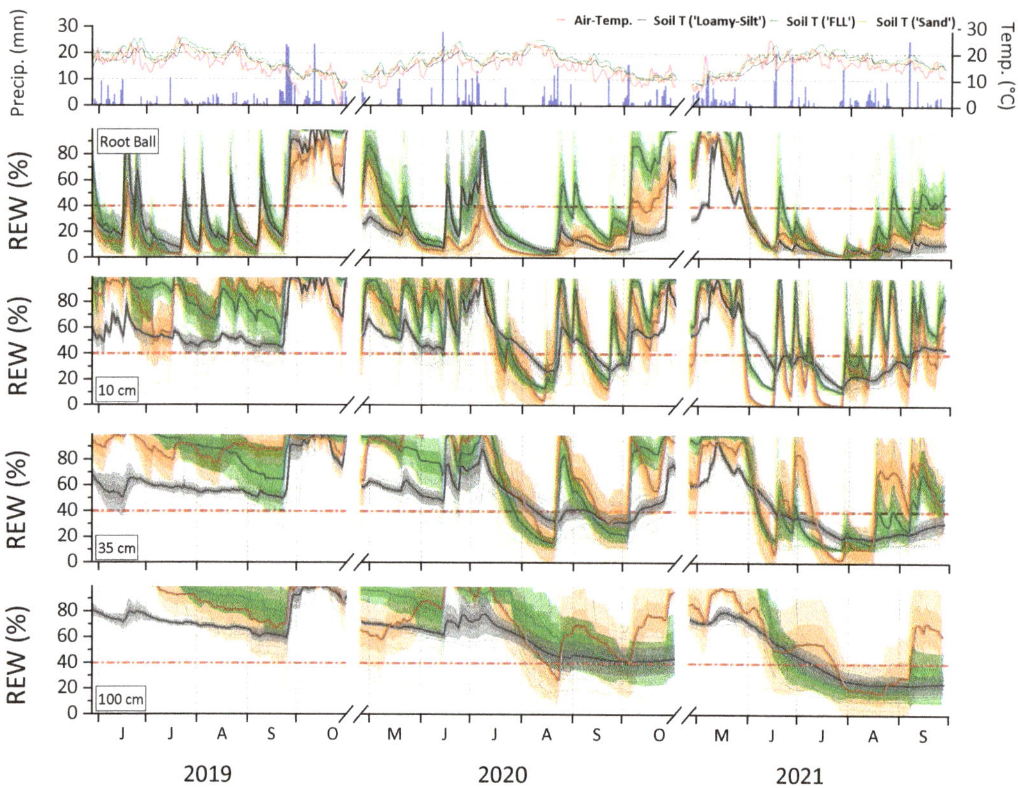

Figure 2. Dynamics of relative extractable water (REW %) at three depths and within the root ball in the three different planting soils ('Sand'(orange), 'FLL'(green), and 'Loamy Silt'(grey)) during the growing seasons of 2019, 2020, and 2021. Bold lines represent the means of all investigated planting pits per planting soil. Shadowed areas indicate the standard error (dark) and the standard deviation (light). Reddish dashed lines show a threshold value of 40% REW. Corresponding precipitation, air, and soil temperatures are given at the top of the graph.

Across all species, the tree DBH growth on 'Sand' and 'FLL' was, on average, −64% and −29%, respectively, lower compared with that on the optimum planting soil. *Kp* was excluded from the comparison. On 'Sand', *Qp* and *Al* were most restricted (−85% and −81%, respectively), whereas *Tc* and *Gt* were the less negatively affected species (−44% and −54%, respectively) compared with the optimum soil. On 'FLL', again, *Qp* and *Cb* were most restricted (−50% and −46%, respectively), whereas *Ls* and *Al* were the less negatively affected species (−10% and −12% respectively) compared with the optimum soil. The species *Al*, *Kp*, *Ls*, and *Oc* showed the same growth patterns found for their DBH growth, and also for their shoot growth (Figure 4b). The remaining species responded differently in terms of shoot growth compared with DBH growth. Across all species, tree shoot growth on 'Sand' and 'FLL' was, on average, −50% and −18%, respectively, lower than that on the optimum planting soil. On 'Sand', *Qp* was, by far, the most restricted (−72%), whereas *Gt* was the less negatively affected species (−27%) compared with the optimum soil. On 'FLL', again, *Qp* was most restricted in terms of growth (−42%) compared with the optimum soil, whereas *Oc*, *Al*, *Kp*, *Ls*, *Qc*, and *Tc* were not significantly affected.

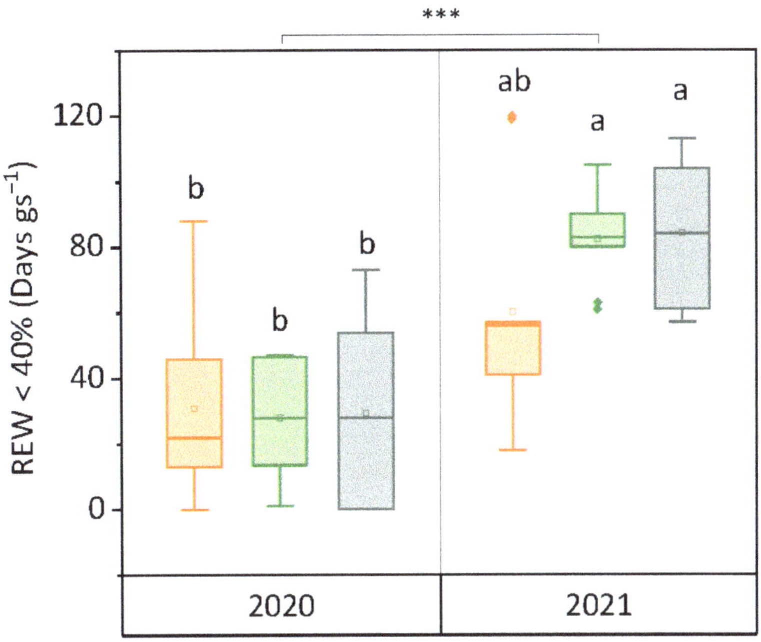

Figure 3. Sum of the number of days with REW < 40% (0–100 cm depth) during the growing season (gs) as a mean of the years 2020 and 2021 in the three planting soils 'Sand' (orange), 'FLL' (green), and 'Loamy Silt' (grey). Statistical analyses were performed by using a two-way ANOVA. Boxes with the same letters indicate no significant differences between the substrates at the $p \leq 0.05$ level. *** indicates significant differences between the years at the $p \leq 0.001$ level. Mean comparisons were performed by Tukey post hoc comparisons.

The assessed vitality scores of the years 2020 and 2021 were, across all species, the highest (i.e., lowered vitality) for the trees growing on 'Sand' (Table 3). In both years, *Al* and *Qp* had, on average, the highest values on 'Sand', whereas *Tc* had the lowest, which was also true for 'FLL' and 'Loamy Silt'. The vitality scores of trees growing on 'FLL' and 'Loamy Silt' were mostly <2. Only *Kp* showed values > 2, which was, in particular, striking for these trees growing in the optimum soil. We also noticed high vitality values for two individuals of *Qc* growing on the optimum soil, which was expressed by the elevated SD (Table 3).

Intra-annual stem diameter measurements for eight tree species (except for *Kp*) in 'Sand' showed a high variation of net growth (Figure 5). Growth onset was, on average, at the beginning of June and ceased, on average, at the beginning of August. During phases with REW < 40%, the daily stem diameter variations were pronounced with amplitudes varying evidently between species—they were weak in *Gs*, *Oc*, and *Qc*, and strong in *Tc* and *Ls*. The tree water deficit (TWD) is defined to occur when the maximum precedent stem diameter is greater than the maximum stem diameter of the actual day. Overall, the TWD was only weakly expressed (<120 µm d^{-1}) and occurred for individuals of almost all species when REW < 40% (data not displayed). However, the TWD during the drought phases was most pronounced for *Al* and *Ls*, with single trees reaching maximum values of 1130 and 627 μm d^{-1}, respectively. Single variable correlations were used to test the impact of the environmental variables REW at 0–35 cm and 0–100 cm, precipitation, and VPD on the tree species growth reactions using the weekly means of relative stem diameter increment (SDI) and the maximum daily shrinkage of the stem diameter (mds). The SDI of all trees investigated, except for *Qc*, was positively correlated with REW at least on the depth level ($p \leq 0.05$), rather than with precipitation or VPD (Table 4). SDI showed the strongest correlations (Spearman's R > 0.6) with REW for the species *Oc*, *Gt*, and *Ls*. The correlations of SDI and REW in 0–35 cm were, for *Gt* and *Al*, more significant than the correlations at 0–100 cm, whereas the growth of *Qp* and *Cb* was correlated more significant with REW 0–100 cm. Mds was correlated with both the soil water status and VPD. Mds and REW at both depth levels were positively correlated in all investigated species ($p \leq 0.001$), with the strongest correlations (Spearman's R > 0.6) for *Tc*, *Qc*, *Cb*, and *Al*. VPD was significantly negatively correlated with mds in all species ($p \leq 0.001$), except for *Gt* ($p > 0.1$) (Table 4).

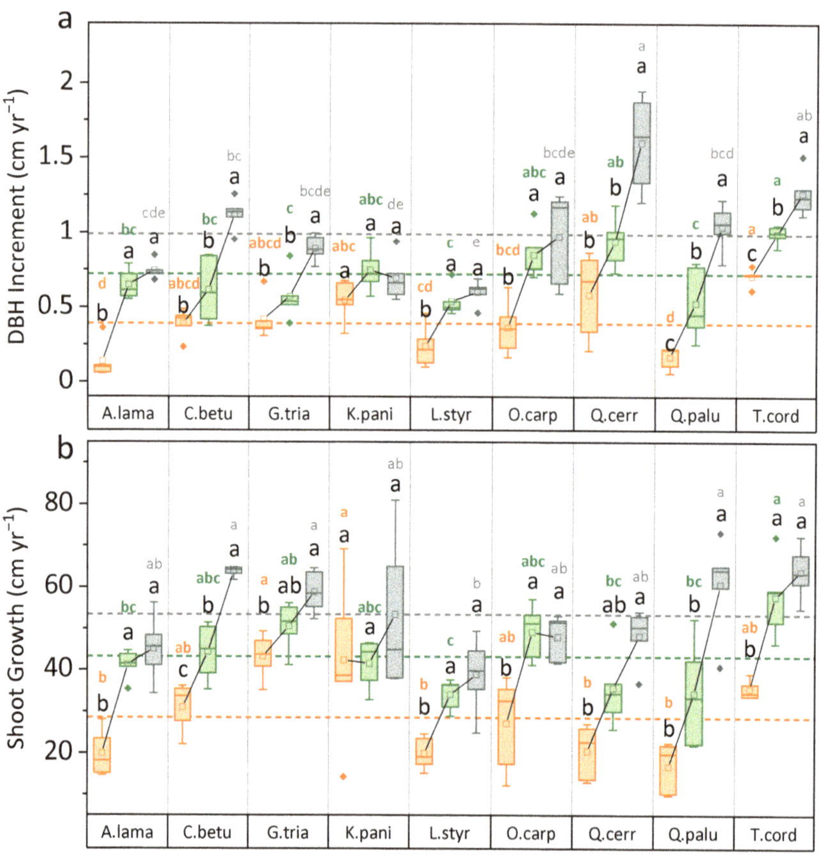

Figure 4. (**a**) Boxplots of annual DBH growth and (**b**) annual shoot growth of all nine tree species and as a mean of the years 2020–2021. Colored dashed lines show the mean growth across all species for the three planting soils 'Sand' (orange), 'FLL' (green), and 'Loamy Silt' (grey). Statistical analyses were performed by using a one-way ANOVA. Boxes with the same letters (black) indicate no significant differences at the $p \leq 0.05$ level within a species between the planting soils. Boxes with the same letters (see color assignment above) indicate no significant differences at the $p \leq 0.05$ level within a planting soil between the species. Mean comparisons were performed by Tukey post hoc comparisons.

Table 3. Tree vitality score assessed by visual inspections of the tree crowns in September 2020 and 2021. Values, ranked from 1 (vital) to 5 (strongly impaired), are presented as the means ± SD for each species x planting soil combination ($n = 5$).

Tree Species 'Cultivar'	Vitality Score [1]					
	'Sand'		'FLL'		'Loamy Silt'	
	2020	2021	2020	2021	2020	2021
Tilia cordata 'Greenspire'	1.6 ± 0.9	1.4 ± 0.5	1.0 ± 0.0	1.0 ± 0.0	1.0 ± 0.0	1.0 ± 0.0
Quercus cerris	2.2 ± 0.4	2.4 ± 0.5	1.4 ± 0.5	1.2 ± 0.4	1.6 ± 0.9	1.8 ± 0.8
Quercus palustris	2.8 ± 0.4	2.8 ± 0.4	1.6 ± 0.5	1.4 ± 0.5	1.0 ± 0.0	1.6 ± 0.5
Carpinus betulus 'Lucas'	2.2 ± 0.4	1.8 ± 0.4	1.8 ± 0.4	1.2 ± 0.4	1.2 ± 0.4	1.0 ± 0.0
Ostrya carpinifolia	2.6 ± 0.5	2.2 ± 0.8	1.8 ± 0.4	1.8 ± 0.4	1.2 ± 0.4	1.2 ± 0.4
Gleditsia triacanthos 'Skyline'	2.4 ± 0.5	1.6 ± 0.5	2.2 ± 0.4	1.6 ± 0.5	1.2 ± 0.4	1.0 ± 0.0
Liquidambar styraciflua	2.2 ± 0.4	2.0 ± 0.0	1.0 ± 0.0	1.0 ± 0.0	1.2 ± 0.4	1.2 ± 0.4
Amelanchier lamarckii	3.0 ± 1.0	3.6 ± 0.9	1.6 ± 0.5	1.6 ± 0.5	1.4 ± 0.5	1.0 ± 0.0
Koelreuteria paniculata	2.4 ± 0.9	2.8 ± 0.8	2.4 ± 0.9	2.0 ± 0.7	2.4 ± 1.1	3.2 ± 1.5

[1] According to vitality assessment procedure from GALK = Deutsche Gartenamtsleiterkonferenz (GALK e.V.; German Garden Agency Directors Conference).

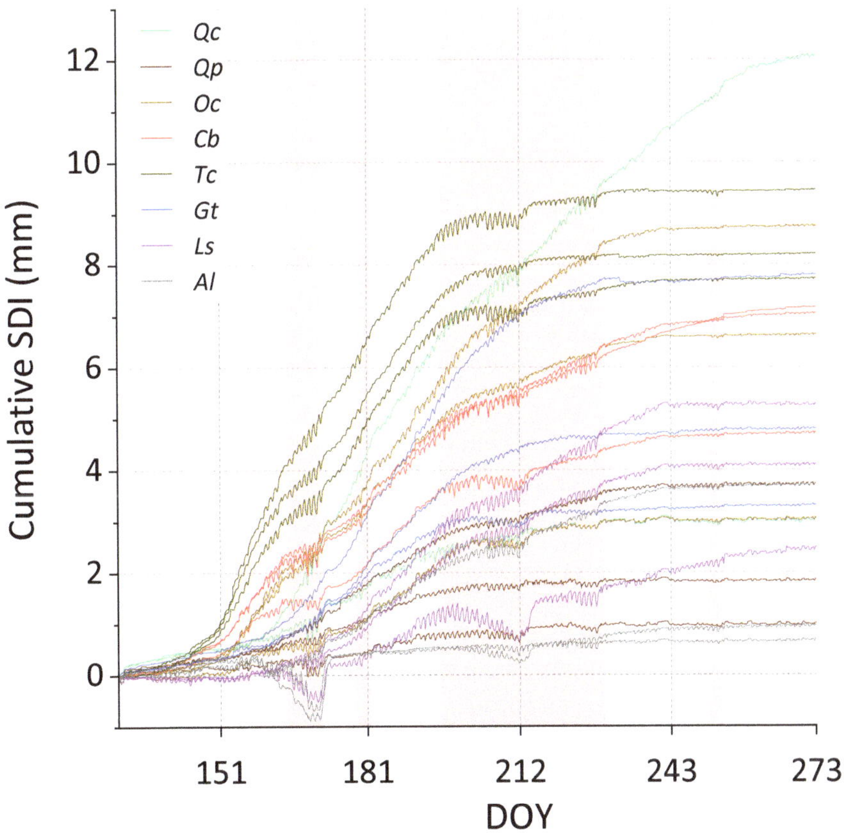

Figure 5. Cumulative stem diameter increment (SDI) of eight tree species grown in 'Sand' during the growing season of 2021. Reddish areas indicate phases when REW was <40% (light red: 0–35 cm; dark red: 0–100 cm).

Table 4. Spearman correlation coefficients® for the correlation of the relative stem diameter increment as weekly means (SDI; $n = 20$ data points per individual tree) and maximum daily shrinkage (mds; $n = 173$ data points per individual tree) of the stem diameter with the hydrological and climatological variables REW (relative extractable water), p (precipitation), and VPD (vapor pressure deficit) in eight species. Investigated period is defined according to the species-specific growing season length (Table 3)).

Weekly SDI	n	$REW_{0–35\ cm}$	$REW_{0–100\ cm}$	p	VPD
Tilia cordata 'Greenspire'	3	0.43 **	0.53 **	ns	ns
Quercus cerris	2	ns	ns	ns	ns
Quercus palustris	3	0.42	0.45 ***	0.43	−0.41
Carpinus betulus 'Lucas'	3	0.32	0.51 **	ns	ns
Ostrya carpinifolia	3	0.69 ***	0.85 ***	0.53 **	ns
Gleditsia triacanthos 'Skyline'	3	0.64 ***	0.37	ns	ns
Liquidambar styraciflua	3	0.61 ***	0.73 ***	ns	ns
Amelanchier lamarckii	3	0.44 **	ns	0.38 **	ns
mds					
Tilia cordata 'Greenspire'	3	0.67 ***	0.72 ***	0.23 **	−0.42 ***
Quercus cerris	2	0.6 ***	0.64 ***	ns	−0.57 ***
Quercus palustris	3	0.47 ***	0.48 ***	ns	−0.43 ***
Carpinus betulus 'Lucas'	3	0.67 ***	0.53 ***	0.16 **	−0.38 ***
Ostrya carpinifolia	3	0.35 ***	0.4 ***	ns	−0.49 ***
Gleditsia triacanthos 'Skyline'	3	0.34 ***	0.39 ***	ns	ns
Liquidambar styraciflua	3	0.6 ***	0.42 ***	0.19 **	−0.4 ***
Amelanchier lamarckii	3	0.63 ***	0.43 ***	0.26 **	−0.51 ***

Significance levels: ** $p \leq 0{,}01$; *** $p \leq 0.001$; ns = not significant.

4. Discussion

4.1. Substrate Characteristics

New street trees are often planted in artificial sandy-textured soil [5,33] or specific load-bearing substrates to resist compaction [23]. Up to now, these substrates were not well characterized in terms of soil hydrological properties [59], and it was unclear how the growth and vitality of specific tree species respond to the individual substrate conditions [9]. We used an experimental field setup with nine selected tree species and two urban tree sites representing planting soils to investigate this response under controlled field conditions.

Soil water availability is the most important parameter controlling tree growth [60–62]. With 10%, the plant available water capacity (PAWC) of 'FLL' was within the range reported for similar technical substrates (7–11%) [63] and exceeded the PAWC of 'Sand' (6%). 'Loamy Silt' had more than twice the amount of plant available water, mainly stored in the medium-sized pores of the dominating silt grains (Figure 1). As expectable, under field conditions, the REW in 'Sand' and 'FLL' was reduced sufficiently and faster than that in the optimum soil with the beginning of the growing season. However, regarding the seasonal average, no difference between the REW negatively exceeding the threshold value of 40% in the artificial soils and 'Loamy Silt' was visible (Figure 3). We assume that the different abilities to transport water caused this finding for the three planting soils. Precipitation water infiltrated the artificial soils faster and more effectively due to the higher infiltration capacity (IC), replenishing the PAWC, particularly at the 10 cm and 35 cm depths regularly. On the other hand, we assumed trees growing in the optimal soil to exploit the soil water stored more effectively, particularly at low soil water potentials, regardless of the absolutely higher quantity. This is reflected by the different hydraulic conductivities of the substrates during the process of drying (Figure 1b).

The large proportion of medium and fine pores in the total pore system of 'Loamy Silt' caused, under moist conditions (<pF 1.8), slower water transport, but that under unsaturated conditions (>pF 1.8–4.2) was almost constant. Compared to the 'Sand' and 'FLL', where large pores became quickly nonconductive with increasing soil water potential [64], a higher amount of water was thus quantitatively available to the trees in 'Loamy Silt'. While the tree roots in the 'Loamy Silt' could be resupplied with soil water with almost no restriction, the development of dry zones around the tree roots [65], causing hydraulically disconnection from surrounding wet soil, was most likely the limiting factor for water supply in sandy and coarse porous soils, causing stomatal closure [66] and thus reduced water consumption. Hence, the drought stress avoidance strategies of trees relying on reducing the plant water potential (anisohydric reaction type) were likely to be not successful in sandy or coarsely porous artificial soils [67]. In order to tap further soil water, a tree able to adapt to sandy and coarse porous soils must invest in the production of fine roots that grow towards available water to bypass dry soil patches. Most likely, the exchange of soil water between the SWP sensor and soil is also affected in a comparative way. This resulted in wide standard deviations of the mean REW in 'Sand' and 'FLL' (Figure 2), suggesting that sufficient soil drying was only detected by sensors close to roots. Thus, a high sensor density would be needed to capture the spatial heterogeneity of the soil water distribution within sandy and coarse porous soils.

Overall, the seasonal water stress conditions determined by using the REW threshold value reported for forests [52] have been comparatively low [14,52,68] (Figure 3). In addition to the explanations mentioned, this may have resulted from an underdeveloped root:shoot ratio within three years after transplanting and regular and intense soil water replenishment from precipitation. The latter, however, would have been lower under actual urban site conditions due to sealing with impermeable pavements and soil compaction. Therefore, the use of rain-out shelters would have been necessary to simulate prolonged drought situations, as used by [41].

4.2. Tree Growth Analysis

So far, data regarding species-specific belowground requirements generated from growth response to ensure the establishment, initial growth, and long-term survival of young street trees in urban environments are scarce. The second- and third-season growth data of the trees planted in artificial soils were in the range of those of other studies for DBH- [24,25,69–71] and shoot growth [32]. In the first growing season, DBH- and shoot growth were similar among species, being low in all soils compared with those in the following years (Figure S3). This growth depression in the first year is in accordance with the finding of [28] that trees need time to recover from a transplanting shock (i.e., reestablishing the root:shoot ratio) and initially mainly profit from the uniform root ball soil conditions and irrigation [19]. This nexus is supported by our observation of high water consumption in the root ball and low consumption in the planting soils at all depths (Figure 2). In the second and third growing seasons, the growth of all species was constrained on the artificial soils compared with the optimal soil [9,24,72]. This is contrary to the findings of [25,29], who reported higher tree growth in the structural planting soils of 1.3–3.2 cm yr^{-1} DBH (*Quercus bicolor*, *Quercus phellos*, and *Pyrus calleryana* 'Chanticleer'), compared with 0.3–0.6 cm yr^{-1} in the artificial soils in our study. In particular, these authors found higher or equal growth of trees in structural soils compared with trees in tree lawns. However, data regarding the soil properties and soil characteristics (i.e., texture, organic matter content, bulk density) are not available and due to the smaller planting pit dimensions and multiple urban environmental constraints in these in situ studies; a comparison of the results with our study should be conducted with caution. Furthermore, it is likely that tree roots developed in the whole planting pit and extended into the surrounding soil a few years after planting [14,73]. This was supported by random excavations at the edges of the planting pits at the end of the third growing season, by the gradual reduction of REW down to 100 cm depth in all substrates until 2021 (Figure 2), and by the elevated growth of trees in 'Sand' and 'FLL' in 2021 compared with that in 2020 (Figure S3). As opposed to our experimental site where natural 'Loamy Silt' surrounded the planting pits, in urban settings, a growth decrease would be most likely when tree roots extend into the surrounding soils comparable to 'Sand' [1,5]. We, therefore, assume that a smaller planting pit and thus an earlier extension of roots into the surrounding soil, when possible would also decrease tree growth and would make long-term comparison between soil conditions inappropriate [29].

Little attention has been paid to the morphological and physiological responses of tree species to artificial urban soils [9,71]. Across all substrates, we found the strongest annual DBH growth rates for *Qc* and the lowest for *Ls*, and the strongest annual shoot growth for *Tc* and the lowest also for *Ls* (Figure 4a,b). The response of the investigated tree species was different [9,24].

Although *Tc* invested strongly in stem growth, even under the unfavorable soil conditions of 'Sand' and 'FLL', the vitality scores, particularly in 'Sand', indicated apparently good performance also in the third growing season. This outcome is in contrast to the findings from [41], where the isohydric *Tc* [74] was highly affected by water scarcity, showing early leaf senescence. Since, in that study, extreme drought situations were caused on sandy loam by using rain-out shelters, it seems plausible that *Tc* is unable to extract water from drying, fine-grained soils [75]. In the sandy and coarse porous artificial soils used in our study, *Tc* showed, contrary to [76], no growth reduction during the phases of low REW in the 'Sand' and, despite the high mds, almost no TWD (Figure 5). These observations indicate that *Tc*, at least the cultivar 'Greenspire', is seemingly well-adapted to artificial urban soils characterized by low hydraulic conductivity. We hypothesize that this adaptation comes with a drought strategy that does not rely on increasing the suction power of roots, but on growing roots towards the water. This hypothetic reaction pattern of *Tc* in urban soils and the long-term effect on the C-balance should be investigated further in order to reliably assess the mortality risk in urban environments under climate change.

Contrary to *Tc*, the overall growth rates of *Ls* were very low [71] (Figure 4). However, compared with the optimal soil, the DBH and shoot growth were equal or above average for all species in 'Sand' and 'FLL', respectively, and trees in 'FLL' maintained the best possible vitality score (Figure 5). During the growing season, the growth of Ls was strongly correlated with REW (Table 4), indicating that investment in above-ground biomass was reduced during phases of low REW. This suggests that, under drought conditions, assimilated and stored C had not been allocated to growth, but rather to mechanisms successfully coping with drought stress. This is in accordance with [71], who found the lowest annual mortality rates for *Ls* among 10 species investigated in a subtropical city in Florida, USA. This indicates, in accordance with our results, that *Ls* is capable of withstanding dry soils [77] and that, in general, high growth rates in urban soils alone are not a categorical identifier for the adaptability of a tree species to the harsh urban environment under future climates.

In contrast to 'FLL', the growth of *Al* in 'Sand' was different to that of *Ls* and almost the lowest of all species, while simultaneously being obviously non-vigorous (Table 3). *Al* was the only species that grew less in 2021 compared with 2020 (data not shown) and where SDI and mds were correlated mainly with REW in the upper soil compartment (Table 4). Thus, we assume that *Al* was neither unable to develop a sufficient, deep-rooting system within the planting pit, nor expand its roots into the favorable surrounding soil.

The *Quercus* species *Qc* and *Qp* showed overall high growth rates in the optimal soil. The DBH growth of *Qc* in the artificial soils was higher than the average of all species. The tree growth in 'Sand' was particularly variable, which suggests different abilities of individuals to react to coarse-textured soil with a low OM content within the species. *Qp*, on the other hand, experienced the largest growth inhibition of all tree species in sandy soils compared with the optimal soil. Currently, *Qp* is, among the studied species, one of the most abundant species in Hamburg's street tree population. However, it appears to be highly reactive to poor soil conditions. Furthermore, care should be taken that soils in *Qp* planting sites have pH values < 6.5 to prevent leaf chlorosis [78], which was visible for the trees in the 'Sand', but not in 'FLL' (Tables 1 and 3), whereas soil compaction and water logging might play minor roles in *Qp* growth [79]. Regardless of the constraints in the artificial soils, heavy precipitation caused prolonged waterlogging conditions during June 2020 and affected the 'Loamy Silt' trees of *Qc* in terms of vitality and *Kp* in terms of growth and vitality (Table 3). This suggests that both tree species need well-aerated, non-compacted soils at sites that do not tend toward waterlogging [80]. Despite the high variability in shoot growth in 'Sand', the DBH growth of *Kp* was comparatively high in the artificial soils. We assumed that *Kp* combined the ability of both, lowering its water potential in fine-grained soils [67], and growing with roots toward the water in coarsely textured and porous soils. However, the comparatively poor vitality of *Kp* in the artificial soils may have also resulted from the low REW conditions during mid-June 2021, since *Kp* is reported to be very sensitive to early growing season drought [67].

The DBH growth rates of *Cb* and *Oc* in the optimal soil were higher than those reported by [81] with similar soil properties. Compared with the optimal soil, *Cb* and *Oc* showed, in our study, similar growth reduction between 60% and 65% in 'Sand'; [81] also reported similar growth reduction for treatment plots where precipitation water infiltration was prevented by rain-out shelters; however, growth was reduced by up to 79% compared with the control plot. Contrary to growth in 'Sand', *Oc* showed in 'FLL' substantially higher growth than the more vital *Cb* compared with the optimum soil; [82] concluded that both species had the lowest resistance of growth under drought conditions in fine-grained soils. However, we found both species to be not affected above average in artificial planting soils, despite the low PAWC.

Considering that *Gt* had above-average growth and the lowest mds rates in 'Sand', it seems most likely that Gt is suitable for harsh urban street tree sites and persists under water-stress conditions [83]. However, the strong and significant correlation of growth with REW at 0–35 cm suggests that the roots are more likely to grow near the surface. This can be problematic and requires further investigation.

Under the slightly dry, but less extreme, meteorological conditions in terms of air/soil temperatures and relative humidity compared with inner cities, the studied trees established within the study period of three years. For the trees, the selected artificial planting soils were thus sufficient for survival in the short term. In the long term, it is most likely that the formation of above ground biomass, the assimilation of carbohydrate reserves, and the provision of ecosystem services may be limited. However, at actual urban street tree sites, trees will face additional constraints affecting tree growth. Further investigations are necessary to understand the mechanistic adaptations of tree species in response to planting soils (permanent stress) and periods of low REW (temporal stress), particularly regarding patterns of carbon allocation under permanent and temporal soil water stress [36]. Our investigations provide a first insight into growth-limiting conditions and tree species-specific differences triggered by 'artificial soils' with different hydrological properties, and the experimental design was proven successful and should be continued in the future. In addition, other relevant tree species not considered in this study should be investigated for their growth behavior under soil conditions representative of urban street sites. Such tree species could include *Robinia pseudoacacia* [84,85], *Quercus robur* [86,87], or *Platanus* spp. [88], which have shown properties suitable for urban road-side conditions.

5. Conclusions

We showed that sandy-textured urban planting soils—one representing structural pit filling and one representing local surrounding soil conditions—have low plant available water capacities and were restricted in terms of hydraulic conductivity when the soil dries. Thus, when the soil water potential decreased and the pore space became non-conductive due to the high percentage of air-filled pores, the amount of water quantitatively available for the tree decreased. Therefore, trees that invest more in the fine root system to bypass soil non-conductivity (e.g., *Tilia cordata* 'Greenspire' and *Liquidambar styraciflua*) might be more successful in sandy and coarse porous soils than trees that lower their water potential, which might be successful in fine-textured drying soils (e.g., *Koelreuteria paniculata*, *Quercus cerris*; [67]). However, the tree growth of all species on the artificial urban soils was significantly constrained. The selected artificial planting soils were sufficient to survive, but most likely did not encourage the trees to build up above ground biomass, to assimilate carbohydrate reserves, or to provide effective ecosystem services in the long term. Thus, improving the hydrological properties of planting soils at street tree sites is crucial to allow newly planted trees to grow and thrive.

Supplementary Materials: The following supporting information can be downloaded at: https://www.mdpi.com/article/10.3390/f13060936/s1 (accessed on). Figure S1: Spatial distribution of planting pits, treatment and control plantation soils, and tree species in a 4 × 35 grid. Figure S2: View of the experimental site in the western direction in June 2019. The three planting soils are clearly visible (light = 'Sand'; dark = 'FLL'; and brown = 'Loamy Silt'). Watering bags were removed after 2019. Figure S3: Annual DBH and shoot growth in each investigated year as a mean of all species (n = 9) in the planting soils (n = 3). Statistical analyses were performed by using a one-way ANOVA. Boxes with the same letters indicate no significant differences at the $p \leq 0.05$ level. Mean comparisons were performed by Tukey post hoc comparisons.

Author Contributions: Conceptualization, A.S., C.R. and A.E.; methodology, A.S.; software, A.S.; validation, J.N.B., C.R. and A.E.; formal analysis, A.S.; investigation, A.S.; resources, A.E.; data curation, J.N.B.; writing—original draft preparation, A.S.; writing—review and editing, J.N.B., C.R., and A.E.; visualization, A.S.; supervision, C.R. and A.E.; project administration, A.E.; funding acquisition, A.E. All authors have read and agreed to the published version of the manuscript.

Funding: This research was funded by the Federal Ministry for the Environment, Nature Conservation, Nuclear Safety and Consumer Protection (BMUV), Zukunft-Umwelt-Gesellschaft (ZUG) GmbH, grant number 67DAS153A, as an initiative under DAS: Förderung von Maßnahmen zur Anpassung an den Klimawandel, and by Deutsche Forschungsgemeinschaft (DFG, German Research Foundation) under Germany's Excellence Strategy—EXC 2037 "CLICCS–Climate, Climatic Change, and Society" Project Number: 390683824—contribution to the Center for Earth System Research and Sustainability (CEN) of Universität Hamburg. Further financial support was given by Freie und Hansestadt Hamburg FHH, and Behörde für Umwelt, Klima, Energie und Agrarwirtschaft (BUKEA).

Institutional Review Board Statement: Not applicable.

Informed Consent Statement: Not applicable.

Acknowledgments: Many thanks to Volker Kleinschmidt for various sensor installations and his professional technical support during planning and measurements. We would like to thank Baumschule Lorenz von Ehren GmbH & Co. KG for providing their ground and the trees, and for their manifold support to realize the experimental site. Many thanks also to K + E (Kompost und Erden GmbH) for providing the structural planting soil and to Annette Wagner and Torsten Melzer from BUKEA (Free and Hanseatic City of Hamburg, Department for the Environment, Climate, Energy and Agriculture) for their cooperation, data support, and tree vitality assessment during 2020 and 2021. We would also like to thank Gerhard Doobe from GALK e.V. AK Stadtbäume and Alexander Gröngröft for fruitful discussions.

Conflicts of Interest: The authors declare no conflict of interest. The funders had no role in the design of the study; in the collection, analyses, or interpretation of data; in the writing of the manuscript, or in the decision to publish the results.

References

1. Jim, C.Y. Urban soil characteristics and limitations for landscape planting in Hong Kong. *Landsc. Urban Plan.* **1998**, *40*, 235–249. [CrossRef]
2. Clark, J.R.; Kjelgren, R. Water as a limiting factor in the development of urban trees. *J. Arboric.* **1990**, *16*, 203–209. [CrossRef]
3. Craul, P.J. A description of urban soils and their desired characteristics. *J. Arboric.* **1985**, *11*, 330–339.
4. Wessolek, G.; Kluge, B.; Toland, A.; Nehls, T.; Klingelmann, E.; Rim, Y.N.; Mekiffer, B.; Trinks, S. Urban Soils in the Vadose Zone. In *Wilfried Endlicher (Hg.): Perspectives in Urban Ecology*; Springer: Berlin/Heidelberg, Germany, 2011.
5. Jim, C.Y.; Ng, Y.Y. Porosity of roadside soil as indicator of edaphic quality for tree planting. *Ecol. Eng.* **2018**, *120*, 364–374. [CrossRef]
6. Armson, D.; Stringer, P.; Ennos, A.R. The effect of street trees and amenity grass on urban surface water runoff in Manchester, UK. *Urban For. Urban Green.* **2013**, *12*, 282–286. [CrossRef]
7. Morgenroth, J.; Buchan, G.; Scharenbroch, B.C. Belowground effects of porous pavements—Soil moisture and chemical properties. *Ecol. Eng.* **2013**, *51*, 221–228. [CrossRef]

8. Cermák, J.; Hruska, J.; Martinková, M.; Prax, A. Urban tree root systems and their survival near houses analyzed using ground penetrating radar and sap flow techniques. *Plant Soil* **2000**, *219*, 103–116. [CrossRef]
9. Pregitzer, C.C.; Sonti, N.F.; Hallett, R.A. Variability in Urban Soils Influences the Health and Growth of Native Tree Seedlings. *Ecol. Restor.* **2016**, *34*, 106–116. [CrossRef]
10. Roetzer, T.; Wittenzeller, M.; Haeckel, H.; Nekovar, J. Phenology in central Europe-differences and trends of spring phenophases in urban and rural areas. *Int. J. Biometeorol.* **2000**, *44*, 60–66. [CrossRef]
11. Schnabel, F.; Purrucker, S.; Schmitt, L.; Engelmann, R.A.; Kahl, A.; Richter, R.; Seele-Dilbat, C.; Skiadaresis, G.; Wirth, C. Cumulative growth and stress responses to the 2018–2019 drought in a European floodplain forest. *Glob. Chang. Biol.* **2021**, *28*, 1870–1883. [CrossRef]
12. Schuldt, B.; Buras, A.; Arend, M.; Vitasse, Y.; Beierkuhnlein, C.; Damm, A.; Gharun, M.; Grams, T.E.E.; Hauck, M.; Hajek, P.; et al. A first assessment of the impact of the extreme 2018 summer drought on Central European forests. *Basic Appl. Ecol.* **2020**, *45*, 86–103. [CrossRef]
13. Hartmann, H. Will a 385 million year-struggle for light become a struggle for water and for carbon?-How trees may cope with more frequent climate change-type drought events. *Glob. Change Biol.* **2011**, *17*, 642–655. [CrossRef]
14. Schütt, A.; Becker, J.N.; Schaaf-Titel, S.; Groengroeft, A.; Eschenbach, A. Soil water stress at young urban street-tree sites in response to meteorology and site parameters. *Urban For. Urban Green. in review.*
15. Schlünzen, K.H.; Hoffmann, P.; Rosenhagen, G.; Riecke, W. Long-term changes and regional differences in temperature and precipitation in the metropolitan area of Hamburg. *Int. J. Climatol.* **2010**, *30*, 1121–1136. [CrossRef]
16. Quigley, M.F. Street trees and rural conspecifics: Will long-lived trees reach full size in urban conditions? *Urban Ecosyst.* **2004**, *7*, 29–39. [CrossRef]
17. Gillner, S.; Vogt, J.; Roloff, A. Climatic response and impacts of drought on oaks at urban and forest sites. *Urban For. Urban Green.* **2013**, *12*, 597–605. [CrossRef]
18. Pretzsch, H.; Biber, P.; Uhl, E.; Dahlhausen, J.; Schütze, G.; Perkins, D.; Rötzer, T.; Caldentey, J.; Koike, T.; Van Con, T.; et al. Climate change accelerates growth of urban trees in metropolises worldwide. *Sci. Rep.* **2017**, *7*, 15403. [CrossRef]
19. Watson, W.T. Influence of Tree Size on Transplant Establishment and Growth. *Horttech* **2005**, *15*, 118–122. [CrossRef]
20. Roman, L.A.; Scatena, F.N. Street tree survival rates: Meta-analysis of previous studies and application to a field survey in Philadelphia, PA, USA. *Urban For. Urban Green.* **2011**, *10*, 269–274. [CrossRef]
21. Roman, L.A.; Battles, J.J.; McBride, J.R. The balance of planting and mortality in a street tree population. *Urban Ecosyst.* **2014**, *17*, 387–404. [CrossRef]
22. Nowak, D.J.; Kuroda, M.; Crane, D.E. Tree mortality rates and tree population projections in Baltimore, Maryland, USA. *Urban For. Urban Green.* **2004**, *2*, 139–147. [CrossRef]
23. Grabosky, J.; Bassuk, N. Testing of structural urban tree soil materials for Use under Pavement to Increase Street Tree Rooting Volumes. *J. Arboric.* **1996**, *22*, 255–263. [CrossRef]
24. Smiley, E.T.; Calfee, L.; Fraedrich, B.R.; Smiley, E.J. Comparison of Structural and Noncompacted Soils for Trees Surrounded by Pavement. *Arboric. Urban For.* **2006**, *32*, 164–169. [CrossRef]
25. Rahman, M.A.; Smith, J.G.; Stringer, P.; Ennos, A.R. Effect of rooting conditions on the growth and cooling ability of Pyrus calleryana. *Urban For. Urban Green.* **2011**, *10*, 185–192. [CrossRef]
26. Grabosky, J. Establishing a common method to compare soil systems designed for both tree growth and pavement support. Research Note. *Soil Sci.* **2015**, *180*, 207–213. [CrossRef]
27. Bühler, O.; Ingerslev, M.; Skov, S.; Schou, E.; Thomsen, I.M.; Nielsen, C.N.; Kristoffersen, P. Tree development in structural soil–An empirical below-ground in-situ study of urban trees in Copenhagen, Denmark. *Plant Soil* **2017**, *413*, 29–44. [CrossRef]
28. Bretzel, F.; Vannucchi, F.; Pini, R.; Scatena, M.; Marradi, A.; Cinelli, F. Use of coarse substrate to increase the rate of water infiltration and the bearing capacity in tree plantings. *Ecol. Eng.* **2020**, *148*, 105798. [CrossRef]
29. Grabosky, J.; Bassuk, N. Seventeen years' growth of street trees in structural soil compared with a tree lawn in New York City. *Urban For. Urban Green.* **2016**, *16*, 103–109. [CrossRef]
30. FLL. Empfehlungen für Baumpflanzungen Teil 2. In *Standortvorbereitungen für Neuanpflanzungen*; Pflanzgruben und Wurzelraumerweiterung, Bauweisen und Substrate; Forschungsgesellschaft Landschaftsentwicklung Landschaftsbau e.V. (FLL): Bonn, Germany, 2010.
31. Nielsen, C.N.; Bühler, O.; Kristoffersen, P. Soil water dynamics and growth of street and park trees. *Arboric. Urban For.* **2007**, *33*, 231–245. [CrossRef]
32. Riikonen, A.; Lindén, L.; Pulkkinen, M.; Nikinmaa, E. Post-transplant crown allometry and shoot growth of two species of street trees. *Urban For. Urban Green.* **2011**, *10*, 87–94. [CrossRef]
33. Schickhoff, U.; Eschenbach, A. Terrestrische und semiterrestrische Ökosysteme. In *Hans von Storch, Insa Meinke und Martin Claußen (Hg.): Hamburger Klimabericht–Wissen über Klima, Klimawandel und Auswirkungen in Hamburg und Norddeutschland*; Springer: Berlin/Heidelberg, Germany, 2018.
34. Somerville, P.D.; Farrell, C.; May, P.B.; Livesley, S.J. Tree water use strategies and soil type determine growth responses to biochar and compost organic amendments. *Soil Tillage Res.* **2019**, *192*, 12–21. [CrossRef]

35. Blackman, C.J.; Creek, D.; Maier, C.; Aspinwall, M.J.; Drake, J.E.; Pfautsch, S.; O'Grady, A.; Delzon, S.; E Medlyn, B.; Tissue, D.T.; et al. Drought response strategies and hydraulic traits contribute to mechanistic understanding of plant dry-down to hydraulic failure. *Tree Physiol.* **2019**, *39*, 910–924. [CrossRef] [PubMed]
36. Trugman, A.T.; Detto, M.; Bartlett, M.K.; Medvigy, D.; Anderegg, W.R.L.; Schwalm, C.; Schaffer, B.; Pacala, S.W. Tree carbon allocation explains forest drought-kill and recovery patterns. *Ecol. Lett.* **2018**, *21*, 1552–1560. [CrossRef] [PubMed]
37. Adams, H.D.; Zeppel, M.J.B.; Anderegg, W.R.L.; Hartmann, H.; Landhäusser, S.M.; Tissue, D.T.; Huxman, T.E.; Hudson, P.J.; Franz, T.E.; Allen, C.D.; et al. A multi-species synthesis of physiological mechanisms in drought-induced tree mortality. *Nat. Ecol. Evol.* **2017**, *1*, 1285–1291. [CrossRef]
38. Mitchell, P.J.; O'Grady, A.P.; Tissue, D.T.; Worledge, D.; Pinkard, E.A. Co-ordination of growth, gas exchange and hydraulics define the carbon safety margin in tree species with contrasting drought strategies. *Tree Physiol.* **2014**, *34*, 443–458. [CrossRef]
39. Roloff, A.; Korn, S.; Gillner, S. The Climate-Species-Matrix to select tree species for urban habitats considering climate change. *Urban For. Urban Green.* **2009**, *8*, 295–308. [CrossRef]
40. Böll, S.; Schönfeld, P.; Körber, K.; Herrmann, J.V. Stadtbäume unter Stress. Projekt »Stadtgrün 2021« untersucht Stadtbäume im Zeichen des Klimawandels. *LWF Aktuell* **2014**, *98*, 4–8.
41. Stratópoulos, L.M.F.; Zhang, C.; Häberle, K.-H.; Pauleit, S.; Duthweiler, S.; Pretzsch, H.; Rötzer, T. Effects of Drought on the Phenology, Growth, and Morphological Development of Three Urban Tree Species and Cultivars. *Sustainability* **2019**, *11*, 5117. [CrossRef]
42. Schönfeld, P. Klimabäume. Welche Arten können in Zukunft gepflanzt werden? *LWG Aktuell.* 2019. Available online: https://www.lwg.bayern.de/mam/cms06/landespflege/dateien/zukunft_klimabaeume.pdf (accessed on 15 March 2022).
43. DWD. CDC (Climate Data Center). Deutscher Wetterdienst. 2020. Available online: https://www.dwd.de (accessed on 15 March 2022).
44. Watson, G.W.; Hewitt, A.M.; Custic, M.; Lo, M. The management of tree root systems in urban and suburban settings II. A Review of Strategies to Mitigate Human Impacts. *Arboric. Urban For.* **2014**, *40*, 249–271. [CrossRef]
45. Roloff, A. Aktualisierte KlimaArtenMatrix 2021 ("KLAM 2.0"). Unter Mitarbeit von Sten Gillner und Ulrich Pietzarka. In *Andreas Roloff (Hg.): Trockenstress bei Bäumen. Ursachen, Strategien, Praxis. Unter Mitarbeit von Anne Dreßler, Britt Kniesel, Doris Krabel, Liu Ming, Ulrich Pietzarka, Andreas Roloff und Lauritz Schrader*; Quelle & Meyer Verlag GmbH & Co.: Wiebelsheim, Germany, 2021; pp. 201–230.
46. Allen, K.S.; Harper, R.W.; Bayer, A.; Brazee, N.J. A review of nursery production systems and their influence on urban tree survival. *Urban For. Urban Green.* **2017**, *21*, 183–191. [CrossRef]
47. UMS GmbH. Manual HYPROP. UMS GmbH. Gmunder Str. 37, 81379 München, Germany (Version 2015-1). 2015. Available online: http://library.metergroup.com/Manuals/UMS/Hyprop_Manual.pdf (accessed on 25 February 2022).
48. Pertassek, T.; Peters, A.; Durner, W. *HYPROP-FIT Software User's Manual*, V. 3.0. Hg. v, UMS GmbH. Gmunder Str. 37, 81379: München, Germany. 2015. Available online: http://www.soil.tu-bs.de/download/downloads/reports/2015.Manual_HYPROP-FIT.pdf (accessed on 25 February 2022).
49. van Genuchten, M.T. A Closed-form Equation for Predicting the Hydraulic Conductivity of Unsaturated Soils. *Soil Sci. Soc. Am. J.* **1980**, *44*, 892–898. [CrossRef]
50. Shock, C.C.; Barnum, J.M.; Seddigh, M. Calibration of Watermark Soil Moisture Sensors for Irrigation Management. 1998. Available online: https://www.researchgate.net/profile/Clinton-Shock/publication/228762944_Calibration_of_W_ermark_Soil_Moisture_Sensors_for_Irrigation_Management/links/55ed971408ae3e12184819e7/Calibration-of-W-ermark-Soil-Moisture-Sensors-for-Irrigation-Management.pdf?_sg%5B0%5D=HUfTqZa51n_ochZlOZEB02YVuEJmDUsBgwCjg5iiy4Sv_CbwSpvfGkQ8yis9DLhS7fHrHlOuGRK4hkLWwK9yWw.z_OgBYeZ6_vc09Hy93oOgqYnxbktt8rNEhCW_0wvFD4FD7NX7icVsJP4hfdPqVNKgnsjCJpB9gWmD-8mzsNqRw&_sg%5B1%5D=J7mKCVEDnOcsiBZC8HLO3AulvwVF5U-xR5w9gkJDwCG2IfsGl9bNH_KVFmXdLEFgW7AKUrn2qQJ3gzGdN98zGopYvHUDdpSHGe6XLszU4x8D.z_OgBYeZ6_vc09Hy93oOgqYnxbktt8rNEhCW_0wvFD4FD7NX7icVsJP4hfdPqVNKgnsjCJpB9gWmD-8mzsNqRw&_iepl= (accessed on 14 February 2022).
51. Allen, R. *Calibration for the Watermark 200SS Soil Water Potential Sensorto Fit the 7-19-96—Calibration #3||Table*; University of Idaho: Kimberly, ID, USA, 2000.
52. Granier, A.; Reichstein, M.; Bréda, N.; Janssens, I.A.; Falge, E.; Ciais, P.; Grünwald, T.; Aubinet, M.; Berbigier, P.; Bernhofer, C.; et al. Evidence for soil water control on carbon and water dynamics in European forests during the extremely dry year: 2003. *Agric. For. Meteorol.* **2007**, *143*, 123–145. [CrossRef]
53. Granier, A.; Breda, N.; Biron, P.; Vilette, S. A lumped water balance model to evaluate duration and intensity of drought constraints in forest stands. *Ecol. Model.* **1999**, *116*, 269–283. [CrossRef]
54. R Core Team. *A Language and Environment for Statistical*; R Foundation for Statistical Computing: Vienna, Austria, 2022.
55. Knüsel, S.; Peters, R.L.; Haeni, M.; Wilhelm, M.; Zweifel, R. Processing and Extraction of Seasonal Tree Physiological Parameters from Stem Radius Time Series. *Forests* **2021**, *12*, 765. [CrossRef]
56. Zweifel, R.; Haeni, M.; Buchmann, N.; Eugster, W. Are trees able to grow in periods of stem shrinkage? *New Phytol.* **2016**, *211*, 839–849. [CrossRef]
57. Köcher, P.; Horna, V.; Leuschner, C. Environmental control of daily stem growth patterns in five temperate broad-leaved tree species. *Tree Physiol.* **2012**, *32*, 1021–1032. [CrossRef]

58. Deslauriers, A.; Rossi, S.; Anfodillo, T. Dendrometer and intra-annual tree growth: What kind of information can be inferred? *Dendrochronologia* **2007**, *25*, 113–124. [CrossRef]
59. Yilmaz, D.; Cannavo, P.; Séré, G.; Vidal-Beaudet, L.; Legret, M.; Damas, O.; Peyneau, P.-E. Physical properties of structural soils containing waste materials to achieve urban greening. *J. Soils Sediments* **2018**, *18*, 442–455. [CrossRef]
60. Allen, C.D.; Macalady, A.K.; Chenchouni, H.; Bachelet, D.; McDowell, N.; Vennetier, M.; Kitzberger, T.; Rigling, A.; Breshears, D.D.; Hogg, E.H.; et al. A global overview of drought and heat-induced tree mortality reveals emerging climate change risks for forests. *For. Ecol. Manag.* **2010**, *259*, 660–684. [CrossRef]
61. Bréda, N.; Granier, A.; Barataud, F.; Moyne, C. Soil water dynamics in an oak stand. I. Soil moisture, water potentials and water uptake by roots. *Plant Soil* **1995**, *172*, 17–27. [CrossRef]
62. Puhlmann, H.; Schmidt-Walter, P.; Hartmann, P.; Meesenburg, H.; von Wilpert, K. Soil Water Budget and Drought Stress. In *Nicole Wellbrock und Andreas Bolte (Hg.): Status and Dynamics of Forests in Germany, Bd. 237*; Springer International Publishing (Ecological Studies): Cham, Switzerland, 2019; pp. 55–91.
63. Grabosky, J.; Haffner, E.; Bassuk, N. Plant available Moisture in Stone-soil Media for Use under pavement while allowing urban tree root growth. *Arboric. Urban For.* **2009**, *35*, 271–278. [CrossRef]
64. Li, K.I.; De Jong, R.; Coe, M.T.; Ramankutty, N. Root-Water-Uptake Based upon a New Water Stress Reduction and an Asymptotic Root Distribution Function. *Earth Interact.* **2006**, *10*, 1–14. [CrossRef]
65. Schulze, E.-D.; Beck, E.; Buchmann, N.; Clemens, S.; Müller-Hohenstein, K.; Scherer-Lorenzen, M. *Plant Ecology*; Springer: Berlin/Heidelberg, Germany, 2019.
66. Abdalla, M.; Carminati, A.; Cai, G.; Javaux, M.; Ahmed, M.A. Stomatal closure of tomato under drought is driven by an increase in soil–root hydraulic resistance. *Plant Cell Environ.* **2021**, *44*, 425–431. [CrossRef] [PubMed]
67. Sjöman, H.; Hirons, A.D.; Bassuk, N.L. Improving confidence in tree species selection for challenging urban sites: A role for leaf turgor loss. *Urban Ecosyst.* **2018**, *21*, 1171–1188. [CrossRef]
68. Schmidt-Walter, P.; Ahrends, B.; Mette, T.; Puhlmann, H.; Meesenburg, H. NFIWADS: The water budget, soil moisture, and drought stress indicator database for the German National Forest Inventory (NFI). *Ann. For. Sci.* **2019**, *76*, 39. [CrossRef]
69. Bühler, O.; Nielsen, C.N.; Kristoffersen, P. Growth and phenology of established Tilia cordata street trees in response to different irrigation regimes. *Arboric. Urban For.* **2006**, *32*, 3–9. [CrossRef]
70. Boukili, V.K.S.; Bebber, D.P.; Mortimer, T.; Venicx, G.; Lefcourt, D.; Chandler, M.; Eisenberg, C. Assessing the performance of urban forest carbon sequestration models using direct measurements of tree growth. *Urban For. Urban Green.* **2017**, *24*, 212–221. [CrossRef]
71. Lawrence, A.B.; Escobedo, F.J.; Staudhammer, C.L.; Zipperer, W. Analyzing growth and mortality in a subtropical urban forest ecosystem. *Landsc. Urban Plan.* **2012**, *104*, 85–94. [CrossRef]
72. Loh, F.C.W.; Grabosky, J.C.; Bassuk, N.L. Growth response of Ficus benjamina to limited soil volume and soil dilution in a skeletal soil container study. *Urban For. Urban Green.* **2003**, *2*, 53–62. [CrossRef]
73. Krieter, M.; Malkus, A. *Untersuchungen zur Standortoptimierung von Straßenbäumen: Ergebnisse eines FLL-Pflanzversuches von Tilia Pallida in 14 Deutschen Städten*; FLL: Bonn, Germany, 1996.
74. Moser, A.; Rahman, M.A.; Pretzsch, H.; Pauleit, S.; Rötzer, T. Inter- and intraannual growth patterns of urban small-leaved lime (Tilia cordata mill.) at two public squares with contrasting microclimatic conditions. *Int. J. Biometeorol.* **2017**, *61*, 1095–1107. [CrossRef]
75. Gillner, S.; Korn, S.; Hofmann, M.; Roloff, A. Contrasting strategies for tree species to cope with heat and dry conditions at urban sites. *Urban Ecosyst.* **2017**, *20*, 853–865. [CrossRef]
76. Moser, A.; Rötzer, T.; Pauleit, S.; Pretzsch, H. The Urban Environment Can Modify Drought Stress of Small-Leaved Lime (*Tilia cordata* Mill.) and Black Locust (*Robinia pseudoacacia* L.). *Forests* **2016**, *7*, 71. [CrossRef]
77. Baraldi, R.; Przybysz, A.; Facini, O.; Pierdonà, L.; Carriero, G.; Bertazza, G.; Neri, L. Impact of Drought and Salinity on Sweetgum Tree (*Liquidambar styraciflua* L.): Understanding Tree Ecophysiological Responses in the Urban Context. *Forests* **2019**, *10*, 1032. [CrossRef]
78. Watson, G.W.; Himelick, E.B. Effects of soil pH, root density and tree growth regulator treatments on pin oak chlorosis. *J. Arboric.* **2004**, *30*, 172–178. [CrossRef]
79. Watson, G.W.; Kelsey, P. The impact of soil compaction on soil aeration and fine root density of Quercus palustris. *Urban For. Urban Green.* **2006**, *4*, 69–74. [CrossRef]
80. Roloff, A.; Gillner, S.; Kniesel, R.; Zhang, D. Interesting and new street tree species for European cities. *Jflr* **2018**, *3*, 1–7. [CrossRef]
81. Stratopoulos, L. "Klimabäume" für die Stadt. Über die Rolle Einer Angepassten Arten- und Sortenwahl für die Kühlleistung von Straßenbäumen. Ph.D. Thesis, Technische Universität München, Weihnenstephan, Germany, 2020.
82. Stratópoulos, L.M.F.; Zhang, C.; Duthweiler, S.; Häberle, K.-H.; Rötzer, T.; Xu, C.; Pauleit, S. Tree species from two contrasting habitats for use in harsh urban environments respond differently to extreme drought. *Int. J. Biometeorol.* **2019**, *63*, 197–208. [CrossRef]
83. Smitley, D.R.; Peterson, N.C. Interactions of Water Stress, Honeylocust Spider Mites (Acari: Tetranychidae), Early Leaf Abscission, and Growth of Gleditsia tnacanthos. *J. Econ. Entomol.* **1996**, *89*, 1577–1581. [CrossRef]

84. Klisz, M.; Puchałka, R.; Netsvetov, M.; Prokopuk, Y.; Vítková, M.; Sádlo, J.; Matisons, R.; Mionskowski, M.; Chakraborty, D.; Olszewski, P.; et al. Variability in climate-growth reaction of Robinia pseudoacacia in Eastern Europe indicates potential for acclimatisation to future climate. *For. Ecol. Manag.* **2021**, *492*, 119194. [CrossRef]
85. Moser-Reischl, A.; Rahman, M.A.; Pauleit, S.; Pretzsch, H.; Rötzer, T. Growth patterns and effects of urban micro-climate on two physiologically contrasting urban tree species. *Landsc. Urban Plan.* **2019**, *183*, 88–99. [CrossRef]
86. Netsvetov, M.; Prokopuk, Y.; Puchałka, R.; Koprowski, M.; Klisz, M.; Romenskyy, M. River Regulation Causes Rapid Changes in Relationships Between Floodplain Oak Growth and Environmental Variables. *Front. Plant Sci.* **2019**, *10*, 96. [CrossRef]
87. Thomsen, S.; Reisdorff, C.; Gröngröft, A.; Jensen, K.; Eschenbach, A. Responsiveness of mature oak trees (*Quercus robur* L.) to soil water dynamics and meteorological constraints in urban environments. *Urban Ecosyst.*. [CrossRef]
88. Dervishi, V.; Poschenrieder, W.; Rötzer, T.; Moser-Reischl, A.; Pretzsch, H. Effects of Climate and Drought on Stem Diameter Growth of Urban Tree Species. *Forests* **2022**, *13*, 641. [CrossRef]

forests

Article

Crown Shapes of Urban Trees-Their Dependences on Tree Species, Tree Age and Local Environment, and Effects on Ecosystem Services

Eleonora Franceschi [1,*], Astrid Moser-Reischl [1,2], Mohammad A. Rahman [2], Stephan Pauleit [2], Hans Pretzsch [1] and Thomas Rötzer [1]

[1] Chair for Forest Growth and Yield Science, Technical University of Munich, Hans-Carl-von-Carlowitz-Platz 2, 85354 Freising, Germany; astrid.reischl@tum.de (A.M.-R.); hans.pretzsch@tum.de (H.P.); thomas.roetzer@tum.de (T.R.)

[2] Chair for Strategic Landscape Planning and Management, Technical University of Munich, Emil-Ramann-Str. 6, 85354 Freising, Germany; ma.rahman@tum.de (M.A.R.); pauleit@tum.de (S.P.)

* Correspondence: eleonora.franceschi@tum.de

Abstract: Crown shapes of common European urban tree species differ from tree species to tree species and are modified by the age of a tree and its local environment. A tree's crown shape has a great influence on the crown volume and thus on the ecosystem service provision of a tree such as the shade area or the shade density. We used the data of 3852 tree individuals from eight German cities and the crown shape data of 528 trees for the species *Acer platanoides*, *Acer pseudoplatanus*, *Aesculus hippocastanum*, *Fraxinus excelsior*, *Platanus × acerifolia*, *Robinia pseudoacacia* and *Tilia cordata* to analyze tree structural dimensions and the crown volume and shade dependency on a tree's crown shapes. Ovoid (57% of all tree individuals) and spherical (24%) crown shapes were mostly observed. However, columnar shape was observed for light-demanding *R. pseudoacacia* in close proximity of objects. The greatest shade areas were measured for spherical shape and the highest shade density for ovoid shape. Logistic regression analysis showed significant effects of age and distance to objects on crown shapes. Significant probability of crown shapes was found for different tree species, e.g., *A. hippocastanum* strongly showed half-ellipsoid crown shapes.

Keywords: urban trees; crown volume; crown shape; climate mitigation; ecosystem services

Citation: Franceschi, E.; Moser-Reischl, A.; Rahman, M.A.; Pauleit, S.; Pretzsch, H.; Rötzer, T. Crown Shapes of Urban Trees-Their Dependences on Tree Species, Tree Age and Local Environment, and Effects on Ecosystem Services. *Forests* **2022**, *13*, 748. https://doi.org/10.3390/f13050748

Academic Editor: Elisabetta Salvatori

Received: 11 March 2022
Accepted: 10 May 2022
Published: 12 May 2022

1. Introduction

Urban trees as a major component of urban green spaces are nature-based solutions to mitigate the intensity of heat islands and to ameliorate the city climate (e.g., [1]) as well as to ensure multiple health benefits (e.g., [2,3]). Tree canopies, most importantly crowns of individual trees, i.e., their size and shape, play a very prominent role in providing these services. They lower the surface temperature by shielding solar radiation, thus reducing the energy that reaches the ground [4]. This lowers the absorption of short-wave radiation, consequently reducing the long-wave emission of the ground to the surrounding environment [5]. Moreover, through the evapotranspiration process, advected heat is absorbed, thus energy is partitioned more as a latent rather than a sensible heat flux [6]. The impact of the cooling effect depends on tree size, leaf amount [7–9] and constitution under prevalent growing conditions and resources [10]. Additional to cooling by shading and evapotranspiration (e.g., [6,11,12]), tree canopies act as a filter and lower air pollution levels [13,14]. While a single tree can have an impact on the surrounding microclimate [15], in parks these effects can extend into the nearby built environment [16]. Tree crown structure, crown density and crown size play a key role in ameliorating the surrounding climate, e.g., the performance of trees in producing shade and filtering solar radiation depends on canopy shape and solidity and overall tree structure [17].

Therefore, changes in crown volume and shape can influence the magnitude of ecosystem service provision. Crown volume can be used as a proxy to estimate leaf area, transpiration and filtration of fine particles [18]. According to Gratani and Varone [19], it is the most significant variable to explain a change in air temperature under the tree canopy. Consequently, the precise calculation of a tree's crown volume is an important prerequisite to accurately estimate such growth parameters and ecosystem services. Of particular importance for green space planning is the development of space occupation over time. Further, the changing rainfall interception with increasing tree size (and therefore plant surface area) has to be taken into account causing changes in a tree's water balance (e.g., [20]).

Data on crown dimensions have been collected for several purposes following different methods; these have been recently reviewed by Zhu et al., [21]. Crown structure assessment with field measurements has shown to be a fast and simple approach. In fact, while recent technologies such as terrestrial laser scanning (TLS) allow partly direct measurements crown-related field measurements offer a straightforward data collection option [21]. At the same time, digital horizontal photography as an alternative to laser scans still requires longer post processing time than field measurements [21]. Recent urban forestry assessment studies (e.g., [18,22]) applied crown radii and crown length estimates for the calculation of tree crown dimensions, assuming the canopies grow in a simplified cylindrical shape. A cylindrical shape of the canopy could be linked to an allocation of biomass in the stem in situations of light competition, as this leads to the tree growing taller to maintain a competitive position where less space for crown extension is available [23]. In urban settings, competition from neighboring trees plays a minor role; consequently, tree crown shapes vary to a greater extent. According to lists for trees recommended for urban plantings (e.g., [24]), urban trees grow their crowns in shapes that widely differ from the cylindrical one and can vary with the tree age. While structure, function and management of urban forests and their relationships are not yet fully understood, traditional forestry sciences have already been considering crown shapes and different canopy growth at a single-tree level for decades. In 1992, Pretzsch et al., [25] set the basis for species-specific crown shapes, ranging from neiloidal to paraboloidal to cylindrical crowns. These considerations have since then been implemented in the single-tree-based stand simulator SILVA [26,27]. There are only a handful of growth models that simulate urban tree growth and estimate ecosystem services independent of spatial and temporal resolution [12]. Among them, the i-Tree growth model group is one of the most prominent ones [28] that can simulate individual tree or trees of single regions. The calculations are based on the average growth rates and biomass allometries as well as canopy structures of urban and forest tree species. In contrast to the generic i-Tree model, CityTree is a physiologically based growth model for single trees [29] and considers carbon and water cycles depending on the environmental conditions.

In this study, we analyzed the crown shapes of common central European urban tree species and their impact on selected ecosystem services. Our overarching hypothesis was that an improved crown shape modeling in urban tree models is essential for a correct estimation of derived crown volume and shade area. We therefore wanted to answer these specific questions:

(1) Is the dependency of the key parameters for calculating the crown volume, i.e., crown radius and crown length, on tree age varying for common urban tree species?
(2) Does the crown shape and thus crown volume of urban trees differ from species to species and change with increasing tree age?
(3) Does the crown shape of urban trees depend on the local environment?
(4) How does the crown shape influence the shade area and the shade density of urban trees (and thereby the cooling potential)?

2. Materials and Methods

2.1. Study Sites

Structural data (see Section 2.2) were collected in eight cities in Germany (Figure 1). The selected cities cover a wide spectrum regarding size of the city and climate. Figure 1 gives an overview of the different precipitation levels of the cities, with the drier sites in Berlin and Wurzburg with less than 600 mm precipitation per year and the highest annual amount for Munich reaching 960 mm [30]. The average annual air temperature values show differences too, highlighting Berlin's drier and warmer climate characteristics compared with the colder cities of Hof, Bayreuth and Augsburg. Additionally, Figure 1 shows the location of each city on a map. Due to the need for high resolution data and the high number of measured trees, local soil conditions of the different cities have not been considered in our analysis; although this plays an important role in tree growth, we expect above-ground conditions to be more decisive for our crown shape analysis.

Name of the city	Air temperature [° C]	Precipitation [mm/year]	Elevation [m]
1 Augsburg	8.5	789	493
2 Bayreuth	8.4	728	351
3 Berlin	9.8	584	40
4 Hof	7.5	716	507
5 Munich	9.6	960	521
6 Nurnberg	9.2	634	312
7 Wurzburg	9.6	599	252

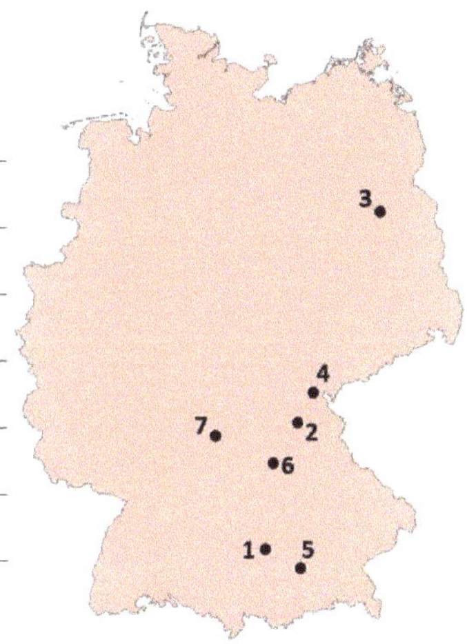

Figure 1. Positions of the eight cities in Germany and averages of annual air temperature and sum of precipitation for the period 1965–2015 (data source: DWD Climate Data Center [30]. Cartography realized with QGIS 3.10 Open Street Map and shapefile of Germany available online on DIVA-GIS [31]).

For this study, we assume that the crown shape of trees is not influenced by the climate or the large-scale environmental conditions of the selected cities. Empirical evidence of the impact of local neighborhood diversity on tree morphological characteristics is rare [32]. Moreover, open-grown trees, as in cities, are less studied even in mixed species forests; rather, they serve as a reference for understanding the response of trees to various biotic competitions [18]. Hasenauer [33] stated that open-grown trees, as is true for many urban trees, can grow larger than comparable forest trees. Of course, these have to be analyzed in a next step for which a broader data set from more cities must be available.

2.2. Tree Species, Structural Data and Tree Age

The present study sets its focus on crown shape analyses, while additional structural data for a more accurate analysis of the first research question were included. Overall, seven tree species were considered: *Acer platanoides* (Norway maple), *Acer pseudoplatanus* (sycamore maple), *Aesculus hippocastanum* (horse chestnut), *Tilia cordata* (small-leaved lime), *Robinia pseudoacacia* (black locust), *Platanus × acerifolia* (London plane), *Fraxinus excelsior* (European ash). The chosen tree species represent considerably different ecological features [34–36]. Moreover, Roloff [37] and Niinemets and Valladares [38] classified the selected species among others regarding their drought tolerance and found different suitability regarding drought stress in urban areas. These tree species are all very abundant in temperate and Mediterranean climate zones [39]. While *T. cordata*, *A. hippocastanum*, *A. platanoides* and *A. pseudoplatanus* are usually shade tolerant species, they are able to adapt to strong light conditions. On the other hand, *R. pseudoacacia*, *F. excelsior* and *P. × acerifolia* are very light demanding but also very drought tolerant [34,37,40].

For all seven species, structural data including diameter at breast height (dbh), tree height (h), crown base (height of the first branch [41,42]), crown radii in eight cardinal directions (north, northeast, east, southeast, south, southwest, west, northwest) as well as the tree locations (geographical coordinates and elevation) were measured. We excluded visually damaged or strongly pruned trees and focused on vital possibly single-standing trees. In addition, for the eight cardinal directions, the distances to neighboring objects (trees or buildings) were measured.

Standard crown dimensions were calculated based on the collected data in the field and formulas from the literature (e.g., [43–45]), i.e., mean crown radius and crown length. While crown length (cl) was defined as the upper segment of tree height from the branch start (i.e., crown base),

$$\mathrm{cl} = \mathrm{h} - \text{crown base} \tag{1}$$

the mean crown radius (cr) was defined as the quadratic mean of the eight crown radii, with i being the eight directions and r the corresponding crown radius value:

$$\mathrm{cr} = \sqrt{\frac{\sum_{i=1}^{8} r_i^2}{8}} \tag{2}$$

The averages of the collected structural data and their standard deviation are shown in Table 1. The number of measured trees per species ranged from 246 individuals for *F. excelsior* to 1065 individuals for *T. cordata*.

Table 1. Characteristics (means and standard deviation) of all measured trees. Abbreviations: *n*— number of samples, dbh— diameter at breast height.

Species	*n*	dbh [cm]	Height [m]	Crown Radius [m]	Crown Length [m]
A. platanoides	356	37.5 ± 17.1	14.1 ± 5.0	4.6 ± 1.6	10.8 ± 4.6
A. pseudoplatanus	245	37.8 ± 14.1	15.6 ± 4.6	4.3 ± 1.2	11.9 ± 4.1
A. hippocastanum	676	50.5 ± 23.9	14.6 ± 4.7	4.9 ± 1.5	11.5 ± 4.3
F. excelsior	246	34.9 ± 18.4	14.3 ± 5.0	4.1 ± 1.6	10.5 ± 4.6
P. × acerifolia	659	42.4 ± 22.3	16.4 ± 5.2	5.8 ± 2.3	12.9 ± 5.1
R. pseudoacacia	605	38.7 ± 18.7	14.4 ± 4.6	4.5 ± 1.4	10.5 ± 3.9
T. cordata	1065	37.1 ± 16.1	14.8 ± 4.6	4.3 ± 1.4	11.1 ± 4.2
	Total = 3852				

All tree species showed a relatively similar range of dbh, with the highest values for *A. hippocastanum* and *P. × acerifolia*. These species showed the biggest crown lengths and crown radii.

For urban tree managers it is important to know the age of the trees, therefore we estimated tree age using the variables dbh and h and species-specific equations (Table 2). From these equations age can be derived and applied as an alternative to dbh. Age was used for the calculation of the leaf area index.

Table 2. Equations and parameters for the calculation of tree age for specific tree species. Sources of the formulas are indicated for every tree species; for *R. pseudoacacia* we applied the formula developed for Gleditschie by Dwyer (2009).

Tree Species	Source	Formula
A. platanoides	Herzog 2021 [46]	$age\ [yrs] = 5.4256 + 1.5741 \cdot dbh$
A. pseudoplatanus	Herzog 2021 [46]	$age\ [yrs] = 9.3524 + 1.7587 \cdot dbh$
A. hippocastanum	Lukaszkiewicz and Kosmala 2008 [47]	$age\ [yrs] = -a + exp(b + c \times dbh/100 + d \times h)$ with $a = 54.2714$, $b = 4.0709$, $c = 0.7988$, $d = 0.0209$
F. excelsior	Lukaszkiewicz and Kosmala 2008 [47]	$age\ [yrs] = -a + exp(b + c \times dbh/100 + d \times h)$ with $a = 210.115$, $b = 5.3523$, $c = 0.2655$, $d = 0.0064$
P. × *acerifolia*	Bühler et al., 2007 [48]	$age\ [\text{yrs}] = 0.996 \cdot \text{dbh}$
R. pseudoacacia	Dwyer 2009 [49]	$age\ [\text{yrs}] = 0.996 \cdot \text{dbh}$
T. cordata	Lukaszkiewicz and Kosmala 2008 [47]	$age[\text{yrs}] = -\text{a} + \exp\left(\text{b} + \text{c} \cdot \frac{\text{dbh}}{100} + \text{d} \cdot \text{h}\right)$ with $\text{a} = 264.073$, $\text{b} = 5.5834$, $\text{c} = 0.3397$, $\text{d} = 0.0026$

2.3. Crown Volume Calculation

Within this study, we analyzed the estimation of crown volume based on shape-specific volume calculations. Therefore, the tree crowns of 528 trees located in Munich and Berlin were assigned to five crown shape types based on Lawrence (1985) [50]; five tree species have been included (Table 3).

Table 3. Number (n) of measured trees included in the crown shape analyses.

	A. platanoides	*A. hippocastanum*	*P.* × *acerifolia*	*R. pseudoacacia*	*T. cordata*
n	91	46	98	65	228

Most of the tree species have specific crown shapes, particularly solitary trees, which are not affected by neighboring trees or buildings. Pyramidal, cylindrical, spherical, ovoid and half-ellipsoidal crown shapes are the most common occurring crown shapes of the analyzed tree species based on the literature and urban tree catalogues [24,50,51] (Figure 2).

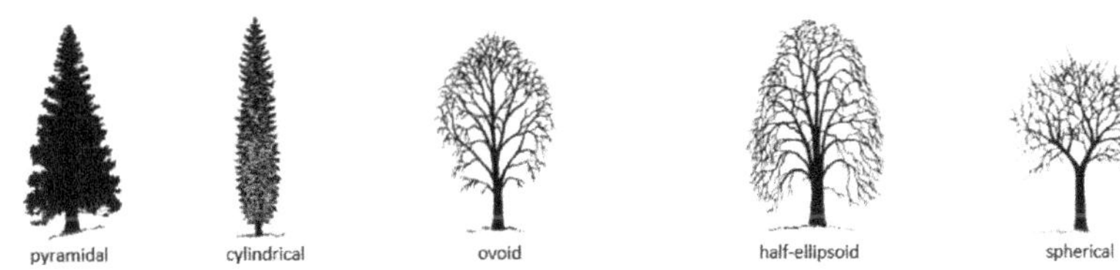

	pyramidal	cylindrical	ovoid	half-ellipsoid	spherical
Tree species	Picea abies Taxus baccata Acer pseudoplatanus	Abies alba Acer platanoides Acer pseudoplatanus Robinia pseudoacacia	Acer platanoides Acer pseudoplatanus Fraxinus excelsior Tilia cordata Platanus x acerifolia Robinia pseudoacacia	Aesculus hippocastanum Betula pendula Fagus sylvatica Tilia cordata	Acer platanoides Acer pseudoplatanus Fraxinus excelsior Tilia cordata Platanus x acerifolia Robinia pseudoacacia
Crown volume	$\frac{\pi \cdot cr^2 \cdot cl}{3}$	$\pi \cdot cr^2 \cdot cl$	$\frac{\frac{4}{3}\pi \cdot cr^2 \cdot \frac{1}{2}cl}{2} + \frac{\frac{4}{3}\pi \cdot cr^3}{2}$	$\frac{\frac{4}{3}\pi \cdot cr^2 \cdot cl}{2}$	$\frac{4}{3}\pi cr^3$
Shade projection area	$\frac{cd \cdot cl}{2}$	$cd \cdot cl$	$\frac{\pi \cdot cr \cdot \frac{1}{2}cl}{2} + \frac{\pi \cdot cr^2}{2}$	$\frac{\pi \cdot cr \cdot cl}{2}$	πcr^2
Surface area	$\pi\, cr^2 + \pi\, cr \cdot \sqrt{cr^2 + cl^2}$	$2\pi \cdot cr^2 + 2\pi \cdot cr \cdot cl$	$\frac{4\pi \sqrt[1.6075]{\frac{(cr^2)^{1.6075} + 2(cr \cdot \frac{1}{2}cl)^{1.6075}}{3}}}{2} + \frac{1}{2} 4\pi cr^2$	$\frac{4\pi \sqrt[1.6075]{\frac{(cr^2)^{1.6075} + 2(cr \cdot cl)^{1.6075}}{3}}}{2} + \pi cr^2$	$4\pi cr^2$

Figure 2. Mathematical equations for the calculation of crown volume, crown shade projection area and surface area for different crown shapes. Crown shapes illustrations from Lawrence, 1985 (Abbreviations as in text and cd— 2·cr).

To calculate a tree's crown volume (cv), crown surface area and crown shade projection area (cspa) the standard geometrical equations shown in Figure 2 were used. Crown radius and crown length were used as the key parameters for these calculations, with cr as radius and cl as crown height (Figure 2). For the half-ellipsoidal shape, the volume of an ellipsoid was calculated and then divided by 2. In this case, the two shorter segments of the ellipsoid were substituted by cr and the longer one by cl. For the ovoid shape, half of an ellipsoid and half of a sphere were summarized to consider both possibilities of a longer upper part of the crown with a flatter lower part and the opposite with an elongated crown shape facing the ground (Figure 2). A bigger cl resulted then in an elongated egg shape, whereas a smaller cl led to a more spherical shape. For the cspa, we used cl and cr to calculate a tree's silhouette. i.e., a circle for the spherical shape, an (half) ellipse for the half-ellipsoid shape and a rectangle for the cylindrical one.

2.4. Statistical Methods

The calculation of the crown volumes (following Figure 2) was applied in Microsoft Office Excel 2013. The visualization of the results was realized using R (software version 3.6.3) basis plot functions. Normal distribution for the main parameters was visually checked (histogram representation in RStudio) before log–log–linear regression was applied to investigate significance and relationship of crown radius and crown length with dbh, following Pretzsch et al., 2012 [52] and Moser et al., (2015) [53].

Dependency of crown shape on the local environment was analyzed applying an index expressing distance levels to objects. The index was equal to 0 when a tree or a building was standing closer than the tree height and equal to 1 when the distance was greater. Logistic regression was then computed in R using the glm function of the package stats. The regression analysis included effects of tree species, diameter at breast height and distance level (0 or 1) for each single crown shape:

$$\text{glm}(\text{crown shape} \sim \text{dbh} + \text{as.factor}(\text{distance level}) + \text{as.factor}(\text{tree species})) \quad (3)$$

2.5. Calculations Based on the Process-Based Growth Model CityTree

Ecosystem services and urban tree growth can be assessed using the process-based model CityTree [32]. For this study, the CityTree model was applied for the estimation of a tree's age-based leaf area index.

The procedure of the CityTree model was also used to calculate the shadow area, shadow density and shadow index of a tree. Hereby, the crown shape was additionally taken into account. The shade area was calculated as the average of the shade area from 8 am to 6 pm for the 21st of June (the longest day in the northern hemisphere). Thus, the shade area was calculated by applying the formulas for cspa, using cd and shade length (instead of cl), with shade length itself calculated using cl and the cotangent for the hour and for the location of the sun height. The single hourly shade areas were averaged into

$$\text{avg. shade area} = \frac{\sum_{i=8}^{18} \text{shade area i}}{11}, \quad (4)$$

with i as the hour of the day and 11 as the number of considered hours. In the calculation of shade density and shade index, the crown shape was considered by applying the shape-specific cv calculation (Figure 2):

$$\text{shade density} = \frac{\text{leaf area } [\text{m}^2]}{\text{cv } [\text{m}^3]} \quad (5)$$

and

$$\text{shade index} = \text{shade area} \cdot \text{shade density}, \quad (6)$$

with

$$\text{leaf area} = \text{LAI} \cdot \text{cpa} \quad (7)$$

In the following results section, examples of the shade area, shade density and shade index were listed to enhance the impact of different crown shapes on those variables. The LAI differences caused by age increase have been considered in the calculations of the variables which include leaf area.

3. Results

3.1. The Crown Structure of Urban Tree Species

The relationship between dbh and crown radius (Figure 3) as well as dbh and crown length (Figure 4) of seven tree species are shown for all 3852 measured trees. Double-logarithmic relationships between the crown dimensions (cr and cl) and the tree diameter were analyzed through linear regression (see Table 4). Their nonlinear least square relationship is represented in Figures 3 and 4 through a black curve. In both analyses, we recognized different species-specific trends, such as a less pronounced increase of crown radius and crown length with increasing dbh for *R. pseudoacacia* (slope equal to 0.53 and 0.56, Table 4) compared with other species such as *F. excelsior* (0.72 and 0.71); whereas, the oldest trees measured were *A. pseudoplatanus* and *A. hippocastanum*.

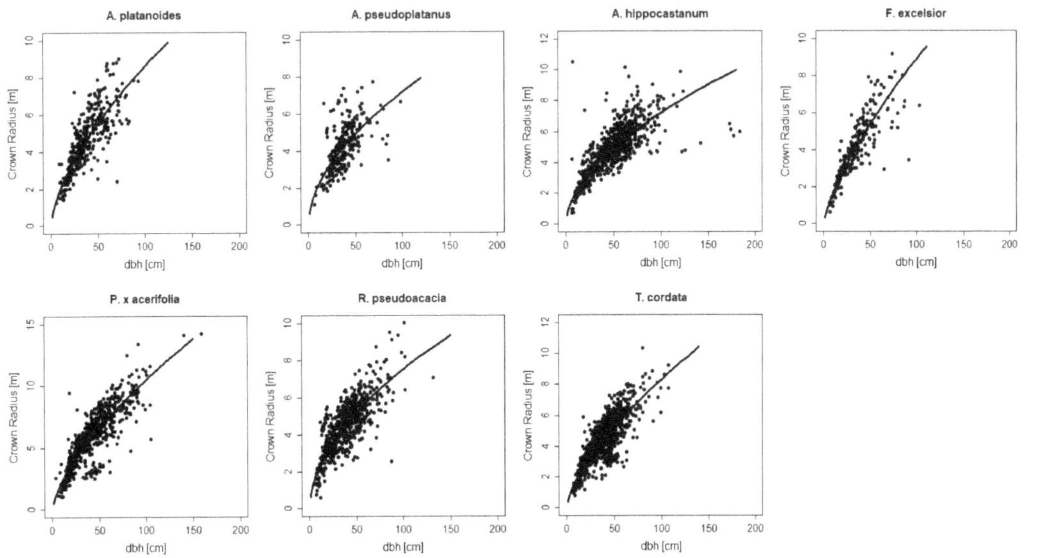

Figure 3. Crown radius and diameter at breast height relationship for seven common urban tree species.

Table 4. Results of the transformed double-logarithmic equation detecting significance of changes in crown radius and crown length related to changes in dbh.

	Crown Radius ln (cr) = a+b·ln (dbh)						
	A. platanoides	*A. pseudoplatanus*	*A. hippocastanum*	*F. excelsior*	*P.* × *acerifolia*	*R. pseudoacacia*	*T. cordata*
a	−0.76	−0.53	−0.55	−1.16	−0.75	−0.44	−0.95
b	0.63	0.54	0.54	0.72	0.67	0.53	0.67
R^2	0.69	0.46	0.68	0.83	0.73	0.63	0.72
p	<0.001	<0.001	<0.001	<0.001	<0.001	<0.001	<0.001
	Crown Length ln (cl) = a + b·ln (dbh)						
a	−0.27	0.03	−0.096	−0.16	0.12	0.29	−0.15
b	0.73	0.67	0.65	0.71	0.65	0.56	0.71
R^2	0.67	0.51	0.65	0.73	0.76	0.58	0.64
p	<0.001	<0.001	<0.001	<0.001	<0.001	<0.001	<0.001

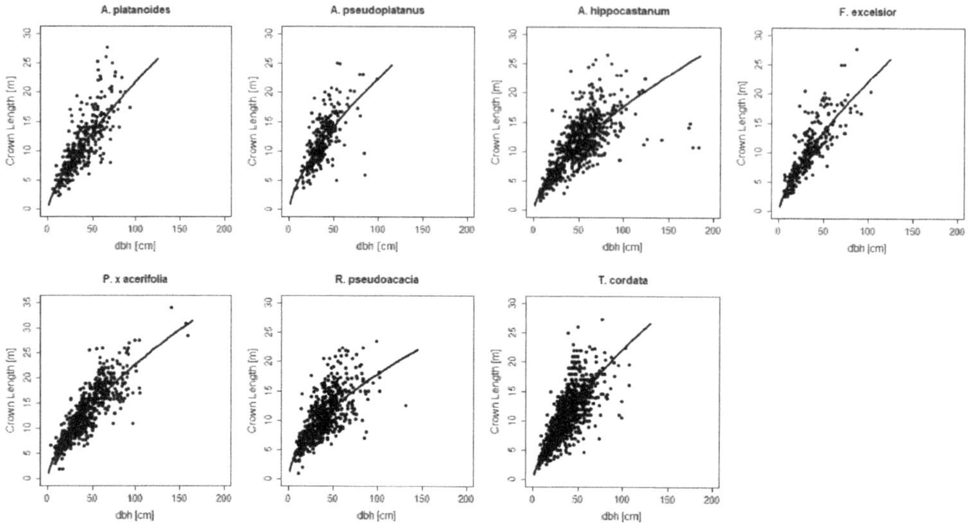

Figure 4. Crown length and diameter at breast height relationship for seven common urban tree species.

The *p*-value for all tree species showed a consistent high significance for the relationship between crown radius and dbh as well as crown length and dbh (Table 4). At the same time, while the R^2 value for *A. pseudoplatanus* was the lowest for both cr and cl dependencies from dbh (0.46 and 0.51), strong dependency was obvious for *F. excelsior* (0.83 and 0.73) and *P.* × *acerifolia* (0.73 and 0.76). For *F. excelsior* an increase of 10% in dbh showed the greatest changes in crown radius and length, with an increase in cr of 7.2% and for cl of 7%. For *R. pseudoacacia* a 10% increase of dbh only led to an increase of 5.2% for cr and 5.5% for cl.

3.2. *Crown Shapes of Urban Tree Species*

The assignment of trees into crown shape categories showed a prevalence of ovoid shapes (57%) for all tree species (Figure 5). The cylindrical shape was seldom (5%) observed in an urban tree context when single-standing trees were analyzed. *R. pseudoacacia* was an exception, as it developed a columnar-shaped crown for 25% of its sample. This light-demanding species is often planted in street canyons, which can be considered as situations of light competition due to the shading buildings.

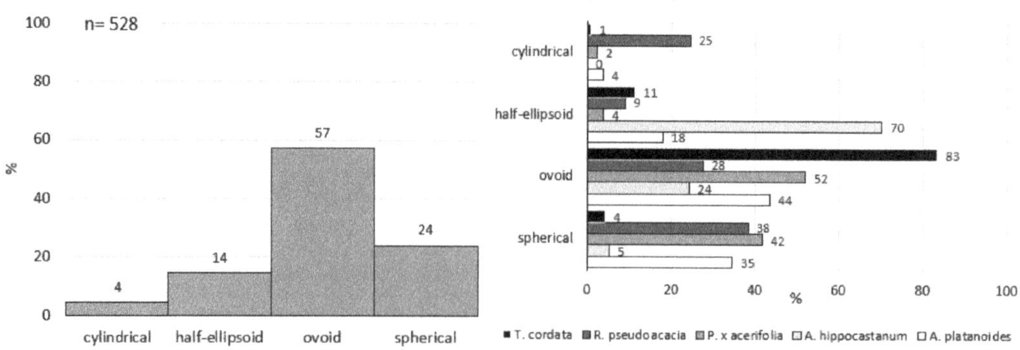

Figure 5. Percentage distribution of crown shapes of all trees measured (**left**) and according to tree species (**right**).

A. hippocastanum showed mostly half-ellipsoidal shaped crowns; however, this was the species with the smallest sample size. Over 50% (52%) of *P.* × *acerifolia*'s observed crowns were ovoid. For *R. pseudoacacia,* there was no clear prevalent shape, whereas 83% of the crowns of *T. cordata* were ovoid shaped. While for *T. cordata* the cylindrical shape was present only for a few trees (1%), 25% of the *R. pseudoacacia* trees showed an elongated crown shape, comparably as often as the ovoid shape. Table 5 highlights the differences of observed crown shapes of small (young) and big (old) trees, where bigger trees were defined as dbh higher than 30 cm.

Table 5. Observed distribution of different crown shapes according to species and age, n—number of observations.

Species	n (%)	Cylindrical n (%)	Half-Ellipsoidal n (%)	Ovoid n (%)	Spherical n (%)
A. platanoides	91 (100)	3 (3)	14 (15)	34 (38)	40 (44)
dbh < 30	21 (100)	1 (5)	1 (5)	7 (33)	12 (57)
dbh > 30	70 (100)	2 (3)	13 (18)	27 (39)	28 (40)
A. hippocastanum	46 (100)	0 (0)	26 (56)	9 (20)	11 (24)
dbh < 30	5 (100)		1 (20)	2 (40)	2 (40)
dbh > 30	41 (100)		25 (61)	7(17)	9 (22)
P. × *acerifolia*	98 (100)	2 (2)	4 (4)	51 (52)	41 (42)
dbh < 30	18 (100)	2 (3)	0 (0)	15 (83)	3 (17)
dbh > 30	80 (100)		4 (5)	36 (45)	38 (47)
R. pseudoacacia	65 (100)	16 (25)	6 (9)	18 (28)	25 (38)
dbh < 30	7 (100)	2 (29)	1 (14)	0 (0)	4 (35)
dbh > 30	58 (100)	14 (24)	5 (9)	18 (31)	21 (36)
T. cordata	228 (100)	2 (1)	26 (12)	190 (83)	10 (4)
dbh < 30	51 (100)	2 (1)	1 (2)	50 (98)	10 (6)
dbh > 30	177 (100)		25 (14)	140 (79)	

While classifying the trees into small and big trees, there were only a few small trees for the species *R. pseudoacacia* and *A. hippocastanum*. In the case of *T. cordata* and *R. pseudoacacia,* the distribution of the most relevant crown shape for the two age classes did not differ considerably. *R. pseudoacacia* showed 1% difference between big and small trees with spherical crowns, while for *T. cordata* 98% of the small trees and 79% of the big trees showed ovoid crowns. For the cylindrical shape, the classes difference was 5% for *R. pseudoacacia* trees. For *P.* × *acerifolia,* we noticed instead a change in the dominant shape; small trees of this species showed prevalently ovoid shaped crowns, while bigger trees had more spherical crowns.

The two most frequent crown shapes overall were ovoid (57%) and spherical (24%), with ovoid as the prevalent shape for *T. cordata* and spherical for *A. platanoides* and *R. pseudoacacia*. Divided in the two classes, we observed an increase of the half-ellipsoidal crowns in big trees (17% instead of 4%), while the cylindrical shape was present in 5% of the trees with dbh bigger than 30 cm and only in 3% of the smaller ones. For ovoid and spherical shapes, the age-distribution differences over the whole sample were not high. The cylindrical shape represented only 4% of the tree crowns in our sample.

3.3. *Influence of Crown Shape on Crown Volume*

As we can see from Figure 6, the distribution and development of crown volume with increasing dbh differed depending on the crown shape and its calculation of the crown volume. While the slope of the exponential increase of crown volume with increasing dbh was the highest for the ovoid and spherical shape, cylindrical and half-ellipsoid crown volumes showed only a small increase per year (slope equal to 0.032 and 0.034). For cylindrical crowns the smallest number of trees was observed; this resulted in a low value of goodness of our model (R^2 = 0.34) in comparison with the ovoid shape (R^2 = 0.68).

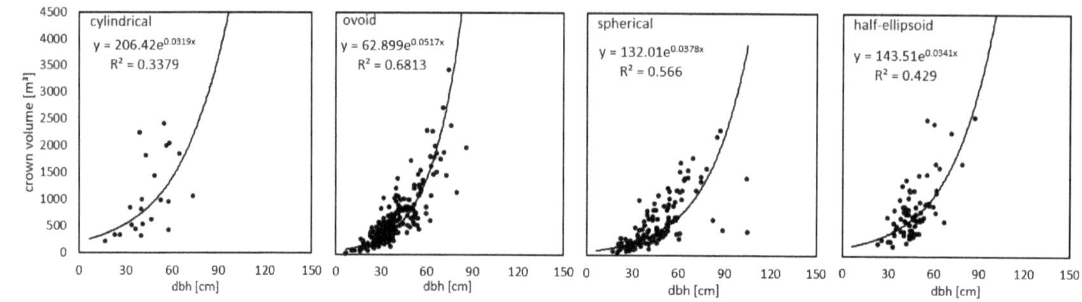

Figure 6. Relationship of crown volume and dbh for the four different crown shapes. Graphs show all 258 trees. The mathematical constant e in the equations indicates the Euler's number.

3.4. Dependency of Crown Shape on the Local Environment

To analyze the dependency of the crown shapes on the local environment we analyzed the influence of the distance to buildings or other trees on the development of the crown shape and volume. For this, two distance classes were defined depending on a tree's height, with one class for the nearest object with a distance smaller or equal to the tree height (h) and one class for the nearest object standing at a distance bigger than the tree height.

For most species, distances from objects such as houses or other trees had little influences on the predominant crown shape. This was distinctly recognizable for *A. hippocastanum*, with spherical-shaped crowns in 79% and 61% (respectively, for shorter and longer distance) of the observations and *T. cordata*, with ovoid shape for 93% and 54%. An exception was represented by *R. pseudoacacia*, with a higher percentage of cylindrically shaped crowns when growing close to objects (44%) and more spherical canopies for wider space conditions (38% spherical and only 17% cylindrical). *P.* × *acerifolia* showed slightly more elongated crowns for small distances (ovoid shape 56%) but spherical crowns for the higher distance level (55%). The change in distribution for *P.* × *acerifolia* was less pronounced than for *R. pseudoacacia*.

3.5. Overarching Analysis of the Crown Shapes of Urban Trees

The results of the logistic regression analysis of the dependencies of the crown shapes on tree species, dbh and distance level can be seen in Table 6.

Table 6. Dependency of the crown shape depending on tree species, dbh and distance level to objects. The used formula is (glm (crown shape)~dbh + as. factor (distance level) + as. factor (tree species)). The distance level 1 refers to the probability in comparison to the distance level with index 0. The reference tree species is *A. platanoides*, to which the tree species variable refers. *p*-value significance follows the RStudio output with "***" for $p < 0.001$, "**" for $p < 0.01$, "*" for $p < 0.1$ and "." for $p < 0.1$.

Variable	Estimate	e (Estimate)	Effect (%)	*p*-Value
		Cylindrical shape		
Intercept	−2.71			**
dbh	0.000037	1.00037	0.0367	
distance level 1	−0.845	0.43	−56.040	.
tree species (*A. hippocastanum*)	−15.52	0.00000018	−99.999	
tree species (*P.* × *acerifolia*)	−0.572	0.565	−43.536	
tree species (*R. pseudoacacia*)	2.154	8.616	761.617	**
tree species (*T. cordata*)	−1.873	0.154	−84.628	.
		Half-ellipsoid shape		
Intercept	−4.659			***
dbh	0.0518	1.0532	5.319	***
distance level 1	1.126	3.083	208.318	***

Table 6. *Cont.*

Variable	Estimate	e (Estimate)	Effect (%)	p-Value
tree species (*A. hippocastanum*)	2.662	14.329	1332.893	***
tree species (*P. × acerifolia*)	−2.143	0.117	−88.267	**
tree species (*R. pseudoacacia*)	−0.977	0.376	−62.416	.
tree species (*T. cordata*)	0.231	1.260	25.994	
		Ovoid shape		
Intercept	2.547			***
dbh	−0.0491	0.952	−4.792	***
distance level 1	−1.119	0.327	−67.345	***
tree species (*A. hippocastanum*)	−0.962	0.382	−61.806	*
tree species (*P. × acerifolia*)	0.263	1.302	30.18	
tree species (*R. pseudoacacia*)	−0.639	0.528	−47.226	.
tree species (*T. cordata*)	1.284	−3.609	260.937	***
		Spherical shape		
Intercept	−2.162			***
dbh	0.0215	1.021	2.174	*
distance level 1	0.853	2.347	134.665	**
tree species (*A. hippocastanum*)	−2.151	0.116	−88.368	**
tree species (*P. × acerifolia*)	0.536	1.709	70.925	
tree species (*R. pseudoacacia*)	0.0981	1.103	10.302	
tree species (*T. cordata*)	−2.009	0.134	−86.596	***

The results of Table 6 are in line with the observed distribution of crown shapes. For most tree crown shapes a significant relationship to dbh and distance to objects was calculated, even if in different levels; the direction of the effect caused by an increase of age and/or distance level varied strongly. For example, with increasing dbh a positive effect on cylindrical, spherical and half-ellipsoid crown shape exists. For each year of increase, a 4.8% lower probability of ovoid crowns occurring can be seen. The direction of the effects for distance to objects also showed different results depending on the observed shape. A bigger distance to objects showed a significant effect with a positive sign for half-ellipsoid and spherical crowns. In contrast, for ovoid shaped crowns the effect is negative and highly significant, but clearly smaller (−67.34% vs. 208.32%). This is consistent with the results of Figure 7, where more ovoid trees were observed at a distance less than the tree height.

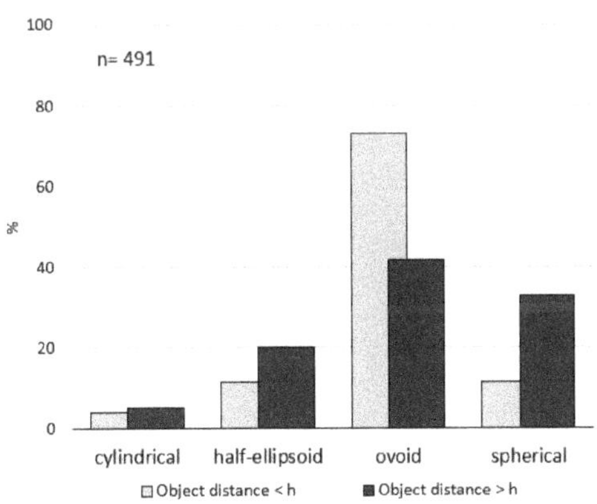

Figure 7. Percentage distribution of crown shapes of all trees for two distance classes of the closest object.

For the different tree species and their relationships to crown shape, we recognize for *R. pseudoacacia* a seven times higher probability of observing cylindrical crowns than for the reference species *A. platanoides*. All other species show lower probabilities. For the half-ellipsoid shape, a strong presence of *A. hippocastanum* was visible with a 13 times higher probability compared with the reference species. Moreover, as shown in Figure 8, *T. cordata* and *A. platanoides* were the species with the highest probability of ovoid shaped crowns, in particular *T. cordata* (2.6 times higher than *A. platanoides*).

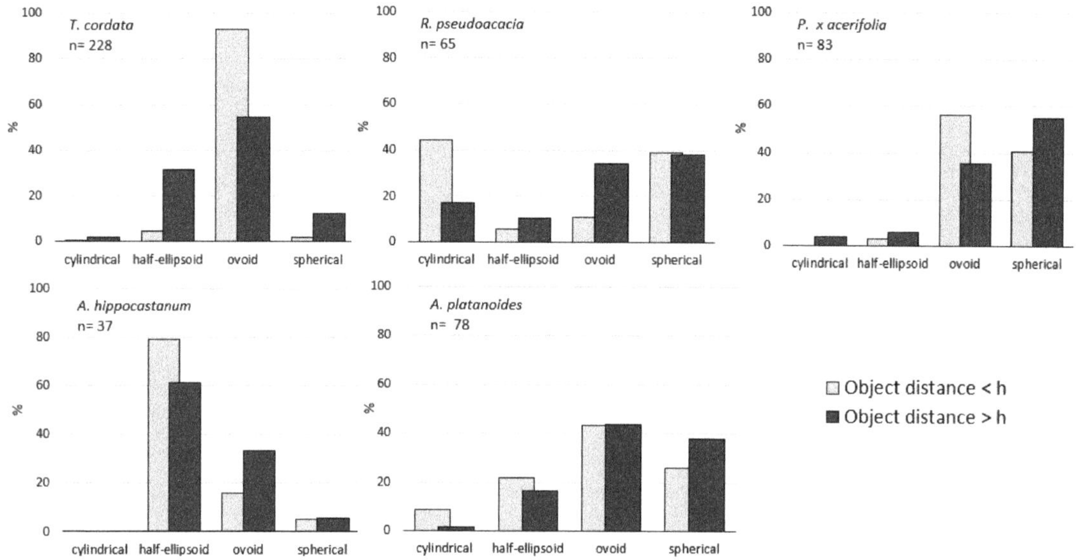

Figure 8. Percentage distribution of crown shapes for the different tree species for two distance classes of the closest object.

3.6. Influence of Crown Shape on the Shade Area and the Shade Density

Resolving the exponential equations for the natural logarithm, an increase of 10% in dbh (x) resulted in an increase of the shade area of 0.43% for ovoid shaped crowns and 0.36% for the half-ellipsoid shape. The shade area for cylindrical shaped crowns did not increase strongly with age (0.2% per 10% increase in stem diameter) (see Figure 9). In contrast, the shade density of the trees showed almost reverse patterns for the different crown forms depending on dbh (Figure 10), with decreasing shade density for every species and the lowest decrease for the cylindrical shape (−0.03% per 10% increase in stem diameter). For example, a tree measuring 60 cm dbh with a spherical shape has a mean shade area of 323.1 m^2, while a cylindrical shaped tree has a mean shade area of 184.6 m^2. A tree with a half ellipsoidal or an ovoid crown, in contrast, has a smaller mean shade area, respectively, of 124.3 m^2 and 114.2 m^2. Smaller trees, measuring 40 cm dbh, also show for the spherical and cylindrical crown the highest shade area values (198.3 m^2 and 120.8 m^2).

Old trees (age > 80) presented ca. 30% to 44% lower shade density than young trees (age < 40), with the highest reduction for ovoid and half-ellipsoid. At the same time, ovoid as well as spherical shaped tree crowns were the densest, with an average shade density of 0.52 m^2/m^3 and 0.42 m^2/m^3, respectively. For a tree with a dbh of 60 cm, the shade density is 0.27 m^2/m^3 for a cylindrical crown shape, 0.32 m^2/m^3 for an ovoid and half-ellipsoid and 0.33 m^2/m^3 for a spherical crown shape. For smaller trees, e.g., with a dbh of 40 cm, the highest shade density is shown by the ovoid (0.47 m^2/m^3) and spherical (0.43 m^2/m^3) crown shapes.

The shade index showed a trend of different levels between crown shapes (Figure 11) similar to Figure 9, with the highest values for spherical and cylindrical crowns having the highest intercepts. At the same time, the highest increment in shade index was shown for the ovoid shape with an increase of 0.25% per 10% increase of dbh. Differently than for the two precedent figures, we noticed the highest R^2 value for the dependency of the shade index from tree stem diameter for cylindrical-shaped trees ($R^2 = 0.54$). For the crown shape half-ellipsoid, the dbh impact was the lowest with an R^2 value of 0.19. For a tree with a dbh of 60 cm, this results in shade indices of 105.5 for spherical and 49.4 for cylindrical crown shape and in smaller shade indices for the half-ellipsoid and ovoid (39.9 and 36.6) shapes.

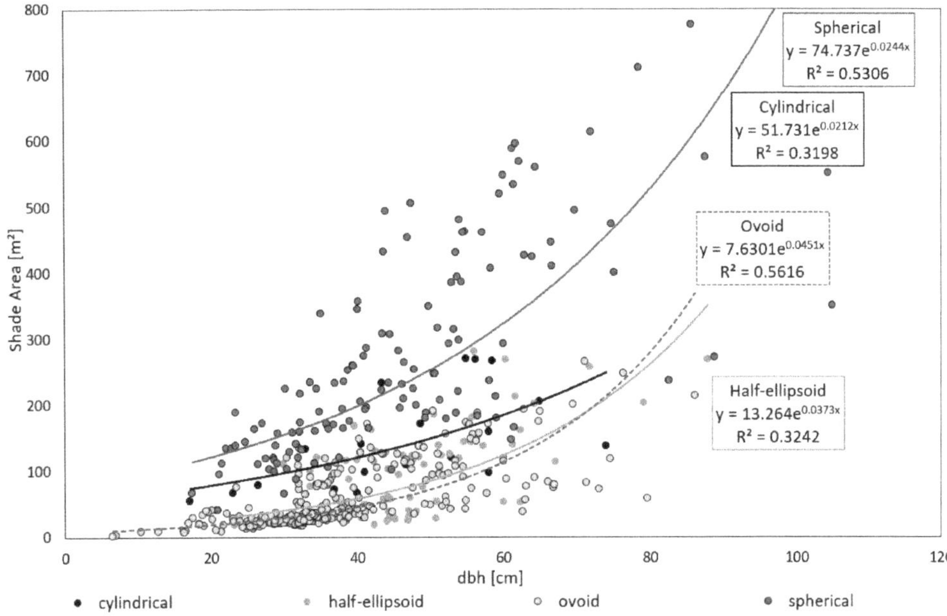

Figure 9. Dependency of shade area on dbh and crown shape. The mathematical constant e in the equations indicates the Euler's number.

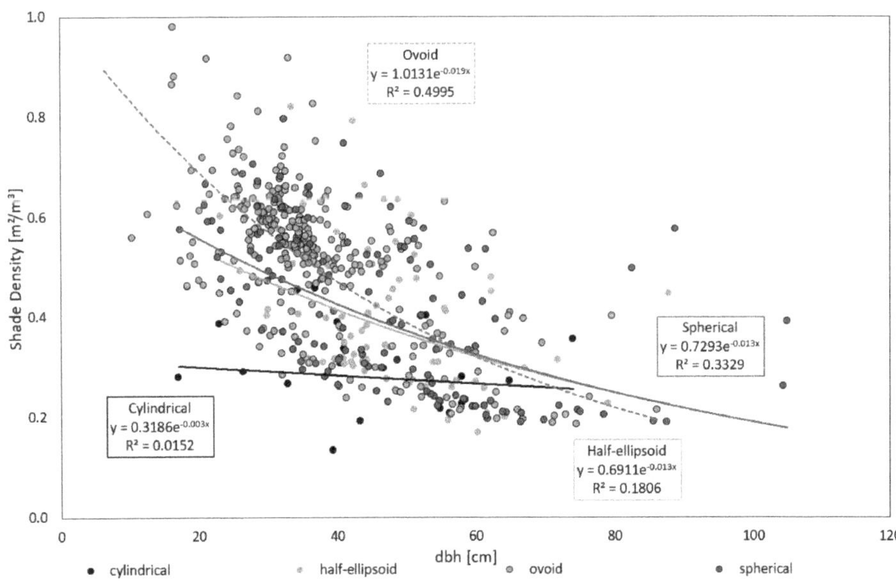

Figure 10. Dependency of shade density on tree dbh and crown shape. The mathematical constant e in the equations indicates the Euler's number.

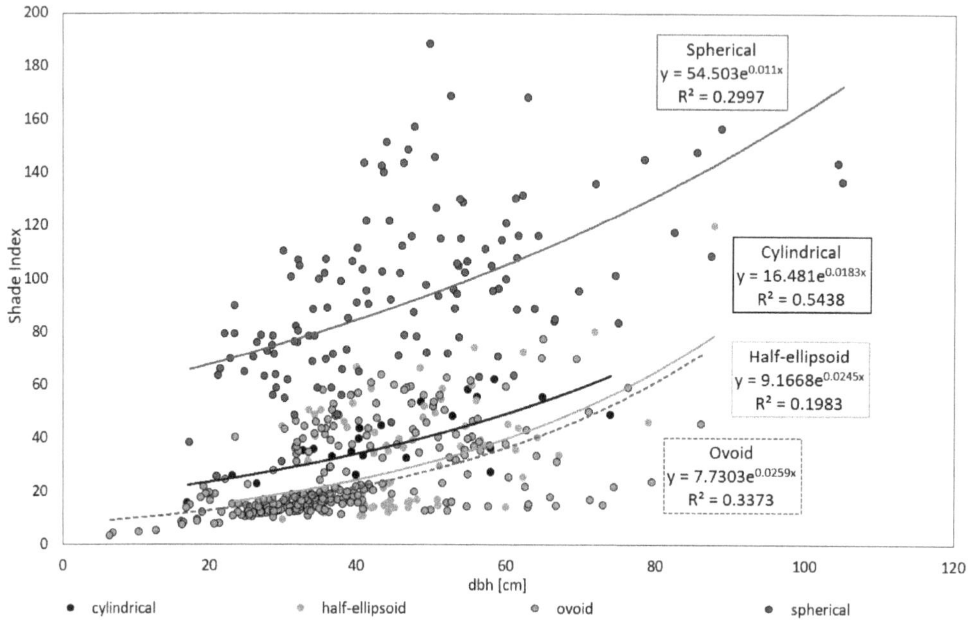

Figure 11. Dependency of shade index on tree dbh and crown shape. The mathematical constant e in the equations indicates the Euler's number.

4. Discussion

The results of this study show the diversity of tree crown dimensions in urban environments in central Europe and possible calculation methods of crown volume and derived ecosystem services. While in the first part we analyzed the development of crown dimensions in dependence on dbh for seven common species, later we analyzed in more depth the effect of different crown shapes on the crown volume and resulting shading parameters. Furthermore, a possible dependency of the crown shape of different tree species on the proximity to neighboring objects was discussed.

The importance of the specific estimation of required space for tree crowns in an urban setting was shown by Pretzsch et al., (2015) [18], who quantified species-specific crown size and its allometry. Our aim was to show the importance of crown shapes for crown volume calculations, to reveal the dependence on tree species, tree dbh and local environment and to quantify effects on ecosystem services, i.e., on the shading effect of urban trees.

This study offers a simple method that can be used in field measurements and be implemented in urban tree models that aim to estimate ecosystem services of urban trees. We are aware that this method cannot offer an analysis linked to growth development without the use of repeated measurements. Moreover, a broader data basis is needed to be able to examine the influence of the climate in the cities on the development of the tree crown shape in more detail. In addition to the classical and simple method of the visual classification of tree crowns as used by us, terrestrial laser scanning (TLS) has been applied more frequently in recent years and also in the field of urban forestry (e.g., [35,54]). The reasons are cheaper and more mobile devices as well as a meanwhile available easy-to-use evaluation software and powerful computers. This method allows, for example, accurate estimation and quantification of individual crown dimensions and structures [55,56]. However, recording a high number of crown shapes—as required for this study—of different tree species of different ages in different locations in several cities using TLS is still very time consuming and costly in contrast to classical observation.

4.1. Influence of the Crown Shape on the Structure and Dimensions of Urban Tree Species

Our results showed ovoid and spherical crown shapes being the most common (Figure 4). This was in line with previous research [20,50,51] reassumed for different species in Table 4. A change of crown shape at different dbh classes could not be seen for most of the species, which could be explained by the small number of small trees. Only for *P.* × *acerifolia* a shift from ovoid to spherical as the prevalent shape was shown with increasing dbh. As Troxel et al., [57] state, the physical dimensions of trees are often highly correlated, but the patterns of growth for individuals of the same species (over time) can vary depending on tree species and site conditions [58].

In several studies, crown volume (cv) is calculated for cylindrical shaped crowns [22,53,57] as well as in urban tree growth models like CityTree [29]. A development into a cylindrical crown shape was linked to light competition that leads to the allocation of biomass in shoots and therefore the tree grows taller [23]. Moreover, obstructions like other close-growing trees, buildings and pruning for traffic safety can result in more dense cylindrical crown shapes. This is also discussed in the following Section 4.2.

However, for single-standing trees the assumption of cylindrical shaped crowns can result in clear under- or overestimation of the real value. For example, for a tree with cr of 4 m, cl 10 m long and a spherical crown, a cylindrical calculation for its cv would result in a crown volume equal to 1.87 times the actual value (87% overestimation), whereas for the same tree dimensions and an ovoid crown the overestimation would be around 67% (1.67 times the actual value). Assuming that a tree with a pyramidal crown shape resembles a cylinder, the actual crown volume and ecosystem services would always be greatly overestimated as the cv for pyramidal crown (cone) is equal to one third of the cylindrical one (see formula for pyramidal shape in Figure 2).

4.2. Influence of the Local Environment on the Crown Shape

Most tree species showed one or two prevalent crown shapes that do not change in relation to the distance to close objects, but instead remain characteristic for the species. An exception is represented by *R. pseudoacacia*, which preferred the cylindrical crown shapes for growing situations with smaller distances to objects like narrow street canyons, where long-elongated crowns permit to escape shading objects. *P.* × *acerifolia*, another light-demanding species like *R. pseudoacacia*, also invested in slightly more elongated crowns on an ovoid shape for small distances and extended its branches into a spherical shape when more space is available. As Troxel et al., [57] state, light-demanding tree species might have weaker dbh–cv relations due to being more susceptible in their growth to their surrounding environment. On the contrary, the shade tolerant tree species seemed to show a more stable crown shape distribution, independent from the distance to objects. This would suggest that the influence of close objects is more or less relevant depending on the tree species character.

4.3. Effects of the Crown Shape/Crown Volume on Ecosystem Services (Shading)

Depending on the analyzed sample and tree species, we recognized different trends and effects of the relationship between crown shape and cooling potential by shading. In terms of shade provision, two important considerations are the shade area and the shade density. The shade area for a taller crown is usually higher, in particular during the mornings and the evenings when the sun is at a lower angle [59], as can be seen for the case of the cylindrical crown in Figure 9. However, the higher shade area from taller crowns comes at the cost of reduced shade density as shown in Figure 10. Shade density is especially important for surface temperature reduction and human thermal comfort [4,60,61]; thus, trees with an ovoid crown shade a specific point on the ground for longer. Similarly, higher canopy density is important in providing higher evapotranspiration cooling [22].

Considering the findings of the insignificant effect of local surroundings over the genetic constituents of species characteristics, it is important to choose species strategically when planting in narrow street canyons or wider avenues to optimize the cooling benefits. It is best to plant trees with broader and denser canopies to maximize the cool "refuge" effect for the city dwellers during hot days. However, the context of underlying surfaces such as higher radiation through light canopies for expediting evapotranspiration from grass lawns [62] or narrow street canyon conditions where light demanding species protrude towards light sources should be taken into consideration.

5. Conclusions

This study answered different questions regarding the crown structure development of seven common urban tree species, crown shape distribution and its link with dbh as well as with the local environment. Additionally, the influence of the crown shape on the crown volume calculation and the resulting differences in shade area and shade density of urban trees were shown. The diverging results depending on the tree crown shapes confirm the need for more specific crown volume assessment methods in urban forestry to better estimate ecosystem services of different tree species.

Species-specific trends for the development of crown dimensions were observed. For all species, the relationship of crown radius and length with tree stem diameter was highly significant, showing dbh as a suitable parameter to predict crown dimension development.

The cylindrical shape, often used in urban tree growth models, was the least observed within this urban sample. Instead, the ovoid and spherical shapes were the prevalent shapes independent of tree species. The highest percentage of cylindrical crowns was found for *R. pseudoacacia*. This light-demanding species responded to the proximity of shading objects with the development of more elongated shapes, while expanding into spherical crowns when more space was available. The influence of nearby buildings was not recognized for the shade tolerant species *T. cordata* and *A. platanoides*. Considering possible reactions of crown development due to different light sensibilities of the tree species could improve green space planning in the varying urban structures and surroundings.

Applying crown shape-specific equations to calculate a tree's crown volume result in more realistic values if only the formula for cylindrical crown shapes is used. According to our study sample, the smallest slope and increase of crown volume with increasing stem diameter was shown for the half-ellipsoid and the cylindrical shape, while the highest increase is recorded for the ovoid and the spherical shape.

Tree crowns with the highest shade area values presented a spherical and a cylindrical shape, with spherical crowns also showing high shade densities (along with ovoid shaped crowns) and shade index. Shade area increased with dbh most strongly for ovoid tree crowns, while cylindrical crowns showed the least decrease of shade density with stem growth. Consideration of ecosystem services such as cooling by shading in urban landscape planning purposes can be improved by applying shape-specific calculations for shade area extension and shade density.

Our findings showed clear over estimations of crown volume (i.e., up to two or three times the actual value depending on the crown shape) when relying on cylindrical shape based calculations, as often applied in urban tree growth models. The methodological approach for this study was chosen as a simple readily available but extensive fast method to be used in field measurements, where TLS scans can hardly be realized. The visual categorization of crown shapes based on reference figures offers a fast, cheap and easy-to-use possibility of obtaining an improved estimation of the crown volume of trees and of deriving ecosystem services that are dependent on the crown shape (e.g., shade area or cooling potential). In the future, however, terrestrial laser scanning may be used more and more, especially if the devices become cheaper and more mobile and easy-to-use analysis software is available.

Author Contributions: Conceptualization, E.F. and T.R.; Methodology, E.F., T.R. and M.A.R.; Formal Analysis, E.F.; Writing–Original Draft Preparation, E.F.; Writing–Review and Editing, E.F., T.R., A.M.-R., M.A.R., H.P. and S.P.; Visualization, E.F.; Supervision, T.R.; Project Administration, T.R., H.P. and S.P.; Funding Acquisition, T.R., H.P., M.A.R. and S.P. All authors have read and agreed to the published version of the manuscript.

Funding: This research was funded by the German Science Foundation (Deutsche Forschungsgemeinschaft), grant number PR 292/21−1 and PA 2626/3−1 and by the Bavarian State Ministry of the Environment and Consumer Protection, grant numbers TUF01UF−64971 and TLK01UFuE69397.

Data Availability Statement: The data are available on request from the corresponding author.

Acknowledgments: We like to thank the DFG for funding the project "Impact of trees on the urban microclimate under climate change: Mechanisms and ecosystem services of urban tree species in temperate, Mediterranean and arid major cities" as well as the Bavarian State Ministry of the Environment and Consumer Protection for funding the project "Urban trees under climate change I + II: their growth, environmental performance, and perspectives" in the frame of the Centre for Urban Ecology and Climate Adaptation. The authors want to thank the municipalities of the cities for the permit to measure urban trees. Thanks also to Martin Honold, Nayanesh Pattnaik and Anna Wenig for their help during the field work.

Conflicts of Interest: The authors declare no conflict of interest.

References

1. Oke, T.R. The micrometeorology of the urban forest. *Phil. Trans. R. Soc. Lond. B* **1989**, *324*, 335–349. [CrossRef]
2. Wolf, K.L.; Lam, S.T.; McKeen, J.K.; Richardson, G.R.A.; van den Bosch, M.; Bardekjian, A.C. Urban Trees and Human Health: A Scoping Review. *Int. J. Environ. Res. Public Health* **2020**, *17*, 4371. [CrossRef] [PubMed]
3. Tzoulas, K.; Korpela, K.; Venn, S.; Yli-Pelkonen, V.; Kaźmierczak, A.; Niemela, J.; James, P. Promoting ecosystem and human health in urban areas using Green Infrastructure: A literature review. *Landsc. Urban Plan.* **2007**, *81*, 167–178. [CrossRef]
4. Rahman, M.A.; Stratopoulos, L.M.; Moser-Reischl, A.; Zölch, T.; Häberle, K.-H.; Rötzer, T.; Pretzsch, H.; Pauleit, S. Traits of trees for cooling urban heat islands: A meta-analysis. *Build. Environ.* **2020**, *170*, 106606. [CrossRef]
5. Rahman, M.A.; Dervishi, V.; Moser-Reischl, A.; Ludwig, F.; Pretzsch, H.; Rötzer, T.; Pauleit, S. Comparative analysis of shade and underlying surfaces on cooling effect. *Urban For. Urban Green.* **2021**, *63*, 127223. [CrossRef]
6. Rahman, M.A.; Moser, A.; Rötzer, T.; Pauleit, S. Within canopy temperature differences and cooling ability of Tilia cordata trees grown in urban conditions. *Build. Environ.* **2017**, *114*, 118–128. [CrossRef]
7. Peters, E.B.; McFadden, J.P.; Montgomery, R.A. Biological and environmental controls on tree transpiration in a suburban landscape. *J. Geophys. Res.* **2010**, *115*. [CrossRef]
8. Armson, D.; Stringer, P.; Ennos, A.R. The effect of tree shade and grass on surface and globe temperatures in an urban area. *Urban For. Urban Green.* **2012**, *11*, 245–255. [CrossRef]
9. Gillner, S.; Vogt, J.; Tharang, A.; Dettmann, S.; Roloff, A. Role of street trees in mitigating effects of heat and drought at highly sealed urban sites. *Landsc. Urban Plan.* **2015**, *143*, 33–42. [CrossRef]
10. Rahman, M.A.; Moser, A.; Rötzer, T.; Pauleit, S. Microclimatic differences and their influence on transpirational cooling of Tilia cordata in two contrasting street canyons in Munich, Germany. *Agric. For. Meteorol.* **2017**, *232*, 443–456. [CrossRef]
11. Gill, S.; Handley, J.; Ennos, A.; Pauleit, S. Adapting Cities for Climate Change: The Role of the Green Infrastructure. *Built Environ.* **2007**, *33*, 115–133. [CrossRef]
12. Rötzer, T.; Moser-Reischl, A.; Rahman, M.A.; Grote, R.; Pauleit, S.; Pretzsch, H. Modelling Urban Tree Growth and Ecosystem Services: Review and Perspectives. In *Progress in Botany Vol. 82*; Cánovas, F.M., Lüttge, U., Risueño, M.-C., Pretzsch, H., Eds.; Springer International Publishing: Cham, Switzerland, 2021; pp. 405–464, ISBN 978-3-030-68619-2.
13. Grote, R.; Samson, R.; Alonso, R.; Amorim, J.H.; Cariñanos, P.; Churkina, G.; Fares, S.; Le Thiec, D.; Niinemets, Ü.; Mikkelsen, T.N.; et al. Functional traits of urban trees: Air pollution mitigation potential. *Front Ecol. Environ.* **2016**, *14*, 543–550. [CrossRef]
14. Nowak, D.J.; Crane, D.E.; Stevens, J.C. Air pollution removal by urban trees and shrubs in the United States. *Urban For. Urban Green.* **2006**, *4*, 115–123. [CrossRef]
15. Yu, C.; Hien, W.N. Thermal benefits of city parks. *Energy Build.* **2006**, *38*, 105–120. [CrossRef]
16. Bowler, D.E.; Buyung-Ali, L.; Knight, T.M.; Pullin, A.S. Urban greening to cool towns and cities: A systematic review of the empirical evidence. *Landsc. Urban Plan.* **2010**, *97*, 147–155. [CrossRef]
17. Shahidan, M.; Jones, P. Plant Canopy Design in Modifying Urban Thermal Environment: Theory and Guidelines. In Proceedings of the 25th Conference on Passive and Low Energy Architecture, Dublin, Ireland, 22–24 October 2008.
18. Pretzsch, H.; Biber, P.; Uhl, E.; Dahlhausen, J.; Rötzer, T.; Caldentey, J.; Koike, T.; van Con, T.; Chavanne, A.; Seifert, T.; et al. Crown size and growing space requirement of common tree species in urban centres, parks, and forests. *Urban For. Urban Green.* **2015**, *14*, 466–479. [CrossRef]
19. Gratani, L.; Varone, L. Carbon sequestration by Quercus ilex L. and Quercus pubescens Willd. and their contribution to decreasing air temperature in Rome. *Urban Ecosyst* **2006**, *9*, 27–37. [CrossRef]
20. Baptista, M.D.; Livesley, S.J.; Parmehr, E.G.; Neave, M.; Amati, M. Variation in leaf area density drives the rainfall storage capacity of individual urban tree species. *Hydrol. Processes* **2018**, *32*, 3729–3740. [CrossRef]
21. Zhu, Z.; Kleinn, C.; Nölke, N. Assessing tree crown volume—A review. *For. Int. J. For. Res.* **2021**, *94*, 18–35. [CrossRef]

22. Rahman, M.A.; Armson, D.; Ennos, A.R. A comparison of the growth and cooling effectiveness of five commonly planted urban tree species. *Urban Ecosyst* **2015**, *18*, 371–389. [CrossRef]
23. Poorter, H.; Niklas, K.J.; Reich, P.B.; Oleksyn, J.; Poot, P.; Mommer, L. Biomass allocation to leaves, stems and roots: Meta-analyses of interspecific variation and environmental control. *New Phytol.* **2012**, *193*, 30–50. [CrossRef] [PubMed]
24. GALK Straßenbaumliste. Available online: https://www.galk.de/arbeitskreise/stadtbaeume/themenuebersicht/strassenbaumliste (accessed on 16 December 2021).
25. Pretzsch, H. Zur Analyse der räumlichen Bestandesstruktur und der Wuchskonstellation von Einzelbäumen. *Forst Und Holz* **1992**, *47*, 408–418.
26. Pretzsch, H. Analysis and modeling of spatial stand structures. Methodological considerations based on mixed beech-larch stands in Lower Saxony. *For. Ecol. Manag.* **1997**, *97*, 237–253. [CrossRef]
27. Pretzsch, H.; Biber, P.; Ďurský, J. The single tree-based stand simulator SILVA: Construction, application and evaluation. *For. Ecol. Manag.* **2002**, *162*, 3–21. [CrossRef]
28. Nowak, D.; Crane, D.; Stevens, J.; Hoehn, R.; Walton, J.; Bond, J. A Ground-Based Method of Assessing Urban Forest Structure and Ecosystem Services. *AUF* **2008**, *34*, 347–358. [CrossRef]
29. Rötzer, T.; Rahman, M.A.; Moser-Reischl, A.; Pauleit, S.; Pretzsch, H. Process based simulation of tree growth and ecosystem services of urban trees under present and future climate conditions. *Sci. Total Environ.* **2019**, *676*, 651–664. [CrossRef]
30. Deutscher Wetterdienst–Climate Data Center. Available online: https://cdc.dwd.de/portal/ (accessed on 16 December 2021).
31. DIVA-GIS. Available online: http://www.diva-gis.org/gdata (accessed on 2 March 2022).
32. Georgi, L.; Kunz, M.; Fichtner, A.; Reich, K.F.; Bienert, A.; Maas, H.-G.; von Oheimb, G. Effects of local neighbourhood diversity on crown structure and productivity of individual trees in mature mixed-species forests. *For. Ecosyst.* **2021**, *8*, 26. [CrossRef]
33. Hasenauer, H. Dimensional relationships of open-grown trees in Austria. *For. Ecol. Manag.* **1997**, *96*, 197–206. [CrossRef]
34. Moser-Reischl, A.; Rötzer, T.; Pauleit, S.; Pretzsch, H. Urban Tree Growth Characteristics of Four Common Species in South Germany. *AUF* **2021**, *47*, 150–169. [CrossRef]
35. Moser, A.; Rötzer, T.; Pauleit, S.; Pretzsch, H. The Urban Environment Can Modify Drought Stress of Small-Leaved Lime (*Tilia cordata* Mill.) and Black Locust (*Robinia pseudoacacia* L.). *Forests* **2016**, *7*, 71. [CrossRef]
36. Kükenbrink, D.; Gardi, O.; Morsdorf, F.; Thürig, E.; Schellenberger, A.; Mathys, L. Above-ground biomass references for urban trees from terrestrial laser scanning data. *Ann. Bot.* **2021**, *128*, 709–724. [CrossRef] [PubMed]
37. Roloff, A. *Bäume in der Stadt. Besonderheiten, Funktion, Nutzen, Arten, Risiken*; Verlag Eugen Ulmer: Stuttgart, Germany, 2013.
38. Niinemets, Ü.; Valladares, F. Tolerance to Shade, Drought, and Waterlogging of Temperate Northern Hemisphere Trees and Shrubs. *Ecol. Monogr.* **2006**, *76*, 521–547. [CrossRef]
39. Pauleit, S.; Jones, N.; Garcia-Martin, G.; Garcia-Valdecantos, J.L.; Rivière, L.M.; Vidal-Beaudet, L.; Bodson, M.; Randrup, T.B. Tree establishment practice in towns and cities–Results from a European survey. *Urban For. Urban Green.* **2002**, *1*, 83–96. [CrossRef]
40. Beck, P.; Caudullo, G.; Tinner, W.; de Rigo, D. Fraxinus Excelsior in Europe: Distribution, Habitat, Usage and Threats. In *San-Miguel-Ayanz*; de Rigo, D.J., Caudullo, G., Durrant, T.H., Mauri, A., Eds.; European Atlas of Forest Tree Species; Publish Office: Luxembourg, 2016.
41. Larsen, F.; Kristoffersen, P. Tilia's physical dimensions over time. *J. Arboric.* **2002**, *28*, 209–214. [CrossRef]
42. Stoffberg, G.H.; van Rooyen, M.W.; van der Linde, M.J.; Groeneveld, H.T. Predicting the growth in tree height and crown size of three street tree species in the City of Tshwane, South Africa. *Urban For. Urban Green.* **2008**, *7*, 259–264. [CrossRef]
43. Pretzsch, H. *Grundlagen der Waldwachstumsforschung*; Springer: Berlin/Heidelberg, Germany, 2019; ISBN 978-3-662-58154-4.
44. Pretzsch, H. *Forest Dynamics, Growth and Yield: From Measurement to Model*; Springer: Berlin/Heidelberg, Germany, 2010; ISBN 978-3-540-88307-4.
45. Moser-Reischl, A.; Uhl, E.; Rötzer, T.; Biber, P.; van Con, T.; Tan, N.T.; Pretzsch, H. Effects of the urban heat island and climate change on the growth of Khaya senegalensis in Hanoi, Vietnam. *For. Ecosyst.* **2018**, *5*, 37. [CrossRef]
46. Herzog, M. Simulation of Growth and Ecosystem Services of Tree Species in Central European Cities under Present and Future Climate Conditions. Master Thesis, Technical University of Munich, Munich, Germany, 2021.
47. Lukaszkiewicz, J.; Kosmala, M. Determining the Age of Streetside Trees with Diameter at Breast Height-based Multifactorial Model. *AUF* **2008**, *34*, 137–143. [CrossRef]
48. Bühler, O.; Kristoffersen, P.; Larsen Søren Ugilt. Growth of Stree Trees in Copenhagen With Empphasis on the Effect of Different Establishment Concepts. *Arboric. Urban For.* **2007**, *33*, 330–337. [CrossRef]
49. Dwyer, J.F. How old is that tree? *Ill. Trees Q. Publ. Q. Publ. Illonis Arborist Assoc.* **2009**, *24*, 13.
50. Lawrence, E. *The Illustrated Book of Trees and Shrubs*; Octopus Books Limited: London, UK, 1985.
51. baumportal.de: Baumportal–Alles über Bäume. Available online: baumportal.de (accessed on 2 March 2022).
52. Pretzsch, H.; Matthew, C.; Dieler, J. Allometry of Tree Crown Structure. Relevance for Space Occupation at the Individual Plant Level and for Self-Thinning at the Stand Level. In *Growth and Defence in Plants*; Matyssek, R., Schnyder, H., Oßwald, W., Ernst, D., Munch, J.C., Pretzsch, H., Eds.; Springer: Berlin/Heidelber, Germany, 2012; pp. 287–310. ISBN 978-3-642-30644-0.
53. Moser, A.; Rötzer, T.; Pauleit, S.; Pretzsch, H. Structure and ecosystem services of small-leaved lime (Tilia cordata Mill.) and black locust (*Robinia pseudoacacia* L.) in urban environments. *Urban For. Urban Green.* **2015**, *14*, 1110–1121. [CrossRef]

54. Bayer, D.; Reischl, A.; Rötzer, T.; Pretzsch, H. Structural response of black locust (*Robinia pseudoacacia* L.) and small-leaved lime (*Tilia cordata* Mill.) to varying urban environments analyzed by terrestrial laser scanning: Implications for ecological functions and services. *Urban For. Urban Green.* **2018**, *35*, 129–138. [CrossRef]
55. Jacobs, M.; Rais, A.; Pretzsch, H. How drought stress becomes visible upon detecting tree shape using terrestrial laser scanning (TLS). *For. Ecol. Manag.* **2021**, *489*, 118975. [CrossRef]
56. Barbeito, I.; Dassot, M.; Bayer, D.; Collet, C.; Drössler, L.; Löf, M.; Del Rio, M.; Ruiz-Peinado, R.; Forrester, D.I.; Bravo-Oviedo, A.; et al. Terrestrial laser scanning reveals differences in crown structure of Fagus sylvatica in mixed vs. pure European forests. *For. Ecol. Manag.* **2017**, *405*, 381–390. [CrossRef]
57. Troxel, B.; Piana, M.; Ashton, M.S.; Murphy-Dunning, C. Relationships between bole and crown size for young urban trees in the northeastern USA. *Urban For. Urban Green.* **2013**, *12*, 144–153. [CrossRef]
58. Quigley, M.F. Street trees and rural conspecifics: Will long-lived trees reach full size in urban conditions? *Urban Ecosyst* **2004**, *7*, 29–39. [CrossRef]
59. Armson, D.; Rahman, M.A.; Ennos, A. A comparison of the shading effectiveness of five different street tree species in Manchester, UK. *Arboric. Urban For.* **2013**, *39*, 157–164. [CrossRef]
60. Lin, B.-S.; Lin, Y.-J. Cooling Effect of Shade Trees with Different Characteristics in a Subtropical Urban Park. *Horts* **2010**, *45*, 83–86. [CrossRef]
61. Rahman, M.A.; Franceschi, E.; Pattnaik, N.; Moser-Reischl, A.; Hartmann, C.; Paeth, H.; Pretzsch, H.; Rötzer, T.; Pauleit, S. Spatial and temporal changes of outdoor thermal stress: Influence of urban land cover types. *Sci. Rep.* **2022**, *12*, 671. [CrossRef]
62. Rahman, M.A.; Moser, A.; Rötzer, T.; Pauleit, S. Comparing the transpirational and shading effects of two contrasting urban tree species. *Urban Ecosyst* **2019**, *22*, 683–697. [CrossRef]

forests

Article

Effects of Climate and Drought on Stem Diameter Growth of Urban Tree Species

Vjosa Dervishi [1,*], Werner Poschenrieder [2], Thomas Rötzer [1], Astrid Moser-Reischl [1] and Hans Pretzsch [1]

1 TUM School of Life Sciences, Chair of Forest Growth and Yield Science, Technical University of Munich, Hans Carl-von-Carlowitz-Platz 2, 85354 Freising, Germany; thomas.roetzer@tum.de (T.R.); astrid.reischl@tum.de (A.M.-R.); hans.pretzsch@tum.de (H.P.)

2 Bavarian Forest Institute (LWF), Hans-Carl-von-Carlowitz-Platz 2, 85354 Freising, Germany; werner.poschenrieder@lwf.bayern.de

* Correspondence: vjosa.dervishi@tum.de

Abstract: Urbanization and climate change are two inevitable megatrends of this century. Knowledge about the growth responses of urban trees to climate is of utmost importance towards future management of green infrastructure with the aim of a sustainable provision of the environmental ecosystem services. Using tree-ring records, this study analyzed growth response to climate by stem diameter at breast height (DBH) of 1178 trees in seven large cities worldwide, including *Aesculus hippocastanum* L. in Munich; *Platanus* × *hispanica* Münchh. in Paris; *Quercus nigra* L. in Houston; *Quercus robur* L. in Cape Town; *Robinia pseudoacacia* L. in Santiago de Chile, Munich, and Würzburg; and *Tilia cordata* Mill. in Berlin, Munich, and Würzburg. Climate was characterized following the de Martonne aridity index (DMI). Overall, trees showed an 8.3% lower DBH under arid than humid climate at the age of 100. Drought-tolerant tree species were overall not affected by climate. However, *R. pseudoacacia* showed a lower diameter when growing in semi-dry than humid climate. In contrast, drought-sensitive tree species were negatively affected by arid climate. Moreover, the effect of drought years on annual diameter increment was assessed. *P.* × *hispanica* and *R. pseudoacacia* appeared as the most drought-resistant species. The highest sensitivity to drought was detected in *T. cordata* and *Q. robur*. *A. hippocastanum* and *Q. nigra* showed a lower diameter growth during drought events, followed by a fast recovery. This study's findings may contribute to a better understanding of urban tree growth reactions to climate, aiming for sustainable planning and management of urban trees.

Keywords: *Aesculus hippocastanum*; diameter growth; drought; *Platanus* × *hispanica*; *Quercus nigra*; *Quercus robur*; *Robinia pseudoacacia*; *Tilia cordata*; urban trees

Citation: Dervishi, V.; Poschenrieder, W.; Rötzer, T.; Moser-Reischl, A.; Pretzsch, H. Effects of Climate and Drought on Stem Diameter Growth of Urban Tree Species. *Forests* **2022**, *13*, 641. https://doi.org/10.3390/f13050641

Academic Editor: Chi Yung Jim

Received: 1 April 2022
Accepted: 20 April 2022
Published: 21 April 2022

1. Introduction

While currently 55% of the world's population lives in urban areas, it is projected that by 2050, almost 70% of the global population will live in urban environments. At the same time, urban settings are frequently exposed to extreme climatic events like dry spells or extreme rainfall events caused by global change [1]. Urban trees, as a crucial part of the green infrastructure in cities, have gained increasing awareness in recent decades. This is mainly due to their role in providing a wide array of environmental ecosystem services [2–5]. Carbon storage, mitigation of heat island effect [6], cooling by transpiration and shading [7–9], reduction of rainwater runoff [10], supporting the biodiversity in urban areas [11], and others are positive environmental impacts of trees. In that sense, due to policy changes and efforts that have been made recently [1], vegetation coverage has increased in many cities worldwide [12]. A recent study [13] showed that more than 10% of built-up areas in 325 large cities globally increased the urban green proportion significantly in the last two decades. However, the quantity of the environmental ecosystem services provided is highly dependent on the species, size, and vitality of trees [6,14].

High impervious ground surface and a higher water loss due to increased evapotranspiration [15,16] may lead to increased water stress in urban areas compared to natural environments. Further, recent studies have shown that drought is the main inciting factor impacting urban trees' health and survival [17,18], and climatic predictions forecast rising temperatures and more frequent heat and drought events. Therefore, there is a need for urban tree management to select tree species according to their future drought response [19]. In order to guarantee a good water supply for vital trees with high environmental ecosystem provision, the requirements of the individual tree species must be known.

As tree species growing in urban areas undergo specific environmental conditions, these conditions may affect the resilience of urban trees [17]. Thus, the growth response in such environments is not guaranteed to correspond to generally expected growth patterns of trees. To improve the strategic management of urban green vegetation, the growth–climate relationship needs to be analyzed in detail. Dendrochronology is a valuable tool to study the relationship between tree growth and climatic factors [20]. Previous studies address how climate change modifies the ecosystem services provided by urban trees [4,11,21]. These studies were based on model predictions and investigated various adaptation actions [22], such as selecting drought-resistant species and various environmental provenances [23,24]. However, less is known about how climate conditions affect the growth of urban trees on a large spatial scale. This study aimed to analyze urban trees' stem diameter growth response depending on the species, their drought tolerance, and urban climate conditions (air temperature, precipitation). The diameter of a tree is a straightforward measurement and can be used as a proxy for calculating other tree parameters, such as tree crown dimension [25–27]. Thus, tree diameter allows one to predict environmental ecosystem services like cooling of urban environments [27,28].

Poschenrieder et al. [29] claimed that tree growth, ecosystem service provision, and stand management strongly depend on a site's climatic conditions, especially precipitation and temperature. Urban tree growth models are proper tools for quantifying environmental ecosystem services depending on tree growth dynamics and changing environments. Prominent models such as UFORE [30,31], i-Tree [32], CITYgreen [33,34], CityTree [6] and UrbTree [35] provide fundamental knowledge for sustainable and future-oriented planning of green infrastructure in cities [28,36]. However, these models do not include tree-related parameters such as mortality, vitality, or nutrient supply, which are important for sustainable planning and management of urban trees [37]. To account for this, Pretzsch et al. [28] recently introduced an urban tree dynamic management model (UTDyn), which was further developed by Poschenrieder et al. [29]. This dynamic model represents an approach of age class balancing and sustainable planning, transferring forest management aspects to urban tree management while including tree growth, mortality, and ecosystem service provision. Pretzsch et al. [28] suggested improving the dynamic model by adding additional parameters.

This study aims to contribute and give additional value to the UTDyn model using tree-ring records of 1178 individual, mature broadleaf trees of six different tree species, distributed across seven cities worldwide as follows: *Aesculus hippocastanum* L. in Munich; *Platanus* × *hispanica* MÜNCHH. in Paris; *Quercus nigra* L. in Houston; *Quercus robur* L. in Cape Town; *Robinia pseudoacacia* L. in Santiago de Chile, Munich, Würzburg; and *Tilia cordata* Mill. in Berlin, Munich, and Würzburg. Climatic conditions (annual precipitation, mean air temperature) were represented using the de Martonne Index (DMI, [38]). The stem diameter growth reaction to climatic conditions and the annual diameter growth reaction to drought events were based on the following research questions:

(Q1) How is the general stem diameter growth reaction of trees depending on climate in urban environments?

(Q2) How does the climate of a city defined by its relative aridity (DMI) affect stem diameter growth of drought-tolerant and drought-sensitive tree species within urban environments?

(Q3) How do *Robinia pseudoacacia* and *Tilia cordata* react to climate within different climates in their stem diameter growth?

(Q4) How do individual drought years affect the annual stem diameter increment of urban tree species?

2. Materials and Methods

2.1. Study Sites and Climate

Seven cities worldwide, in which tree cores of six tree species were recorded [20,39], form the base of this study. Table 1 gives an overview of the selected cities, geography, and climate conditions. The climate dataset consists of total annual precipitation data and long-term mean temperature data for the period of 1981–2010 for each city. Annual precipitation data for the cities Cape Town, Houston, and Paris were provided by the National Climatic Data Center (NOAA NCEI, [40]). For the German cities Berlin, Munich, and Würzburg, data was provided by the Open Data Server of the German Metrological Service (DWD, [41]). A historical climate dataset, as described by González-Reyes [42], was used for Santiago de Chile.

The cities within this study are distributed across several climate zones (Table 1). According to the climate classification by Köppen-Geiger [43], the cities Berlin, Munich, Würzburg, and Paris are located in the temperate zones of Central Europe. They are characterized by an oceanic climate (Cfb), featuring mild summers and cool winters. However, the amount of precipitation and mean temperature differs between the sites. While Berlin and Würzburg show similar climatic patterns with an annual precipitation sum of around 600 mm and a mean temperature of around 9 °C, Paris is characterized by a higher mean temperature of 12.3 °C and a higher precipitation sum of 632 mm. Compared to other European cities within this study, precipitation is highest in Munich (948 mm). Santiago de Chile and Cape Town are characterized by Mediterranean climate (Csb) with hot, dry summers and mild, wet winters. However, the long-term mean temperature and the amount of precipitation in Cape Town are higher (16.7 °C, 544 mm) compared to Santiago de Chile (14.7 °C, 325 mm). Houston has a humid, subtropical climate (Cfa). The annual precipitation sums up to 1091 mm and is ample throughout the year. The temperature average in Houston is 21 °C.

Table 1. Overview of the geographical position and climate conditions (1981–2010) of the study sites. Climate zone refers to the Köppen-Geiger climate classification [43].

Location, Geographic Position of Study Sites						
Berlin, Germany	Munich, Germany	Würzburg, Germany	Paris, France	Santiago de Chile, Chile	Cape Town, South Africa	Houston, USA
52.31 °N 13.24 °E	48.14 °N 11.58 °E	49.46 °N 9.57 °E	48.51 °N 2.21 °E	33.27 °S 70.40 °W	33.55 °S 18.25 °E	29.46 °N 95.23 °W
Climate zone, total annual precipitation [mm a^{-1}] ± SD, mean temperature [°C]						
Temperate (Cfb)	Temperate (Cfb)	Temperate (Cfb)	Temperate (Cfb)	Mediterra-nean (Csb)	Mediterra-nean (Csb)	Subtropical (Cfa)
591 ± 100	948 ± 141	601 ± 108	632 ± 116	325 ± 156	544 ± 118	1091 ± 249
9.5	9.7	9.6	12.3	14.7	16.7	21
DMI Index (mean value) and climatic classification following Baltas [44]						
32	53	32	32	15	23	34
humid	very humid	humid	humid	semi-dry	mediterranean	humid

2.2. Aridity Index according to de Martonne

Aridity is a long-term, climate phenomenon characterized by a shortage of water supply/precipitation. In this study, the aridity at a given location was quantified using the de Martonne aridity index (DMI). This index was developed by de Martonne in 1926 [38] and describes the combined effect of temperature and precipitation as follows:

$$\text{DMI} = \frac{\text{Prec}}{(\text{Temp} + 10)} \quad (1)$$

In this analysis, Prec is described as total annual precipitation (mm) and Temp (°C) as the long-term mean temperature (1981–2010) per city. DMI shows the annual value for the climate, while DMI avg presents the mean DMI value for each tree, which was calculated for the entire lifespan of a given tree. Higher DMI values indicate cool and moist climatic conditions, while lower values express warm and dry climatic conditions. According to Baltas [44] classification, the DMI index of the seven cities represents a range from semi-dry climatic conditions (Santiago de Chile, DMI = 15) to very humid conditions (Munich, DMI = 53) (see Table 1). Due to restricted data availability, growth comparisons for single tree species across different climatic conditions were only possible for the tree species *R. pseudoacacia* and *T. cordata*. Thus, tree species within this study were categorized into drought sensitivity classes (drought-sensitive, drought-tolerant) following the study of Niinemets and Valladares [45]. Figure A1 shows the climatic conditions for each city throughout the years.

2.3. Selection of Tree Species and Tree Individuals

The analyzed trees within this study represent medium-sized to large, long-lived deciduous tree species. Table 2 lists information on the tree species regarding drought tolerance, wood anatomy, and maximal height. Most of the species—*A. hippocastanum*, *P.* × *hispanica*, *Q. robur*, *R. pseudoacacia*, *T. cordata*—are widely present in the urban landscapes of Central Europe [46–49]. *Q. nigra* and *R. pseudoacacia* are native to North America [50]. However, due to its invasiveness, *R. pseudoacacia* is now widespread across Europe and partially across Asia, South America, Africa, Canada, and Australia [51].

Table 2. Description of sampled tree species and their characteristics on drought tolerance (following Niinemets & Valladares, 2006), wood anatomy, and maximal achievable height.

Tree Species	Drought Tolerance	Wood Anatomy	Height [m]
Aesculus hippocastanum L.	sensitive	diffuse-porous	39
Platanus × *hispanica* Münchh.	tolerant	diffuse-porous	35
Quercus nigra L.	tolerant	ring-porous	30
Quercus robur L.	tolerant	ring-porous	30
Robinia pseudoacacia L.	tolerant	ring-porous	35
Tilia cordata Mill.	sensitive	diffuse-porous	30

2.4. Tree Data Collection

The recorded tree data includes dendrochronological information taken from increment cores of 1178 individual trees worldwide. Trees were selected along transects from the city center to the four cardinal directions, thus covering an urbanization gradient from highly paved urban areas to more open green suburban parts of a city, comprising solitary street trees, as well as trees from city parks and urban forests. The sampled trees were free from any biotic or abiotic damages. Two increment cores were taken from each tree at a height of 1.3 m in perpendicular directions (North, East), using an increment borer with an inner diameter of 5 mm (Haglöf, Sweden). The sampled tree species are *A. hippocastanum*, *T. cordata and R. pseudoacacia* in Munich, *T. cordata* in Berlin, *P.* × *hispanica* in Paris, *R. pseudoacacia* in Santiago de Chile, *Q. robur* in Cape Town, *T. cordata* and *R. pseudoacacia* in Würzburg, and *Q. nigra* in Houston. Afterwards, the increment cores were glued on wooden slides and polished on a sanding machine using 120 to 1200 grit, depending on the tree species, in order to enhance the optimum visibility of the growth rings. This preparation allowed tree-ring width measurements with a precision of 0.01 mm using a digital positioning table Digitalpositiometer [52]. Crossdating of the observed tree-ring data was based on the methods provided with the R-library dplR [53] and performed up to a tree age of one year. The age of all sampled trees was computed with age formulas based on tree structures. Individual tree ages could be estimated by combining city administration records with increment core series. From the increment cores, the stem diameter growth of each tree on an annual basis was traced, resulting in a total of 66,000 observations. The most extensive observation years were covered in Berlin and ranged from 1876 to 2013.

The tree-based characteristics of the data are presented in Table 3. The number of sampled trees ranged from 69 *Q. robur* individuals in Cape Town to 251 *T. cordata* trees in Berlin. *Q. robur* showed the highest mean diameter at breast height (67.9 cm), while *T. cordata* in Würzburg showed the lowest mean DBH (33.2 cm). Hence, *T. cordata* in Würzburg showed the lowest mean height of 12.5 m, while *P.* × *hispanica* in Paris were the highest trees with a mean of 18.8 m. *A. hippocastanum* were on average the oldest trees sampled in this study with a mean age of 118 years, while *R. pseudoacacia* in Würzburg and Santiago de Chile had an average age of 52 years. *T. cordata* in Berlin showed the lowest mean annual diameter increment (5.0 mm), while *Q. nigra* in Houston showed the highest mean annual diameter increment (11.3 mm).

Table 3. Statistical characteristics of the tree ring series from the sampled trees. *N* = Number of sampled trees: drought-tolerant 571 (species *P.* × *hispanica*, *Q. nigra*, *Q. robur*, and *R. pseudoacacia*), drought-sensitive 607 (species *A. hippocastanum*, *T. cordata*), DBH = Diameter at Breast Height.

City	Sampling Year	*N*	Mean Tree Age	Mean Tree Height [m]	DBH [cm]	Diameter Increment [mm a^{-1}]
			Aesculus hippocastanum L.			
Munich	2013	193	118	16.1	63.3 (19.6–117.0)	5.4 (2.0–15.8)
			Platanus × *hispanica* Münchh.			
Paris	2013	133	105	18.8	64.8 (40.3–144.0)	7.3 (1.4–23.0)
			Quercus nigra L.			
Houston	2014	179	53	16.2	59.9 (34.2–98.0)	11.3 (3.8–29.2)
			Quercus robur L.			
Cape Town	2011	69	103	15.6	67.9 (40.3–112.9)	6.5 (2.2–17.3)
			Robinia pseudoacacia L.			
Munich	2014	30	57	15.7	44.5 (14.0–101.9)	8.5 (3.7–20.7)
Würzburg	2014	31	52	15.1	44.3 (11.0–102.2)	9.0 (4.8–18.3)
Santiago de Chile	2012	129	52	15.3	41.4 (19.8–56.1)	8.5 (3.6–19.2)
			Tilia cordata Mill.			
Berlin	2010–2013	251	85	16.9	44.2 (16.5–81.1)	5.0 (1.2–12.2)
Munich	2014, 2018	133	81	13.1	34.2 (12.0–86.7)	5.6 (2.1–10.3)
Würzburg	2014	30	62	12.5	33.2 (14.0–71.5)	6.2 (3.0–11.0)

2.5. Data Analysis

Statistical evaluations were conducted using the software R-4.0.5 [54]. Linear mixed effect models were applied using lme4 package [55] to analyze the effects of climate on tree diameter growth. Fixed effects included the stem diameter (DBH), tree age (Age), and a climate parameter characterized by the annual de Martonne Index (DMI) or the trees' average de Martonne Index (DMI avg) during each tree's lifespan. Random effects for City,

Species, Individual tree, and Calendar year were used for correcting temporal and spatial autocorrelations. Furthermore, to check for differences in the means, post-hoc tests were performed by using emmeans package with Tukey correction [56]. Model-based figures were produced using the package sjplot [57]. The climate is expressed using the 2.5th and 97.5th quantile of the DMI for the arid and humid climate conditions, respectively. This was necessary as DMI values for drought-sensitive and drought-tolerant species differed strongly, and considering relative droughts was thus possible. For analyzing the general long-term stem diameter growth reaction to climate for all trees within this study, the following model was applied:

$$\ln DBH_{ijkt} = a_0 + a_1 \times \ln Age_{ijkt} + a_2 \times DMI_{it} + a_3 \times \ln Age_{ijkt} \times DMI_{it} + b_i + b_{ij} + b_{ijk} + \varepsilon_{ijkt} \quad (2)$$

Hereby, DBH is the response variable and refers to the stem diameter of the trees in breast height; Age represents the tree age. The indices represent (i) city, (j) species, (k) individual tree, and (t) calendar year.

For analyzing the stem diameter growth reaction of drought-tolerant (DT) and drought-sensitive (DS) tree species in urban environments, the following model was applied separately for each group:

$$\ln DBH_{ij} = a_0 + a_1 \times \ln Age_{ij} + a_2 \times DMI\ avg_j + b_i + \varepsilon_{ij} \quad (3)$$

Hereby, DMI avg refers to an average value of the Martonne Index (DMI) of an individual tree's lifespan. The indices i refer to i^{th} tree individual, and j refers to the calendar year. The variable Species was not included in the random effects due to a low range of species for the drought-tolerant (DT) group (2 levels: *A. hippocastanum, T. cordata*).

The species-specific long-term stem diameter growth with age was calculated by the following model equation separately for the tree species *R. pseudoacacia* and *T. cordata*:

$$\ln DBH_{ij} = a_0 + a_1 \times \ln Age_{ij} + a_2 \times City + a_3 \times \ln Age_{ij} \times City + b_i + \varepsilon_{ij} \quad (4)$$

using Age and City as fixed effect and the indices i for individual tree and j for calendar year as random effects.

To quantify the influence of drought years on stem diameter increment, the Superposed Epoch Analysis (SEA) [39,58] was applied on the ring-width dataset using the R package dplR [53]. The SEA function assesses the significance of departures in tree ring-width indices (RWI) from the mean for a given set of key event years (e.g., drought years) and lagged years (superposed epoch) by comparing the value of the superposed epoch to randomly selected epochs, which are selected from the tree-ring dataset using random sets of 11 years (five years before and after the drought year) from 1000 bootstrapped sets.

3. Results

An overview of the diameter-age relationship for the analyzed species in each city is shown in Figure 1. For this, simple linear models were applied for each tree of a given species and location. Based on the steepness of the slope, *Quercus nigra* L. (*Q. nigra*) showed—in the subtropic climate of Houston—the highest diameter-age relationship (1.131, Figure 1c), followed by *Robinia pseudoacacia* L. (*R. pseudoacacia*) growing in the temperate climate cities of Würzburg (0.931, Figure 1g) and Munich (0.888, Figure 1f), and the Mediterranean climate of Santiago de Chile (0.827, Figure 1e). The slopes for *Platanus* × *hispanica* MÜNCHH. (*P.* × *hispanica*) in the humid climate of Paris and *Quercus robur* L. (*Q. robur*) in the Mediterranean climate of Cape Town were 0.723 (Figure 1b) and 0.638 (Figure 1d), respectively. *Aesculus hippocastanum* L. (*A. hippocastanum*) showed a lower diameter-age growth relationship (0.522, Figure 1a) under humid climate conditions in Munich. Similar trends were observed for *Tilia cordata* Mill. (*T. cordata*) growing in Würzburg (0.629, Figure 1j), Munich (0.566, Figure 1i), and Berlin (0.495, Figure 1h).

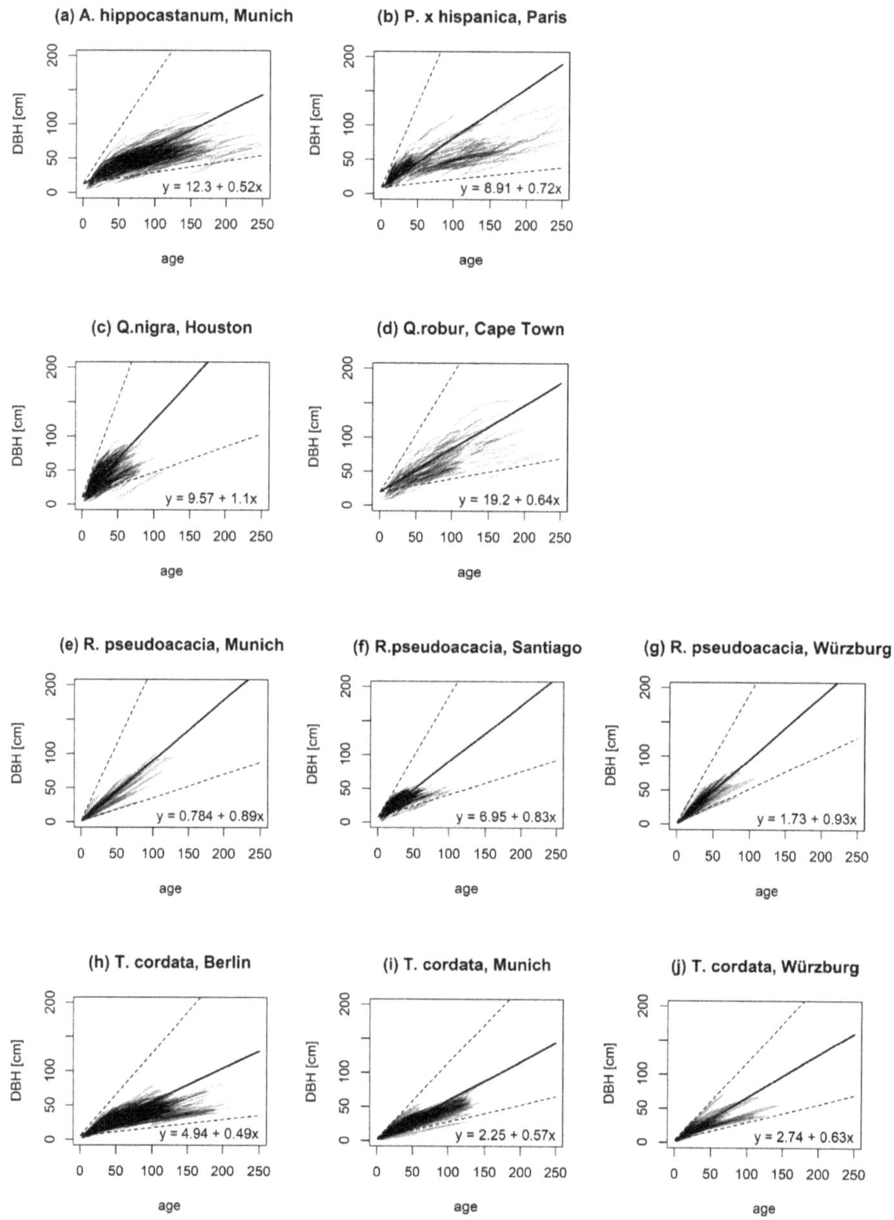

Figure 1. Diameter-age relationship for the tree species *Aesculus hippocastanum* L. in Munich (**a**); *Platanus* × *hispanica* Münchh. in Paris (**b**); *Quercus nigra* L. in Houston (**c**); *Quercus robur* L. in Cape Town (**d**); *Robinia pseudoacacia* L. in Santiago de Chile (**e**), Munich (**f**), and Würzburg (**g**); and *Tilia cordata* Mill. in Berlin (**h**), Munich (**i**), and Würzburg (**j**). The solid line represents the mean stem diameter growth, while the dashed lines show the minimum and maximum.

3.1. Stem Diameter Growth Reaction of Urban Trees in Dependence on the Aridity

The results of the long-term tree diameter growth reaction depending on the climate within this study are presented in Figure 2 and listed in Table A1. Overall, the estimate for the diameter–age relationship was 0.784 ($p < 0.001$). This effect was stronger with increased DMI value (interaction ln Age and DMI, $p < 0.001$), showing that trees had a higher DBH growth with increased tree age and DMI value. On average, trees at the age of 60 showed a 2.9% lower DBH (45.7 cm) when growing under arid climate conditions than trees growing under humid climate (DBH = 47 cm). This effect increases with increasing tree age. Consequently, at 80, 100 and 140, the difference is by −5.9%, −8.3%, and −12.1%, respectively.

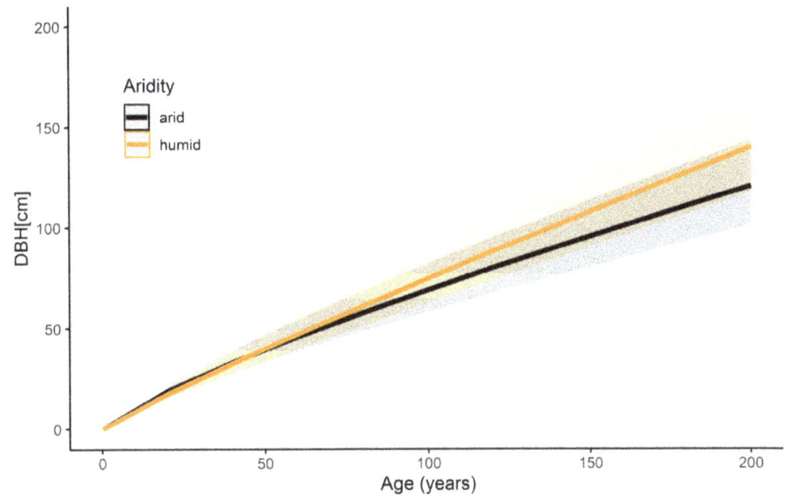

Figure 2. Long-term stem diameter growth reaction of all trees within this study (N = 1178) in dependence on tree age and DMI aridity index based on Equation (2) (see Table A1 for model parameters). The black curve visualizes the stem diameter growth under arid climatic conditions (DMI = 11, corresponding to the 2.5th quantile of annual DMI of all cities). The orange curve visualizes the diameter growth under humid climatic conditions (DMI = 64, corresponding to 97.5th quantile of annual DMI of all cities). Shaded bands visualize the predicted confidence interval of the curves.

3.2. Drought Effects on Stem Diameter Growth of Drought-Tolerant and Drought-Sensitive Tree Species

The statistical model results regarding the diameter growth reaction of drought-tolerant and drought-sensitive tree species are presented in Figure 3. Age has a significant effect on the diameter growth for both drought-tolerant and drought-sensitive tree species ($p < 0.001$). Averaged DMI values of each tree (DMI avg) have a significant effect on the diameter growth of drought-sensitive tree species ($p < 0.001$, Figure 3b) but no effect on the diameter growth of drought-tolerant tree species ($p = 0.456$, Figure 3a). On average, drought-sensitive tree species growing under arid climatic conditions showed a reduced diameter growth of 16.6% compared to the growth under humid climate conditions. At 100 years, drought-sensitive tree species had an average diameter of 50.4 cm when growing under arid climate conditions and 58.8 cm when growing under humid climate conditions.

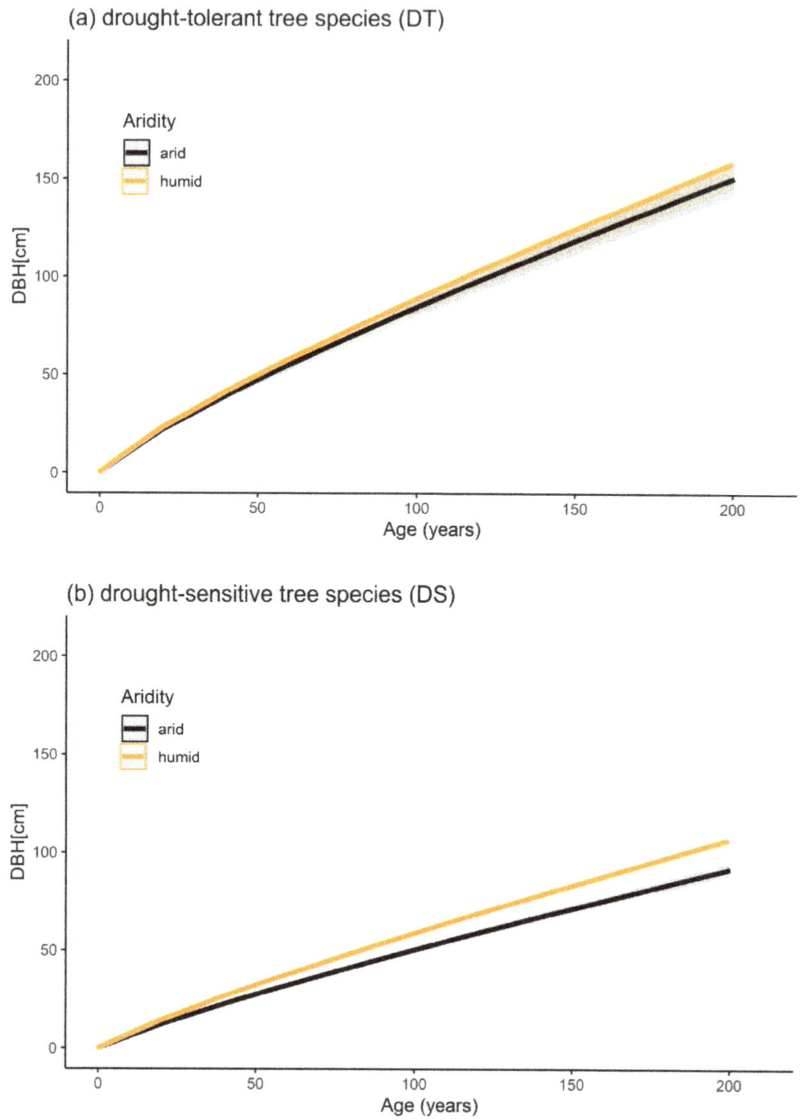

Figure 3. Stem diameter growth reaction of drought-tolerant (DT, (**a**)) and drought-sensitive (DS, (**b**)) urban tree species depending on the climate, based on model Equation (3) (see Table A1 for model parameters). Black curves visualize the diameter growth under arid climatic conditions (corresponding to the 2.5th quantile of a DMI avg-level, DMI avg = 13 for DT and DMI avg = 30 for DS). Orange curves visualize the diameter growth in humid climatic conditions (corresponding to the 97.5th quantile level of the DMI avg, 52 and 53 for DT and DS, respectively). The shaded bands visualize the prediction confidence interval of the curves.

3.3. Species Specific Reaction of Stem Growth on Aridity

Model results for the stem diameter in dependence of age for *R. pseudoacacia* in three cities, Santiago de Chile (DMI = 15), Munich (DMI = 53), and Würzburg (DMI = 32), are presented in Figure 4 and listed in Table A2. The post-hoc test revealed significant differences in the slopes of all three cities, whereby the differences between Santiago de Chile and both Munich and Würzburg were significantly higher ($p < 0.001$) than between Munich and Würzburg ($p < 0.0223$). Furthermore, the lower slope for Santiago de Chile (0.746 ± 0.003), compared with Munich (1.015 ± 0.004) and Würzburg (1.001 ± 0.004), indicate that the trees in Santiago de Chile tend to have a lower diameter growth with age, compared with Munich ($p < 0.001$) and Würzburg ($p < 0.001$). At the age of 40, *R. pseudoacacia* in Santiago de Chile and Würzburg had a similar DBH of around 38 cm, while the trees in Munich showed a 10% lower DBH (34.3 cm). However, at 60, the trees showed a similar DBH in Santiago de Chile and Munich (51.5 and 51.8, respectively), while the trees in Würzburg had a 10.9% higher DBH (56.6 cm). At 80, trees in Würzburg showed a 15% higher DBH than Santiago de Chile and 8% to Munich. At 100 years, trees in Santiago de Chile had a 20% lower DBH compared to Würzburg and a 13% lower than Würzburg.

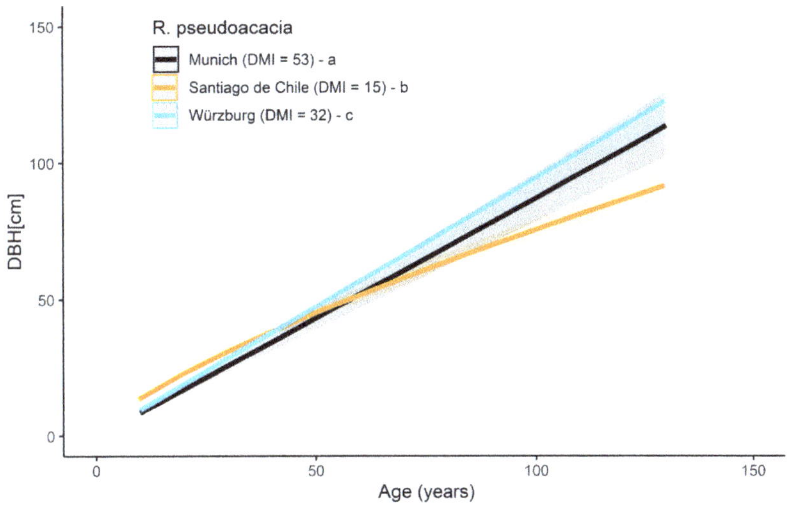

Figure 4. Stem diameter growth of *R. pseudoacacia* in Santiago de Chile (black curve, *N* = 129 trees), Munich (orange curve, *N* = 30 trees), and Würzburg (blue curve, *N* = 31 trees). Letters a, b, and c show significant differences between the slopes. Shaded bands visualize the prediction confidence interval of the curves.

T. cordata in Berlin, Munich, and Würzburg revealed significant differences in tree growth in all three cities (see Figure 5 and Table A2). The highest slope was observed for Munich (0.9401 ± 0.0019), followed by Würzburg (0.9031 ± 0.0036) and Berlin (0.8519 ± 0.0021). On average, *T. cordata*, at the age of 40, showed a lower DBH in both cities, Berlin (−13%) and Munich (−10%), compared to Würzburg. Even though Würzburg and Berlin have similar climatic patterns (DMI = 32), growth differences of the DBH get more substantial with increasing age. At 80 years, compared with the trees in Würzburg (BHD = 48.1 cm), T. cordata had a 17% lower DBH in Berlin (41.2 cm), and a 7% lower DBH in Munich (DBH = 45 cm).

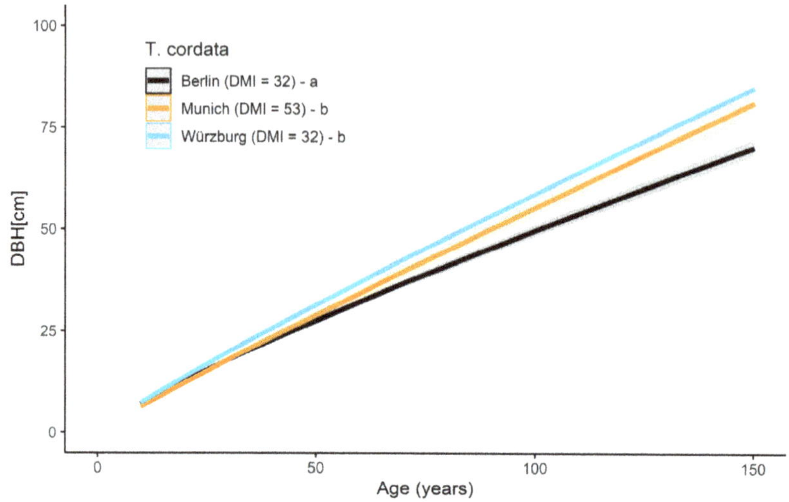

Figure 5. Stem diameter growth in dependence of age of *T. cordata* in Berlin (black curve, N = 251 trees), Munich (orange curve, N = 133 trees), and Würzburg (blue curve, N = 30 trees). Letters a and b show significant differences between the slopes. The shaded bands visualize the prediction confidence interval of the curves.

3.4. Effect of Drought Years on Stem Diameter Increment of Common Urban Tree Species

The superposed epoch analysis (SEA) was used to investigate the impact of drought events on the stem diameter increment. For this, three to five extreme drought years were selected as key event years for each city based on the climate dataset (see Figure A1 in Appendix A). Depending on the tree-data timeframe, the drought years with the lowest DMI value were chosen for which the tree-dataset covered at least 50% of the sampled trees of a given species in each city. The selected drought years for each tree species in each city are listed in Table 4.

Table 4. Selected drought years of the seven cities for the analysis of the reaction of the annual diameter increment to drought by using superposed epoch analysis (SEA).

Tree Species	City	Drought Years				
A. hippocastanum	Munich	1976	1982	2003		
P. × *hispanica*	Paris	1963	1976	2005		
Q. nigra	Houston	1988	1998	2011		
Q. robur	Cape Town	1973	2003	2010		
R. pseudoacacia	Santiago de Chile	1968	1988	1998	2007	
	Munich	1976	1982	2003		
	Würzburg	1976	1991	2003		
T. cordata	Berlin	1982	1989	2003		
	Munich	1947	1976	1982	2003	2015
	Würzburg	1991	1996	2003		

The SEA results reveal varying reactions for tree species to drought events within the analyzed cities (Figures 6 and 7, and Table A3). *A. hippocastanum* trees in Munich showed a negative stem diameter increment only during the drought years (Figure 6a). *P.* × *hispanica* in Paris had the highest resistance to droughts and showed no significant reaction (Figure 6b). *Q. nigra* in Houston experienced a negative ring-width indices (RWI) development prior to and during the drought year, followed by a fast recovery, and thus showed a significant positive RWI two years after the drought (Figure 6c). *Q. robur* trees in Cape Town had negative RWIs in the first and the fourth year after a drought event (Figure 6d). *R. pseudoacacia* (Figure 7a) was resistant to drought in Munich and Würzburg, while it experienced a decreased growth in Santiago de Chile explicitly during the year of a drought event. *T. cordata* (Figure 7b) growing in Munich showed the highest sensitivity during and after drought years. On the other hand, *T. cordata* in Berlin was resistant to drought years. In Würzburg, *T. cordata* experienced a positive RWI four years before the drought and a negative RWI in the second year after the drought.

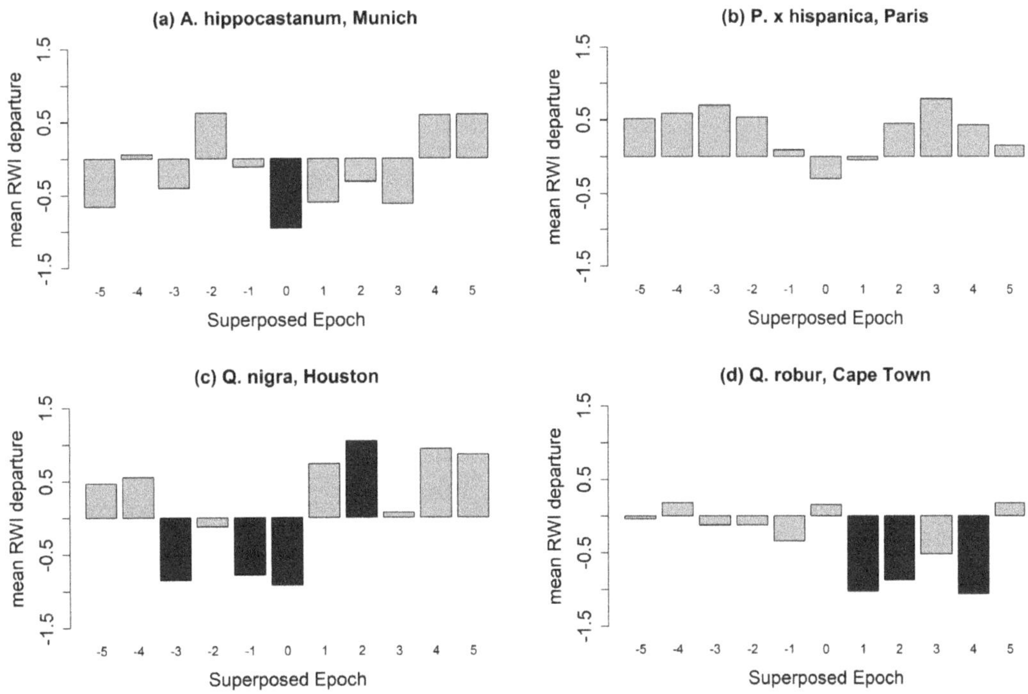

Figure 6. Superposed epoch analysis (SEA) on the influence of drought years on the ring width index (RWI) of *A. hippocastanum* in Munich (**a**), *P.* × *hispanica* in Paris (**b**), *Q. nigra* in Houston (**c**), and *Q. robur* in Cape Town (**d**). RWI is shown for the drought years (0), the pre-drought (−5 to −1), and post-drought years (1 to 5). Input drought years for each city are shown in Table 4. Significant RWIs ($p < 0.05$) are shown in dark grey.

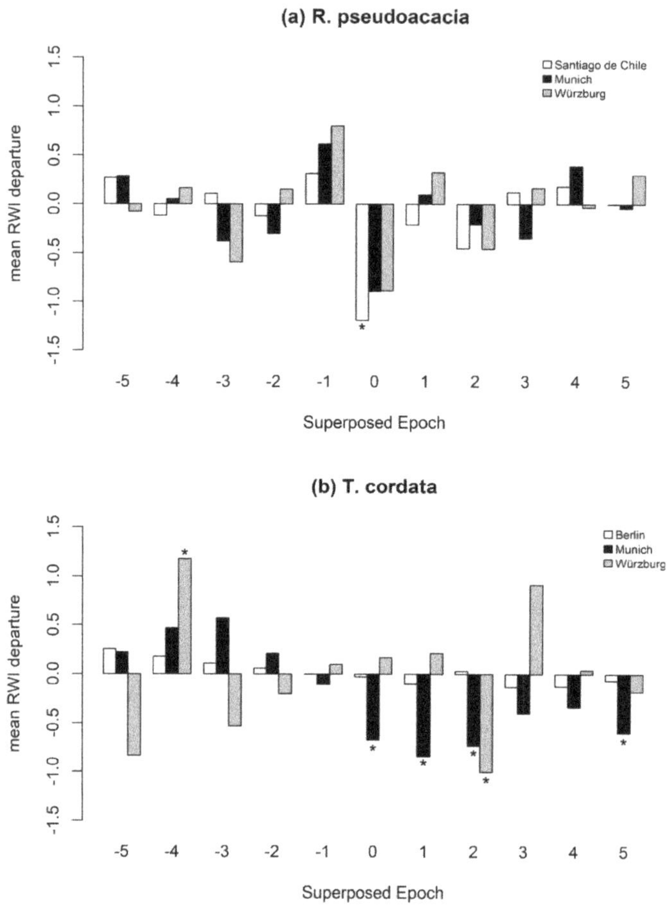

Figure 7. Superposed epoch analysis on the influence of drought years in the RWI of *R. pseudoacacia* (**a**) in Santiago de Chile (white bars), Munich (black bars) and Würzburg (grey bars), and *T. cordata* (**b**) in Berlin (white bars), Munich (black bars) and Würzburg (grey bars). Shown are RWIs during drought years (0), pre-drought (−5 to −1), and post-drought years (1 to 5). Significant RWIs ($p < 0.05$) are marked by *.

4. Discussion

4.1. Impact of Urban Climate on the Long-Term Growth Response of the Stem Diameter

The effect of drought on the stem diameter growth of all trees was notable and led to a suppression of the diameter at breast height (DBH) by 8.3% at the age of 100 years, presuming a change from the current Central European long term average of humid (DMI = 32 in Berlin, Paris, and Würzburg to 53 in Munich) to semi-dry conditions as in Santiago de Chile, where the average DMI is 15 (see Table 1). Such a strong shift is not likely to occur in Central Europe until 2050 [59]. However, on average, the climate of cities in the northern hemisphere is expected to shift towards warmer conditions in the next decades [59]. That shift corresponds to an average speed of 20 km a^{-1} and will undoubtedly persist beyond 2050 [60].

Considering the fact that the sampled tree species within this study grow in one climate zone or region explicitly (except for *R. pseudoacacia*), an intra-comparison for tree species growing in different climate zones was not possible. Therefore, the study at hand classified tree species into drought-tolerant tree species (including *P.* × *hispanica, Q. nigra Q. robur, R. pseudoacacia*) and drought-sensitive tree species (*A. hippocastanum, T. cordata*). This classification was done following the study of Niinemets and Valladares [45]. However, the general sensitivity of the DBH to arid climate conditions was due to the drought reaction of the drought-sensitive species group (see Figure 3), showing a 16.6% decline in the diameter growth at the age of 100 years. In contrast, the predicted DBH over time in the drought-tolerant group was proven to be tolerant for a DMI range between 13 to 52 (corresponding to semi-dry to very humid climate).

Due to the broad range of data for the species *R. pseudoacacia* and *T. cordata*, analyses of the DBH were possible at a species-specific level. Considering *R. pseudoacacia* as a prominent example for the drought-tolerant group, the diameter growth was strongly dependent on tree age and showed a different growth course in Santiago de Chile, compared with Munich and Würzburg. In Santiago de Chile, where the climate is classified as semi-dry (DMI = 13), the diameter-age growth course marked a higher diameter growth in the younger age, followed by a lower diameter growth in a mature stage (see Figure 4) compared with Munich and Würzburg. On the contrary, the diameter-age growth course in the humid climate of Munich (DMI = 53) and Würzburg (DMI = 32) was similar and marked a quasi-linear growth. On average, *R. pseudoacacia* showed a higher DBH in Würzburg, followed by Munich and Santiago de Chile. For example, at the age of 100, *R. pseudoacacia* showed an average diameter of 94.3 cm in Würzburg, 87 cm in Munich, and 75.4 cm in Santiago de Chile. However, the steeper slopes for *R. pseudoacacia* in humid climate cities indicate stronger growth–age relationships with increasing water supply. The differing long-term growth reaction of *R. pseudoacacia* in Santiago de Chile compared to Munich and Würzburg can generally be explained by the typical climate of these locations implying the phenotypic acclimation or the species adaption to the respective local conditions to which the species is exposed to. It seems that, in the long-term, *R. pseudoacacia* profits from the higher precipitation in the humid climate of Munich. Thus, in terms of a shift of future climate conditions in central European cities, urban planning needs to consider markedly slower tree growth in the long term, even for well-known drought-tolerant species. However, the persistence of drought-tolerant tree species under notably dry conditions will support their suitability in the future urban environment.

Furthermore, the drought sensitivity of *R. pseudoacacia* may be affected by the physiological adaptation of this species to a more favorable water supply. Due to its isohydric stomatal behavior strategy, *R. pseudoacacia* closes the stomata early during drought events to minimize water loss through transpiration [61]. This allows *R. pseudoacacia* to resist short drought periods [39]. However, the isohydric strategy may lead to carbon starvation and tree death. Recent findings suggest that *R. pseudoacacia*, as a leguminous species, is drought-resistant due to its symbiosis with nitrogen-fixing bacteria [62]. The symbiotic nitrogen (N) fixation may play a role in drought tolerance and drought avoidance by supplying N to leaves for acclimation and facilitating compensatory growth following drought. Recent findings on forest stands of Eastern Europe have shown that *R. pseudoacacia* will benefit from climate change [63,64] and that it is a species that reacts very plastically with radial increments to thermal and precipitation conditions [65]. However, limitation of water supply may lead to morphological changes such as reduction of individual leaf dimensions of *R. pseudoacacia*, resulting in a highly reduced leaf area [66] and thus, in combination with the isohydric stomatal behavior strategy, providing lower cooling of the urban environment.

T. cordata showed higher DBH in Würzburg compared with Berlin regardless of the similar long-term climate patterns in Berlin and Würzburg (DMI = 32) (see Figure 5). On average, *T. cordata* showed, at the age of 100 years, a diameter of 58.9 cm in Würzburg, 55.5 cm in Munich, and 49.8 cm in Berlin. Moreover, a higher slope was estimated for Munich (0.9401) and Würzburg (0.9031), indicating a higher DBH in these two cities compared to Berlin (0.8519). Possible explanations for the differing reaction of *T. cordata* in Würzburg and Berlin rely on climate aspects that may affect the tree growth during the vegetation period and winter. These aspects refer to the distribution of precipitation amount throughout the year, sunshine duration, temperature, relative air humidity, and winter frost. Furthermore, microclimatic aspects in the nearby environment of the trees may impact tree growth. Many studies refer to *T. cordata* as a tree species sensitive to the environmental site conditions within a city. In a recent study by Rötzer et al. [18], *T. cordata* developed a higher biomass growth in the suburban parts of Würzburg compared to the highly paved parts of the city center. The authors state that due to different microclimatic conditions, individual environmental sites may impact tree growth. A study by Moser-Reischl et al. [67] on urban tree growth in southern Germany reported a lower diameter growth for *T. cordata* trees growing in city squares compared to other growing sites such as parks. Other studies conducted in temperate climate cities of Central Europe recorded declined growth of *T. cordata* due to high environmental stress along roadside areas [68] and higher rates of necrotic leaves along heavy traffic roads [69]. This leads to the conclusion, that beside climatic tolerances, other environmental aspects such as traffic emissions need to be considered when analyzing influences on urban tree growth. In a study of temperate climate conducted in different cities, greater root development was observed for *Tilia* ssp. within irrigated areas and in areas where coarser fractions of gravel and debris were mixed with finer materials such as clay and silt [70]. Moser et al. [39] analyzed the growth reaction of *T. cordata* and *R. pseudoacacia* during the extremely dry and warm summer of 2015 in two different urban sites in Munich, namely highly paved squares and more open, green squares. The study showed that trees growing at open green sites had a higher diameter growth than trees growing in highly paved squares. Following the anisohydric stomatal regulation strategy, *T. cordata* keeps an ongoing leaf gas exchange even under drought, leading to water loss and tree mortality as a result. Together with the mentioned studies, the observed results suggest the avoidance of future plantings of *T. cordata* in highly paved urban areas and along roadsides.

Other factors such as genotype adaptation or provenance play an important role in the long-term growth of trees. Moreover, phenotypic acclimation or genetic adaptation, or a balance of both, may explain the growth responses of tree species to the changing climate [71]. Broadmeadow et al. [72] recommend using provenances from origins, where the prevailing climatic conditions follow the projected future conditions of the target site.

4.2. Impact of Drought Years on the Stem Diameter Increment

The SEA results revealed that *P.* × *hispanica* and *R. pseudoacacia* growing in temperate climate cities were the most tolerant tree species during drought events (see Figures 6a and 7a, and Table A3). These results reflect the importance of drought-tolerant and fast-growing tree species for urban forestry, as shown in numerous studies on urban tree growth. For example, a dendrochronological study [73] on five different urban tree species in temperate climate cities showed that the diameter increment of *P.* × *hispanica* was not significantly affected by drought years. Furthermore, analyses on leaf-gas exchange of five tree species in the temperate climate resulted in a species-specific response to urban sites, with *P.* × *hispanica* resulting as a tree species with a high water use efficiency. Even in the selected periods of high atmospheric drought and low soil moisture, the water use efficiency remained high [74]. Other studies reported *P.* × *hispanica* as the most tolerant tree species to environmental stress for roadsides in cities of temperate climate [68]. Therefore, under the prevailing environmental and climatic conditions, *P.* × *hispanica* seems well adapted to urban environments. Moser-Reischl et al. [67] reported for *P.* × *hispanica* trees growing in three urban environments, namely parks, public squares and roadsides at six cities with a temperate climate, a similar diameter and height increment regardless of the growing site. However, other studies on *P.* × *hispanica* conducted in cities with temperate climate showed opposite results for this tree species. These studies reported a negative influence for *P.* × *hispanica* by higher sums of long-term precipitation and higher numbers of freezing days in winter on the vitality of this species [75] and reduced growth by high temperatures during the growing season [76].

For *R. pseudoacacia*, the observed SEA results (see Figure 7a, Table A3) revealed a high resistance of this tree species to drought years when growing in temperate climate cities. *R. pseudoacacia* growing in the Mediterranean climate of Santiago de Chile showed negative growth reactions during the drought years, followed by a fast recovery in the following years. Due to its great ability to cope with environmental drought, *R. pseudoacacia* is being seen as an important tree species in both ecosystems, forests, and urban landscapes. Thus, there has been enormous research investigating the adaptation and avoidance strategy of *R. pseudoacacia* to environmental drought in recent decades. In a study conducted in German rural areas, young trees of *R. pseudoacacia* showed ecophysiological and morphological adaptation to prolonged drought conditions by reducing water loss through transpiration and leaf size. Under drought conditions, the stomatal down-regulation of transpiration was reduced by 50% compared to growth under sufficient soil moisture supply [66]. In a recent study by Rötzer et al. [18] on the growth reaction of mature *R. pseudoacacia* and *T. cordata* in Würzburg, *T. cordata* showed a significant lower diameter increment compared to *R. pseudoacacia* and thus a lower provision of environmental ecosystem services such as carbon fixation and cooling by transpiration during the extreme drought years 2018 and 2019. Nola et al. [77] analyzed the response of xylem anatomy of mature *R. pseudoacacia* and *Q. robur* to climate variability in temperate forests of Pavia, Italy. Despite the same wood anatomy (ring-porous) of these species, the authors found differences in the vessel size and distribution within tree rings, with. *Q. robur* showing a typical ring-porous reaction by building a uniform vessel number and size depending on the previous summer precipitation and autumn-winter temperatures. Conversely, *R. pseudoacacia* was able to modulate its response to climate by varying the number and size of vessels. The xylem plasticity enables *R. pseudoacacia* to cope better with both inter-annual climate variations and extreme drought events than *Q. robur*. SEA results for *Q. robur* in Cape Town showed a high sensitivity of this species to drought events by showing a significantly lower diameter increment in the years after the drought events (see Figure 6d). *T. cordata* growing in Munich showed the highest sensitivity during and after drought years by experiencing reduced growth in the following years after drought events (see Figure 7b). Similar results were shown for the diameter increment growth of mature *T. cordata* in Munich for drought years during the period of 1985 to 2015 [78]. On the other hand, the observed results show that *T. cordata* growing in Würzburg and Berlin were resistant when facing drought events. In Würzburg, this tree species showed a significantly lower diameter growth in the second year after the drought, while in Berlin, the diameter increment was resistant to drought years. A possible reason for this reaction could be the local climate of the different cities, especially the different annual precipitation sums and the distribution over the year. While Berlin and Würzburg are characterized by a drier and warmer climate with a long-term mean annual precipitation sum of 590 mm a^{-1}, Munich is characterized by an almost twice as high precipitation (950 mm a^{-1}). This growth reaction may be explained by a possible adaptation of *T. cordata* to the dry and warm local climate conditions of Berlin and Würzburg.

A. hippocastanum trees in Munich experienced an overall reduced increment growth during drought years (Figure 6a). However, *A. hippocastanum* showed the ability to recover rapidly after drought years. Studies on *A. hippocastanum* have shown that this tree species is very sensitive to the growing site. Moser-Reischl et al. [67] analyzed the growth reactions of almost 600 trees of *A. hippocastanum* in South Germany. The study's results show different growth reactions of *A. hippocastanum* depending on the sites of the cities. Trees growing in park sites had the highest average diameter increment compared to trees on street sites and highly paved public squares.

For *Q. nigra* in Houston, a significantly reduced growth before and during drought years was assessed, followed by an immediate recovery leading to significant, positive mean ring-width-index departure after two years (Figure 6c). Results of a study by Moser et al. [79] in Houston showed that young trees of *Q. nigra* grew better in the city center than in suburban and rural surroundings of Houston. However, this effect was reversed with increasing age.

Other aspects of urban green management, such as pruning operations, irrigation, or leaf litter removal, may affect the short-term growth reaction of urban trees. A study conducted on poplar trees (*Populus* spp.) [80] recorded higher leaf carbon isotopic ratios ($\delta^{13}C$) for pruned trees than unpruned trees, indicating a higher stomatal conductance of leaves, thus suggesting that pruned trees could have higher drought stress resistance than unpruned trees. Due to a reduced transpiration surface, the pruning treatment improved the trees' water status. Another crucial aspect is the soil biota, especially the mycorrhizal composition in the soil, which may increase the water uptake and thus protect the tree during drought events due to the symbiosis with the host plant [81]. For forest trees, mycorrhizal fungi play an important role in the natural regeneration of forests [82]. Rusterholz et al. [83] analyzed the mycorrhizal composition on juvenile sycamore maple (*Acer pseudoplatanus*) along an urbanization gradient from rural surroundings to the city center of a temperate climate city. The study's results show that the mycorrhizal fungal colonization was 15–45% lower in areas of higher urbanization than in rural areas.

Due to already mentioned and other biotic and abiotic environmental conditions, trees in urban areas show strong stress responses in their growth reactions. Further, increasing urbanization may negatively influence the growing sites of trees, raising the mortality rate of urban trees. The life expectancy of trees within urban sites is relatively low compared to forest trees [20] and depends on factors such as growing sites and proper management, amongst others. Analyzing the growth of trees and other components of the urban green infrastructure, such as shrubs and lawns, plays a crucial role towards sustainable planning and management of green infrastructure in future cities. As shown in this study, combined tree and climate datasets and permanent monitoring sites in urban environments provide the background for in-depth analyses of the provision of ecosystem services by a city's tree cover. Such data and climate-growth relationship of urban trees can further serve as a base for the development and improvement of growth models such as the dynamic Urban Tree growth model UTDyn (Pretzsch et al. [28], Poschenrieder et al. [29]) or the process-based urban tree growth model CityTree (Rötzer et al. [6]).

5. Conclusions

This study focused on analyzing the long-term stem diameter growth reaction of 1178 trees in seven cities worldwide depending on climate conditions (air temperature, precipitation). The observed results indicate that the effect of climate on the long-term stem diameter growth at any age was outstanding and that dry and warm climate, i.e., a low de Martonne Index, led to a reduction of diameter growth. The growth decline of drought-sensitive tree species such as *A. hippocastanum* and *T. cordata* was notably higher compared to the growth of drought-tolerant tree species such as *P. hispanica*, *Q. nigra*, *Q. robur*, and *R. pseudoacacia*. It is known that ecosystem environmental services provided by trees strongly depend on the growth and dimensions of these. At the same time, tree growth depends on species-related tolerance against drought events and climatic conditions. This dependency will most likely be amplified under expected future climatic conditions. Thus, future urban green management needs to focus on a tree selection adapted to ongoing climatic changes to ensure a sustainable and further provision of environmental ecosystem services. This study's findings and datasets may contribute to a better understanding of urban tree growth reactions to climate and to further development of concepts such as the dynamic Urban Tree Growth model UTDyn (see Pretzsch et al. [28] and Poschenrieder et al. [29]), aiming for sustainable planning and management of urban trees.

Due to the long lifespan of the trees, and their enhanced provision of ecosystem services in mature and old stages of growth in particular, urban management needs thought-out and long-term designs aiming for a sustainable management and vital growth of trees. In this context and towards future climatic impacts, there is a need for interdisciplinarity within urban green planning, including architecture, forestry, ecology, water management, new technologies, and the urban communities as important drivers.

Author Contributions: V.D.: Conceptualization; data curation; formal analysis; investigation; methodology; visualization; writing—original draft; W.P.: data curation; formal analysis; writing—original draft; T.R.: conceptualization; funding acquisition; writing—review and editing; A.M.-R.: Data curation; writing—review and editing; H.P.: Conceptualization; Project administration; Funding acquisition. All authors have read and agreed to the published version of the manuscript.

Funding: This research was funded by the Audi Environmental Foundation, grant number 5101954, the German Science Foundation (DFG), grant number PR 292/21-1 and by the Bavarian State Ministry of the 504 Environment and Consumer Protection, grant numbers TUF01UF-64971, TKP01KPB-71924 and TEW01CO2P-75382.

Institutional Review Board Statement: Not applicable.

Informed Consent Statement: Not applicable.

Data Availability Statement: Not applicable.

Acknowledgments: The authors gratefully acknowledge all funding institutions and municipalities that supported the study through funding and permissions of tree sampling. These are the AUDI Environmental Foundation for funding the project: "Reaction kinetics of trees under climate change" (5101954), the German Science Foundation (DFG) for providing funds for the project PR 292/21-1 "Impact of trees on the urban microclimate under climate change: Mechanisms and ecosystem services of urban tree species in temperate, Mediterranean and arid major cities", the Bavarian State Ministry of the Environment and Consumer Protection for funding the study through projects: "Urban Trees under Climate Change: their growth, environmental performance, and perspectives" (TUF01UF-64971), "Climate experience Würzburg: Influence of trees on the microclimate of the city of Würzburg" (TKP01KPB-71924), and "Ecosystem services of urban green at public squares in Munich" (TEW01CO2P-75382). The authors also thank the responsible municipal authorities of Munich, Würzburg, Berlin, Paris, Houston, Cape Town, and Santiago de Chile for the support and permission to measure and core the trees. The authors thank Álvaro González-Reyes for providing the historical climate dataset for Santiago de Chile. Finally, the authors would like to thank Pascal Edelmann and Peter Biber for the constructive discussion.

Conflicts of Interest: The authors declare no conflict of interest.

Appendix A

Table A1. Statistical results of the linear mixed effect model as described by Equations (2) and (3). DT = Drought tolerant, DS = Drought sensitive. Significant p values ($p < 0.05$) are shown in bold.

Model Equation	**Equation (2)**			**Equation (3) on DT**			**Equation (3) on DS**		
Predictors	*Estimates*	*SE*	*p*	*Estimates*	*SE*	*p*	*Estimates*	*SE*	*p*
Intercept	**0.604**	0.090	**<0.001**	**0.576**	0.051	**<0.001**	**−0.265**	0.048	**<0.001**
ln Age	**0.784**	0.002	**<0.001**	**0.835**	0.002	**<0.001**	**0.865**	0.001	**<0.001**
DMI	**−0.007**	0.000	**<0.001**	0.001	0.002	0.456	**0.007**	0.001	**<0.001**
ln Age:DMI	**0.002**	0.000	**<0.001**						
Random Effects									
σ2		0.01			0.01			0.01	
$\tau_{00\ treeID:(species:city)}$		0.11							
$\tau_{00\ species:city}$		0.03							
$\tau_{00\ city}$		0.03							
$\tau_{00\ treeID}$					0.17			0.09	
$\tau_{00\ treeID:city}$									
ICC		0.93			0.92			0.88	
N_{treeID}		1178			571			607	
$N_{species}$		6							
N_{city}		7							
Observations		66,000			25,219			41,023	
R^2		0.701			0.718			0.795	

Table A2. Statistical results of the linear mixed effect model described by Equation (4) on the diameter growth for *R. pseudoacacia* in Santiago de Chile, Munich, and Würzburg and *T. cordata* in Berlin, Munich, and Würzburg. Significant p values ($p < 0.05$) are shown in bold.

Model Equation	**Equation (4) on *R. pseudoacacia***			**Equation (4) on *T. cordata***			
Predictors	*Estimates*	*SE*	*p*	*Predictors*	*Estimates*	*SE*	*p*
Intercept	0.889	0.027	**<0.001**	Intercept	−0.015	0.018	0.395
ln Age	0.746	0.003	**<0.001**	ln Age	0.852	0.002	**<0.001**
Munich	−1.096	0.06	**<0.001**	Munich	−0.298	0.029	**<0.001**
Würzburg	−0.953	0.06	**<0.001**	Würzburg	−0.068	0.05	0.174
ln Age:Munich	0.269	0.005	**<0.001**	ln Age:Munich	0.088	0.003	**<0.001**
ln Age:Würzburg	0.256	0.005	**<0.001**	ln Age:Würzburg	0.051	0.004	**<0.001**
Random Effects							
σ^2		0.01		σ^2		0.01	
$\tau_{00\ treeID}$		0.08		$\tau_{00\ treeID}$		0.06	
ICC		0.87		ICC		0.84	
N_{treeID}		190		N_{treeID}		414	
Observations		8025		Observations		27,229	
R^2		0.828		R^2		0.842	

Table A3. Superposed epoch analysis (SEA) for the single urban tree species in seven cities. p values are listed for the pre-drought years (−5 to −1), the drought years (0), and the post-drought years (1–5). Significant p values ($p < 0.05$) are printed in bold.

		Years										
		Pre-Drought Years					**Drought Years**	**Post-Drought Years**				
Tree Species	**City**	**−5**	**−4**	**−3**	**−2**	**−1**	**0**	**1**	**2**	**3**	**4**	**5**
A. hippocastanum	Munich	0.078	0.438	0.226	0.111	0.456	**0.016**	0.117	0.254	0.079	0.135	0.136
P. × *hispanica*	Paris	0.174	0.151	0.093	0.182	0.577	0.112	0.238	0.303	0.083	0.306	0.604
Q. nigra	Houston	0.196	0.162	**0.046**	0.324	**0.048**	**0.034**	0.094	**0.028**	0.484	0.055	0.077
Q. robur	Cape Town	0.498	0.324	0.405	0.395	0.264	0.368	**0.013**	**0.028**	0.149	**0.018**	0.320
R. pseudoacacia	Santiago de Chile	0.280	0.383	0.415	0.364	0.233	**0.019**	0.353	0.217	0.283	0.247	0.556
	Munich	0.332	0.425	0.253	0.327	0.132	0.066	0.428	0.380	0.279	0.200	0.471
	Würzburg	0.449	0.431	0.131	0.377	0.082	0.061	0.267	0.221	0.314	0.513	0.269
T. cordata	Berlin	0.336	0.402	0.426	0.464	0.488	0.466	0.413	0.500	0.372	0.373	0.426
	Munich	0.304	0.104	0.081	0.318	0.396	**0.050**	**0.020**	**0.039**	0.154	0.167	**0.050**
	Würzburg	0.074	**0.022**	0.193	0.387	0.402	0.383	0.380	0.026	0.060	0.451	0.374

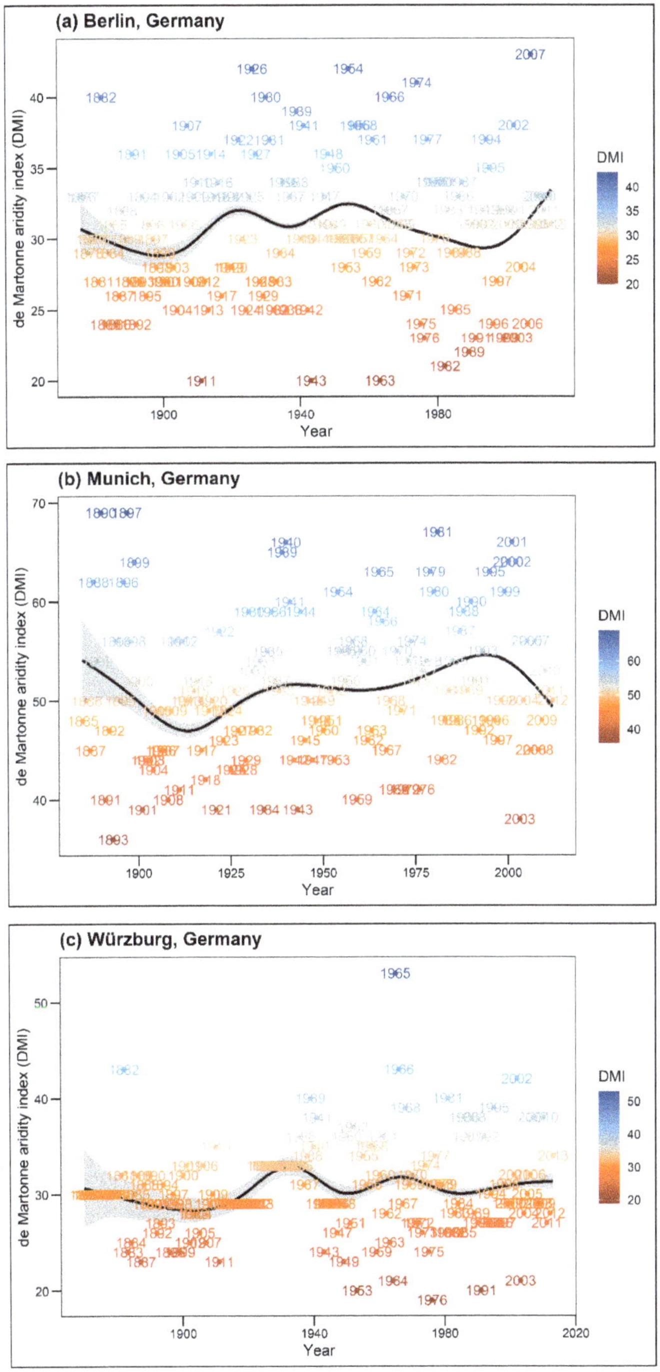

Figure A1. *Cont.*

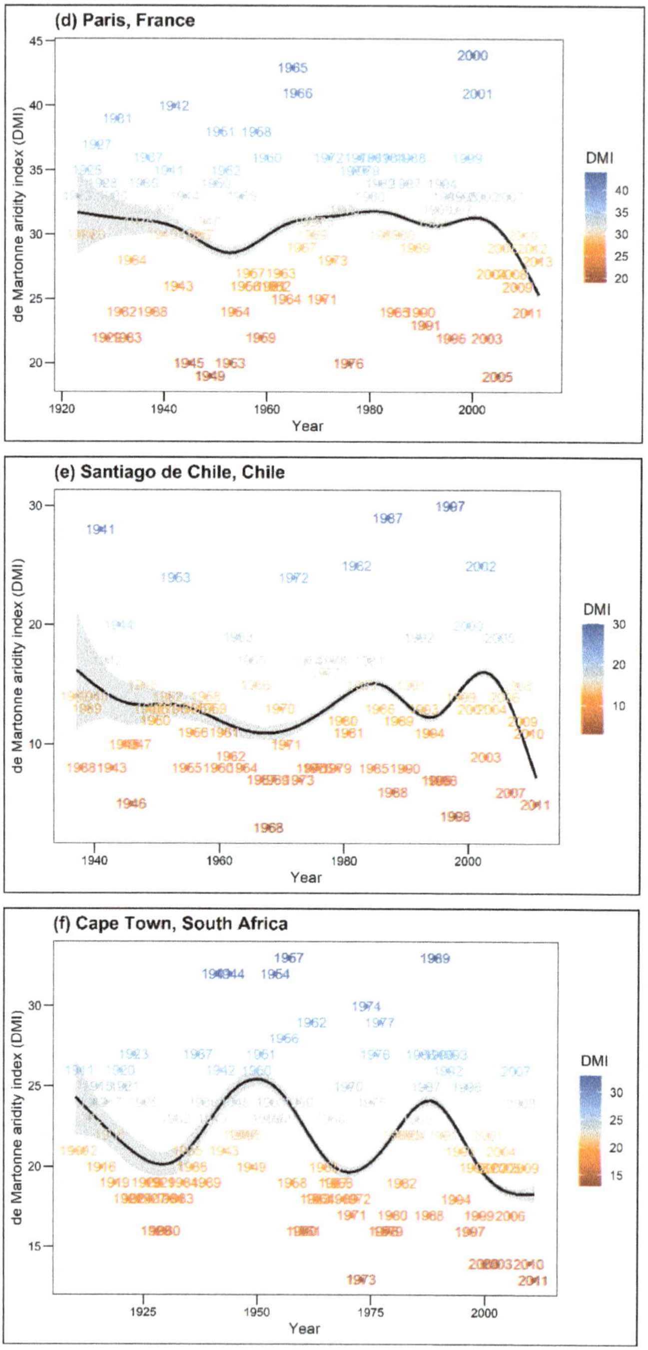

Figure A1. *Cont.*

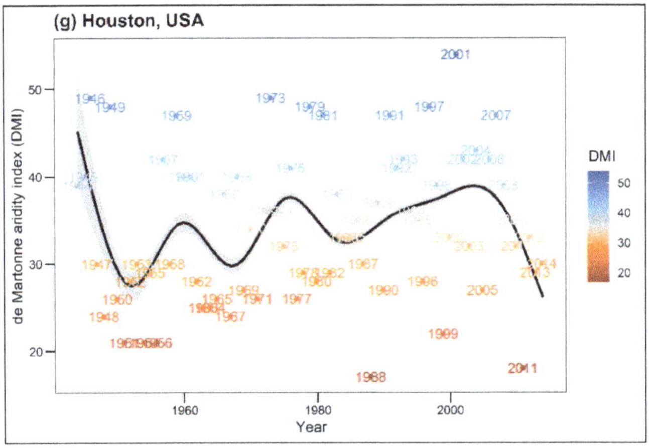

Figure A1. Annual de Martonne Indices (DMI) over the entire periods for each city. High DMI values indicate humidity, i.e., cool and moist conditions, low DMI values indicate aridity, i.e., dry and warm conditions. The black line shows the smoothed mean DMI course through the years using gam algorithm.

References

1. United Nations. *World Urbanization Prospects*; United Nations: New York, NY, USA, 2018; Volume 12, ISBN 9789211483192.
2. Roeland, S.; Moretti, M.; Amorim, J.H.; Branquinho, C.; Fares, S.; Morelli, F.; Niinemets, Ü.; Paoletti, E.; Pinho, P.; Sgrigna, G.; et al. Towards an integrative approach to evaluate the environmental ecosystem services provided by urban forest. *J. For. Res.* **2019**, *30*, 1981–1996. [CrossRef]
3. Moser, A.; Rötzer, T.; Pauleit, S.; Pretzsch, H. Structure and ecosystem services of small-leaved lime (*Tilia cordata* Mill.) and black locust (*Robinia pseudoacacia* L.) in urban environments. *Urban For. Urban Green.* **2015**, *14*, 1110–1121. [CrossRef]
4. Nowak, D.J.; Crane, D.E. Carbon storage and sequestration by urban trees in the USA. *Environ. Pollut.* **2002**, *116*, 381–389. [CrossRef]
5. Roy, S.; Byrne, J.; Pickering, C. A systematic quantitative review of urban tree benefits, costs, and assessment methods across cities in different climatic zones. *Urban For. Urban Green.* **2012**, *11*, 351–363. [CrossRef]
6. Rötzer, T.; Rahman, M.A.; Moser-Reischl, A.; Pauleit, S.; Pretzsch, H. Process based simulation of tree growth and ecosystem services of urban trees under present and future climate conditions. *Sci. Total Environ.* **2019**, *676*, 651–664. [CrossRef] [PubMed]
7. Rahman, M.A.; Stratopoulos, L.M.F.; Moser-Reischl, A.; Zölch, T.; Häberle, K.-H.; Rötzer, T.; Pretzsch, H.; Pauleit, S. Traits of trees for cooling urban heat islands: A meta-analysis. *Build. Environ.* **2020**, *170*, 106606. [CrossRef]
8. Rahman, M.A.; Hartmann, C.; Moser-Reischl, A.; von Strachwitz, M.F.; Paeth, H.; Pretzsch, H.; Pauleit, S.; Rötzer, T. Tree cooling effects and human thermal comfort under contrasting species and sites. *Agric. For. Meteorol.* **2020**, *287*, 107947. [CrossRef]
9. Rahman, M.A.; Dervishi, V.; Moser-Reischl, A.; Ludwig, F.; Pretzsch, H.; Rötzer, T.; Pauleit, S. Comparative analysis of shade and underlying surfaces on cooling effect. *Urban For. Urban Green.* **2021**, *63*, 127223. [CrossRef]
10. Berland, A.; Shiflett, S.A.; Shuster, W.D.; Garmestani, A.S.; Goddard, H.C.; Herrmann, D.L.; Hopton, M.E. The role of trees in urban stormwater management. *Landsc. Urban Plan.* **2017**, *162*, 167–177. [CrossRef]
11. MacGregor-Fors, I.; Escobar, F.; Rueda-Hernández, R.; Avendaño-Reyes, S.; Baena, M.L.; Bandala, V.M.; Chacón-Zapata, S.; Guillén-Servent, A.; González-García, F.; Lorea-Hernández, F.; et al. City "green" contributions: The role of urban greenspaces as reservoirs for biodiversity. *Forests* **2016**, *7*, 146. [CrossRef]
12. Richards, D.R.; Fung, T.K.; Belcher, R.N.; Edwards, P.J. Differential air temperature cooling performance of urban vegetation types in the tropics. *Urban For. Urban Green.* **2020**, *50*, 126651. [CrossRef]
13. Sun, L.; Chen, J.; Li, Q.; Huang, D. Dramatic uneven urbanization of large cities throughout the world in recent decades. *Nat. Commun.* **2020**, *11*, 1–9. [CrossRef] [PubMed]
14. Moser-Reischl, A.; Rahman, M.A.; Pauleit, S.; Pretzsch, H.; Rötzer, T. Growth patterns and effects of urban micro-climate on two physiologically contrasting urban tree species. *Landsc. Urban Plan.* **2019**, *183*, 88–99. [CrossRef]
15. Sand, E.; Konarska, J.; Howe, A.W.; Andersson-Sköld, Y.; Moldan, F.; Pleijel, H.; Uddling, J. Effects of ground surface permeability on the growth of urban linden trees. *Urban Ecosyst.* **2018**, *21*, 691–696. [CrossRef]
16. Brune, M. Urban trees under climate change. *Clim. Serv. Cent. Ger.* **2016**, *24*, 123.
17. Brandt, L.A.; Lewis, A.D.; Scott, L.; Darling, L.; Fahey, R.T.; Iverson, L.; Nowak, D.J.; Bodine, A.R.; Bell, A.; Still, S. *Chicago Wilderness Region Urban Forest Vulnerability Assessment and Synthesis: A Report from the Urban Forestry Climate Change Response*

Framework Chicago Wilderness Pilot Project; Gen. Tech. Rep. NRS-168; U.S. Department of Agriculture, Forest Service, Northern Research Station: Newtown Square, PA, USA, 2017; 142p.
18. Rötzer, T.; Moser-Reischl, A.; Rahman, M.A.; Hartmann, C.; Paeth, H.; Pauleit, S.; Pretzsch, H. Urban tree growth and ecosystem services under extreme drought. *Agric. For. Meteorol.* **2021**, *308–309*, 108532. [CrossRef]
19. Burley, H.; Beaumont, L.J.; Ossola, A.; Baumgartner, J.B.; Gallagher, R.; Laffan, S.; Esperon-Rodriguez, M.; Manea, A.; Leishman, M.R. Substantial declines in urban tree habitat predicted under climate change. *Sci. Total Environ.* **2019**, *685*, 451–462. [CrossRef]
20. Pretzsch, H.; Biber, P.; Uhl, E.; Dahlhausen, J.; Schütze, G.; Perkins, D.; Rötzer, T.; Caldentey, J.; Koike, T.; Van Con, T.; et al. Climate change accelerates growth of urban trees in metropolises worldwide. *Sci. Rep.* **2017**, *7*, 15403. [CrossRef]
21. Iverson, L.R.; Prasad, A.M. *Predicting Abundance of 80 Tree Species Following Climate Change in the Eastern United States*; Wiley: Hoboken, NJ, USA, 1998; Volume 68, pp. 465–485.
22. Gill, D.; Magin, G.; Bertram, E. *Trees and Climate Change. A Guide to the Factors that Influence Species Vulnerability and a Summary of Adaptation Options*; FAUNA FLORA Int.: Cambridge, UK, 2013; pp. 1–16.
23. Churkina, G.; Grote, R.; Butler, T.M.; Lawrence, M. Natural selection? Picking the right trees for urban greening. *Environ. Sci. Policy* **2015**, *47*, 12–17. [CrossRef]
24. Roloff, A.; Korn, S.; Gillner, S. The Climate-Species-Matrix to select tree species for urban habitats considering climate change. *Urban For. Urban Green.* **2009**, *8*, 295–308. [CrossRef]
25. Coombes, A.; Martin, J.; Slater, D. Defining the allometry of stem and crown diameter of urban trees. *Urban For. Urban Green.* **2019**, *44*, 126421. [CrossRef]
26. Pretzsch, H. Tree growth as affected by stem and crown structure. *Trees–Struct. Funct.* **2021**, *35*, 947–960. [CrossRef]
27. Pretzsch, H.; Biber, P.; Uhl, E.; Dahlhausen, J.; Rötzer, T.; Caldentey, J.; Koike, T.; van Con, T.; Chavanne, A.; Seifert, T.; et al. Crown size and growing space requirement of common tree species in urban centres, parks, and forests. *Urban For. Urban Green.* **2015**, *14*, 466–479. [CrossRef]
28. Pretzsch, H.; Moser-Reischl, A.; Rahman, M.A.; Pauleit, S.; Rötzer, T. Towards sustainable management of the stock and ecosystem services of urban trees. From theory to model and application. *Trees–Struct. Funct.* **2021**, 1–20. [CrossRef]
29. Poschenrieder, W.; Rötzer, T.; Biber, P.; Uhl, E.; Dervishi, V.; Pretzsch, H. Sustainable management of urban tree stocks based on multi-criteria scenario modelling. *Urban For. Urban Green.* **2022**. forthcoming.
30. Nowak, D.; Crane, D. The Urban Forest Effects (UFORE) Model: Quantifying urban forest structure and functions', in Integrated tools for natural resources inventories in the 21st century. *Integr. Tools Nat. Resour. Invent.* **2000**, *212*, 714–720.
31. Nowak, D.J.; Crane, D.E.; Stevens, J.C.; Hoehn, R.E.; Walton, J.T.; Bond, J. A ground-based method of assessing urban forest structure and ecosystem services. *Arboric. Urban For.* **2008**, *34*, 347–358. [CrossRef]
32. Tree i-Tree. Available online: https://www.itreetools.org/ (accessed on 12 July 2021).
33. Longcore, T.; Li, C.; Wilson, J.P. Applicability of citygreen urban ecosystem analysis software to a densely built urban neighborhood. *Urban Geogr.* **2004**, *25*, 173–186. [CrossRef]
34. Peng, L.; Chen, S.; Liu, Y.; Wang, J. Application of CITYgreen model in benefit assessment of Nanjing urban green space in carbon fixation and runoff reduction. *Front. For. China* **2008**, *3*, 177–182. [CrossRef]
35. Kramer, H.; Oldengarm, J. Urbtree: A Tree Growth Model for the Urban Environment. *Photogramm. Remote Sens.* **2010**, *38*, 4/C7.
36. Pauleit, S.; Zölch, T.; Hansen, R.; Randrup, T.B.; Konijnendijk van den Bosch, C. Nature-Based Solutions and Climate Change—Four Shades of Green. In *Nature-Based Solutions to Climate Change Adaptation in Urban Areas*; Kabisch, N., Korn, H., Stadler, J., Bonn, A., Eds.; Theory and Practice of Urban Sustainability Transitions; Springer: Cham, Switzerland, 2017; pp. 29–49. [CrossRef]
37. Rötzer, T.; Moser-Reischl, A.; Rahman, M.A.; Grote, R.; Pauleit, S.; Pretzsch, H. Modelling Urban Tree Growth and Ecosystem Services: Review and Perspectives. *Prog. Bot.* **2020**, *82*, 405–464. [CrossRef]
38. De Martonne, E. Une nouvelle function climatologique: L'indice d'aridite. *Meteorologie* **1926**, *2*, 449–458.
39. Moser, A.; Rötzer, T.; Pauleit, S.; Pretzsch, H. The urban environment can modify drought stress of small-leaved lime (*Tilia cordata* Mill.) and black locust (*Robinia pseudoacacia* L.). *Forests* **2016**, *7*, 71. [CrossRef]
40. NOAA. NCEI Climate Data Online. Available online: https://www.ncdc.noaa.gov/cdo-web/datasets (accessed on 1 July 2020).
41. Deutscher Wetterdienst (DWD). Open Data Server of the German Meteorological Service (DWD). Available online: https://opendata.dwd.de/climate_environment/CDC/observations_germany/climate/annual/more_precip/historical/ (accessed on 15 July 2020).
42. González-Reyes, Á. Ocurrencia de eventos de sequías en la ciudad de Santiago de Chile desde mediados del siglo XIX. *Rev. Geogr. Norte Gd.* **2016**, *32*, 21–32. [CrossRef]
43. Peel, M.C.; Finlayson, B.L.; McMahon, T.A. Updated world map of the Köppen-Geiger climate classification. *Hydrol. Earth Syst. Sci.* **2007**, *11*, 1633–1644. [CrossRef]
44. Baltas, E. Spatial distribution of climatic indices in northern Greece. *Meteorol. Appl.* **2007**, *14*, 69–78. [CrossRef]
45. Niinemets, Ü.; Valladares, F. Tolerance to shade, drought, and waterlogging of temperate northern hemisphere trees and shrubs. *Ecol. Monogr.* **2006**, *76*, 521–547. [CrossRef]
46. Roloff, A. *Bäume in der Stadt: Besonderheiten, Funktion, Nutzen, Arten, Risiken*; Ulmer Eugen Verlag: Stuttgart, Germany, 2013.

47. Eaton, E.; Caudullo, G.; Oliveira, S.; de Rigo, D. *Quercus robur* and *Quercus petraea* in Europe: Distribution, habitat, usage and threats. In *European Atlas of Forest Tree Species*; San-Miguel-Ayanz, J., Rigo, D.d., Caudullo, G., Durrant, T.H., Mauri, A., Eds.; Publication Office of the European Union: Luxembourg, 2016; pp. 160–163.
48. Sitzia, T.; Cierjacks, A.; de Rigo, D.; Caudullo, G. *Robinia pseudoacacia* in Europe: Distribution, habitat, usage and threats. In *European Atlas of Forest Tree Species*; Publication Office of the European Union: Luxembourg, 2016; pp. 166–167.
49. Radoglou, K.; Dobrowolska, D.; Spyroglou, G.; Nicolescu, V.N. A review on the ecology and silviculture of limes: (*Tilia cordata* Mill., *Tilia platyphyllos* Scop, and *Tilia tomentosa* Moench.) in Europe. *Bodenkultur* **2009**, *60*, 9–19.
50. FNA Flora of North America. Available online: http://floranorthamerica.org/Main_Page (accessed on 21 July 2021).
51. Martin, G.D. South African Journal of Botany Addressing geographical bias: A review of *Robinia pseudoacacia* (black locust) in the Southern Hemisphere. *South Afr. J. Bot.* **2019**, *125*, 481–492. [CrossRef]
52. Biritz GmbH. Digitalpositiometer Typ 2. Available online: https://www.biritz.at/digitalpositiometer/digitalpositiometer-typ-2/ (accessed on 13 April 2022).
53. Bunn, A.; Korpela, M.; Biondi, F.; Campelo, F.; Mérian, P.; Qeadan, F.; Zang, C. R Package Version 1.7.2. dplR: Dendrochronology Program Library in R. 2021. Available online: https://CRAN.R-project.org/package=dplR (accessed on 27 August 2021).
54. R Core Team R: A Language and Environment for Statistical Computing. 2020. Available online: https://www.R-project.org/ (accessed on 21 July 2021).
55. Bates, D.; Maechler, M.; Bolker, B.; Walker, S. Lme4: Linear Mixed-Effects Models Using Eigen and S4. Available online: https://cran.r-project.org/web/packages/lme4/index.html (accessed on 9 August 2021).
56. Russell, V.L. R Package Version 1.7.2. Emmeans: Estimated Marginal Means, aka Least-Squares Means. 2022. Available online: https://CRAN.R-project.org/package=emmeans (accessed on 2 January 2022).
57. Lüdecke, D. R Package Version 2.8.10. sjPlot: Data Visualization for Statistics in Social Science. 2021. Available online: https://CRAN.R-project.org/package=sjPlot (accessed on 15 September 2021).
58. Lough, J.M.; Fritts, H.C. An assessment of the possible effects of volcanic eruptions on North American climate using tree-ring data, 1602 to 1900 A.D. *Clim. Chang.* **1987**, *10*, 219–239. [CrossRef]
59. Bastin, J.F.; Clark, E.; Elliott, T.; Hart, S.; van den Hoogen, J.; Hordijki, I.; Ma, H.; Majumder, S.; Manoli, G.; Maschler, J.; et al. Understanding climate change from a global analysis of city analogues. *PLoS ONE* **2019**, *14*, e0217592. [CrossRef]
60. IPCC. *Assessment Report 6 Climate Change 2021: The Physical Science Basis*; IPCC: Geneva, Switzerland, 2021.
61. Klein, T. The variability of stomatal sensitivity to leaf water potential across tree species indicates a continuum between isohydric and anisohydric behaviours. *Funct. Ecol.* **2014**, *28*, 1313–1320. [CrossRef]
62. Minucci, J.M.; Miniat, C.F.; Teskey, R.O.; Wurzburger, N. Tolerance or avoidance: Drought frequency determines the response of an N2-fixing tree. *New Phytol.* **2017**, *215*, 434–442. [CrossRef] [PubMed]
63. Dyderski, M.K.; Paź, S.; Frelich, L.E.; Jagodziński, A.M. How much does climate change threaten European forest tree species distributions? *Glob. Chang. Biol.* **2018**, *24*, 1150–1163. [CrossRef] [PubMed]
64. Puchałka, R.; Dyderski, M.K.; Vítková, M.; Sádlo, J.; Klisz, M.; Netsvetov, M.; Prokopuk, Y.; Matisons, R.; Mionskowski, M.; Wojda, T.; et al. Black locust (*Robinia pseudoacacia* L.) range contraction and expansion in Europe under changing climate. *Glob. Chang. Biol.* **2021**, *27*, 1587–1600. [CrossRef].
65. Klisz, M.; Puchałka, R.; Netsvetov, M.; Prokopuk, Y.; Vítková, M.; Sádlo, J.; Matisons, R.; Mionskowski, M.; Chakraborty, D.; Olszewski, P.; et al. Variability in climate-growth reaction of *Robinia pseudoacacia* in Eastern Europe indicates potential for acclimatisation to future climate. *For. Ecol. Manage* **2021**, *492*, 119194. [CrossRef]
66. Mantovani, D.; Veste, M.; Freese, D. Black locust (*Robinia pseudoacacia* L.) ecophysiological and morphological adaptations to drought and their consequence on biomass production and water-use efficiency. *New Zealand J. For. Sci.* **2014**, *44*, 1–11. [CrossRef]
67. Moser-Reischl, A.; Rötzer, T.; Pauleit, S.; Pretzsch, H. Urban Tree Growth Characteristics of Four Common Species in South Germany. *Arboric. Urban For.* **2021**, *47*, 150–169. [CrossRef]
68. Swoczyna, T.; Kalaji, H.M.; Pietkiewicz, S.; Borowski, J.; Zaraś-Januszkiewicz, E. Photosynthetic apparatus efficiency of eight tree taxa as an indicator of their tolerance to urban environments. *Dendrobiology* **2010**, *63*, 65–75.
69. Nebesnyi, V.B.; Grodzinskaya, A.A.; Gonchar, A.Y. The use of *Tilia cordata* Mill. as bioindicator for the evaluation of the ecological state of Kyiv urbanized areas (Ukraine). *J. Med. Plants Stud.* **2016**, *4*, 277–282.
70. Day, S.D.; Wiseman, P.E.; Dickinson, S.B.; Harris, J.R. Tree Root Ecology in the Urban Environment and Implications for a Sustainable Tree Root Ecology in the Urban Environment and Implications for a Sustainable. *Rhizosphere* **2010**, *36*, 193–205. [CrossRef]
71. Canham, C.D.; Murphy, L.; Riemann, R.; McCullough, R.; Burrill, E. Local differentiation in tree growth responses to climate. *Ecosphere* **2018**, *9*, 9. [CrossRef]
72. Broadmeadow, M.S.J.; Ray, D.; Samuel, C.J.A. Climate change and the future for broadleaved tree species in Britain. *Forestry* **2005**, *78*, 145–161. [CrossRef]
73. Gillner, S.; Bräuning, A.; Roloff, A. Dendrochronological analysis of urban trees: Climatic response and impact of drought on frequently used tree species. *Trees–Struct. Funct.* **2014**, *28*, 1079–1093. [CrossRef]
74. Gillner, S.; Korn, S.; Roloff, A. Leaf-gas exchange of five tree species at urban street sites. *Arboric. Urban For.* **2015**, *41*, 113–124. [CrossRef]

75. Gregorová, B.; Černý, K.; Holub, V.; Strnadová, V. Effects of climatic factors and air pollution on damage of London plane Effects of climatic factors and air pollution on damage of London plane (*Platanus hispanica* Mill.). *Hortic. Sci.* **2010**, *37*, 109–117. [CrossRef]
76. Cedro, A.; Nowak, G. Effects of climatic conditions on annual tree ring growth of the *Platanus* × *hispanica 'Acerifolia'* under urban conditions of Szczecin. *Dendrobiology* **2006**, *55*, 11–17.
77. Nola, P.; Bracco, F.; Assini, S.; von Arx, G.; Castagneri, D. Xylem anatomy of *Robinia pseudoacacia* L. and *Quercus robur* L. is differently affected by climate in a temperate alluvial forest. *Ann. For. Sci.* **2020**, *77*, 12–16. [CrossRef]
78. Moser, A.; Rahman, M.A.; Pretzsch, H.; Pauleit, S.; Rötzer, T. Inter- and intraannual growth patterns of urban small-leaved lime (*Tilia cordata* Mill.) at two public squares with contrasting microclimatic conditions. *Int. J. Biometeorol.* **2017**, *61*, 1095–1107. [CrossRef]
79. Moser, A.; Uhl, E.; Rötzer, T.; Biber, P.; Dahlhausen, J.; Lefer, B.; Pretzsch, H. Effects of Climate and the Urban Heat Island Effect on Urban Tree Growth in Houston. *Open J. For.* **2017**, *07*, 428–445. [CrossRef]
80. Maurin, V.; DesRochers, A. Physiological and growth responses to pruning season and intensity of hybrid poplar. *For. Ecol. Manage* **2013**, *304*, 399–406. [CrossRef]
81. Augé, R.M.; Stodola, A.J.W.; Tims, J.E.; Saxton, A.M. Moisture retention properties of a mycorrhizal soil. *Plant Soil* **2001**, *230*, 87–97. [CrossRef]
82. Kadowaki, K.; Yamamoto, S.; Sato, H.; Tanabe, A.S.; Hidaka, A.; Toju, H. Mycorrhizal fungi mediate the direction and strength of plant–soil feedbacks differently between arbuscular mycorrhizal and ectomycorrhizal communities. *Commun. Biol.* **2018**, *1*, 1–5. [CrossRef] [PubMed]
83. Rusterholz, H.P.; Studer, M.; Zwahlen, V.; Baur, B. Plant-mycorrhiza association in urban forests: Effects of the degree of urbanisation and forest size on the performance of sycamore (*Acer pseudoplatanus*) saplings. *Urban For. Urban Green.* **2020**, *56*, 126872. [CrossRef]

forests

Article

Does Vertical Greening Really Play Such a Big Role in an Indoor Thermal Environment?

Jiayu Li and Bohong Zheng *

School of Architecture and Art, Central South University, Changsha 410083, China; J.Y.Li@csu.edu.cn
* Correspondence: Zhengbohong@csu.edu.cn

Abstract: Little attention has been paid to indoor cooling compared with the surface cooling of vertical greening. The few studies on the indoor cooling of vertical greening are almost all conducted in a hot climate area with windowless building models, which is suspected to exaggerate the role of vertical greening in an indoor thermal environment. Through two improvements, this paper explored the realistic impact of vertical greening on an indoor thermal environment. First, we built models according to the actual window-to-wall ratio rather than a fictitious model without windows. Second, an annual cycle evaluation, considering both hot summer and cold winter, was used to replace the typical hot day. With the support of Envi-met and Kriging models, the results revealed that the existing research not only exaggerated vertical greening's positive effects on an indoor thermal environment in hot seasons but also ignored its potential harms to thermal perception in cold seasons. These exaggerated results could easily cause the abuse of vertical greening in cities. In actual windowed buildings, the role of vertical greening in indoor temperatures is not always positive, and the positive effect is not as strong as previous studies suggest.

Keywords: vertical greening; indoor thermal environment; annual cycle; windowed model

Citation: Li, J.; Zheng, B. Does Vertical Greening Really Play Such a Big Role in an Indoor Thermal Environment? *Forests* **2022**, *13*, 358. https://doi.org/10.3390/f13020358

Academic Editors: Thomas Rötzer, Stephan Pauleit, Mohammad A Rahman and Astrid Reischl

Received: 20 January 2022
Accepted: 18 February 2022
Published: 20 February 2022

Publisher's Note: MDPI stays neutral with regard to jurisdictional claims in published maps and institutional affiliations.

1. Introduction

Since the twentieth century, investigations on urban greening have substantially increased [1–3]. Among those studies, more and more attention has been paid to the use of vertical greening, because building facades occupy a higher and higher proportion of the city [4–6]. Vertical greening has been confirmed as an effective contributor to cooling in the built environment [7–9]. Although the wall-surface temperature and indoor air temperature are closely linked [10,11], regarding vertical greening, the majority of concern is given to wall-surface cooling rather than to indoor air cooling [12–14]. Only a few studies have investigated indoor air cooling of vertical greening [15,16]. Olivieri et al. pointed out that vegetal façades reduced a building's average interior air temperature by 4 °C [17]. Similarly, another experiment also claimed that an indoor mean air temperature decrease of 4 °C was achieved, with a maximum decrement of 6 °C observed [18]. In Coma's study, the mean reduction of indoor air temperature was only 1 °C [19]. In addition, Haggag et al. measured indoor air temperatures during the hottest month of July in a hot and arid climate zone, where the green facades maintained a reduction of 5 °C in indoor air temperature [20]. Existing studies have concluded that room size and weather conditions are two factors affecting the cooling performance of vertical greening [21], and another experiment with a similar room size and climatic conditions only reached a maximum indoor cooling of 1.5 °C [22]. Comparing the settings, in Chen's experiment [22], the building model had windows, but for the other studies mentioned above, they were all conducted in windowless buildings. Glasses have a higher solar heat gain coefficient (SHGC) and a higher coefficient of heat conductivity [23,24], which is the main channel of heat exchange. Besides this, previous investigations have found that the cooling effect of vertical greening in summer was stronger than of that in winter [25,26]. Almost all of the existing articles about indoor cooling, in terms of vertical greening, choose a typical summer day to evaluate [27,28]. These deficiencies all contribute to the exaggeration of vertical greening's performance on indoor cooling. These magnified performances could lead to the abuse of vertical greening.

This research aimed to pragmatically evaluate the indoor cooling performance of vertical greening. Compared with the previous studies, two improvements were made in this research. First, this study no longer took the typical summer day as the research cycle, and the research area was not limited to the tropical region; instead, the annual evaluation cycle was used to explore vertical greening in hot summer and cold winter climate zones. Second, this research abandoned the windowless model and constructed the research model according to the actual window-to-wall ratio.

The typical building model in Changsha, China, was built according to the actual layout of the urban environment. Subsequently, the Envi-met model was employed as the research tool after the field measurement was conducted to validate its accuracy [29,30]. The average meteorological data of the 12 months in 2020 were reviewed and used as the boundary conditions. Finally, the annual distributions of indoor temperature for five vertical greening scenarios were fitted with the Kriging model [31].

2. Materials and Methods

2.1. Analytical Scenarios

The study was carried out in Changsha, located at 28.19° N, 113.22° E, in the subtropical monsoon climate zone. Changsha has become one of the "four furnaces" in China [32], whose mean temperature ranges from 16.8 to 17.3 °C. Changsha experiences the lowest temperature between 4.4 and 5.1 °C in January, while the highest temperature above 30 °C happens in July [33]. The extremely hot summer brings threats to energy consumption and citizens' health [34,35]. Cooling the city, therefore, is an urgent issue in climate-responsive urban planning [36]. Urban planners concentrate on the total amount of greening in the form of indicators, including vertical greening ratio and ground greening ratio. Meanwhile, ground grass has no shading effect on indoor thermal environments compared with vertical greening, which is essential to indoor cooling [37]. Therefore, this model explored the changes of an indoor thermal environment during the transformation from ground greening to vertical greening when the total greening area was fixed. Considering the mean window-to-wall ratios of Changsha, we adopted a 20% window-to-wall ratio in this research [38]. The research model covered an area of 7744 m^2 and was composed of 22 × 22 grids. The buildings were 20 m high and 8 m wide. The research model is presented in Figure 1, where V is the ratio of vertical greening, and G represents the ratio of ground greening.

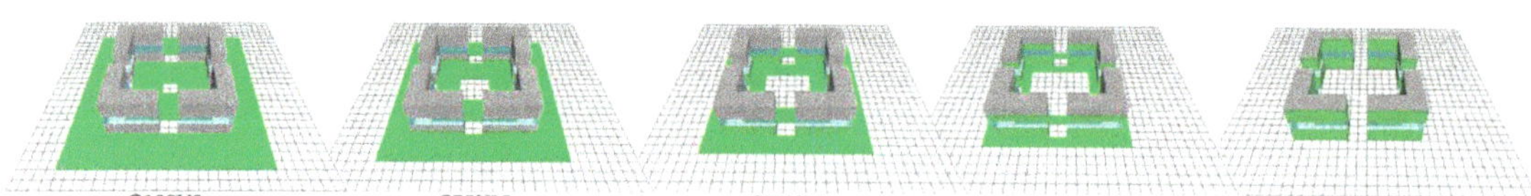

Figure 1. Research models. G represents the ratio of ground greening to total greening, and V represents the ratio of vertical greening to total greening.

The area of the wall in each model was 7680 m^2, where the window area was 1536 m^2. The areas of each wall component in every scenario are displayed in Table 1.

In this research, we adopted grass as the plant of ground greening and took modular greening as the vertical greening. The attributes of ground greening and vertical greening are listed in Table 2.

Table 1. The areas of each wall component.

Scenarios	Ground Greening (m^2)	Vertical Greening (m^2)	Window (m^2)
G100V0	6144	0	1536
G75V25	4608	1536	1536
G50V50	3072	3072	1536
G25V75	1536	4608	1536
G0V100	0	6144	1536

Table 2. The attributes of ground greening and vertical greening.

Greening Type	Elements	Parameters	Value
Ground greening	Plant	Leaf type	Grass
		Albedo	0.20
		Transmittance (Frac)	0.30
		Plant height (m)	0.25
		Root zone depth (m)	0.20
		LAD	0.30
		RAD	0.10
Vertical greening	Plant	LAI (m^2/m^2)	1.5
		Leaf angle distribution	0.5
	Substrate	Emissivity of Substrate (Frac)	0.95
		Albedo of Substrate (Frac)	0.30
		Water Coefficient of Substrate for plant	0.50
		Air Gap between Substrate and wall (m)	0.0

Clear float glass is commonly used in Changsha's residential buildings. Here, we employed clear float glass as the material of the window in this research. The parameters of the clear float glass are shown in Table 3 [39].

Table 3. The attributes of clear float glass.

Parameters	Value
Thickness (mm)	20.00
Absorption (Frac)	0.05
Transmission (Frac)	0.90
Solar heat gain coefficient	0.80
Reflection (Frac)	0.05
Emissivity (Frac)	0.90
Specific Heat (J/(kg·K)	750.00
Thermal conductivity (W/(m·K)	1.05
Density (kg/m^3)	2500.00

The walls of the buildings had a thickness of 310 mm, and were composed of concrete, with a roughness length of 0.02. The attributes of concrete are presented in Table 4 [40].

Table 4. The attributes of the concrete in this simulation.

Parameters	Value
Absorption	0.50
Transmission (Frac)	0.00
Reflection (Frac)	0.50
Emissivity (Frac)	0.90
Specific heat (J/kg*k)	0.90
Thermal conductivity (W/m*k)	1.60
Density (kg/m^3)	2220.00

2.2. Simulation Tools

The Envi-met model was employed as the research tool in this research. Unlike other simulation tools, Envi-met can model vegetated surfaces and their substrates, including heat and moisture transfer, as well as their interactions with the building [41]. It calculates the leaf temperature individually for each model grid box and can also simulate the photosynthetic rate, the evapotranspiration rate, and the water availability in the soil, all of which make Envi-met preferable to the other simulation tools in the simulation of facade greening [42]. Envi-met derives the indoor air temperature from the heat convection on the interior surface of the associated walls and roofs and the energy transmitted through the transparent glass [29]. The indoor temperature is calculated by Equation (1) [43].

$$T_i^* = T_i + \frac{1}{C_p V} \int_{e=1}^{E} A(e)\big(Q_{sw}^{tr}(e) + h_{c,i}(T_3^*(e) - T_i)\big)dt \quad (1)$$

In the formula, T_i means the original air temperature in zone i, V is the volume of zone i, C_p represents the specific heat capacity of air, and T_i^* indicates the updated air temperature after time dt. E shows the number of façades in zone i, and $A(e)$ is the surface area of zone i. Q_{sw}^{tr} indicates the shortwave radiation transmitted into zone i though the transparent façade e, and $h_{c,i}$ means the heat convection coefficient between the inner walls and ambient air.

Many documents have validated the accuracy of the Envi-met model [44,45]. For example, it was reported that the square correlation coefficients (R^2) of measurement and simulation were between 0.52 and 0.97 for all the variables, and those of air temperature ranged from 0.91 to 0.97 [46]. This study also conducted a survey to confirm the accuracy of the Envi-met model in terms of simulating the effects of greening on indoor temperature.

The validation survey was conducted in Changsha (113.109° E, 28.235° N) from 22 to 24 August 2020. A HOBO Data Logger (MX2302), which is a compact, battery-powered device equipped with an internal microprocessor, data storage, and sensors, was employed to record the temperatures. The field experiment is shown in Figure 2A, where two HOBO Data Loggers were set to record the air temperature both inside and outside the vertical greening building. The Envi-met model of the experiment is presented in Figure 2B. The variation of the outdoor air temperature recorded by the Data Logger is presented in Figure 3A, and the indoor air temperatures as measured and simulated are shown in Figure 3B.

Statistical results revealed that the Pearson coefficient of the simulated and measured indoor air temperatures was 0.969, which confirmed that the Envi-met model was reliable for this research.

(A) (B)

Figure 2. Measured and simulated models: (**A**) measurement of indoor and outdoor air temperatures; (**B**) simulation model of the measured building.

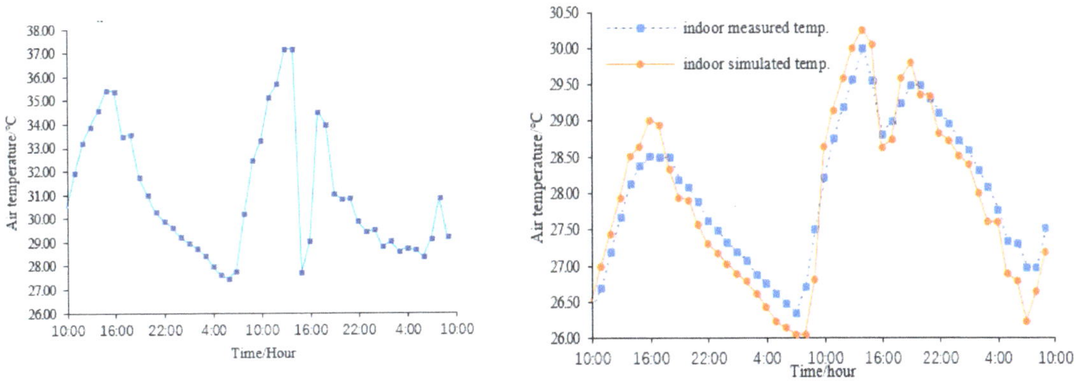

(A) (B)

Figure 3. Variations of outdoor temperatures and indoor air temperatures: (**A**) variation of outdoor air temperatures; (**B**) measured indoor air temperatures and the simulated indoor air temperature.

2.3. Data Resources

Meteorological data applied in this research were obtained from the Changsha historical weather website (http://lishi.tianqi.com/changsha/, accessed on 20 April 2021). The annual temperatures of 2020 indicated that August was the hottest month of Changsha. The average meteorological data within August at each hour in 2020 were intensively studied. Analytical results showed that the mean high air temperature in August was 35 °C, and that of the average low air temperature was 26 °C. The mean wind velocity was 3.5 m/s, and the relative humidity ranged from 78% to 88%. Similarly, the remaining 11 months were also intensively reviewed. The averages of the meteorological data of each month are shown in Table 5 and are used as the boundary conditions in the following simulations.

Table 5. The averages of the meteorological data in every month of 2020.

Month	Initial Meteorological Condition	Data	Month	Initial Meteorological Condition	Data
January	Wind velocity (m/s) (10 m off the ground)	3	February	Wind velocity (m/s) (10 m off the ground)	3
	Wind angle (°)	315		Wind angle (°)	135
	Mean max—air temperature (°C)	8		Mean max—air temperature (°C)	14
	Mean min—air temperature (°C)	4		Mean min—air temperature (°C)	8
	Mean max—relative humidity (%)	95		Mean max—relative humidity (%)	91
	Mean min—relative humidity (%)	85		Mean min—relative humidity (%)	81
March	Wind velocity (m/s) (10 m off the ground)	3	April	Wind velocity (m/s) (10 m off the ground)	3
	Wind angle (°)	315		Wind angle (°)	315
	Mean max—air temperature (°C)	17		Mean max—air temperature (°C)	22
	Mean min—air temperature (°C)	11		Mean min—air temperature (°C)	13
	Mean max—relative humidity (%)	88		Mean max—relative humidity (%)	89
	Mean min—relative humidity (%)	78		Mean min—relative humidity (%)	79
May	Wind velocity (m/s) (10 m off the ground)	3	June	Wind velocity (m/s) (10 m off the ground)	3.5
	Wind angle (°)	315		Wind angle (°)	315
	Mean max—air temperature (°C)	28		Mean max—air temperature (°C)	31
	Mean min—air temperature (°C)	20		Mean min—air temperature (°C)	25
	Mean max—relative humidity (%)	88		Mean max—relative humidity (%)	89
	Mean min—relative humidity (%)	78		Mean min—relative humidity (%)	79
July	Wind velocity (m/s) (10 m off the ground)	3	August	Wind velocity (m/s) (10 m off the ground)	3.5
	Wind angle (°)	270		Wind angle (°)	135
	Mean max—air temperature (°C)	32		Mean max—air temperature (°C)	35
	Mean min—air temperature (°C)	26		Mean min—air temperature (°C)	26
	Mean max—relative humidity (%)	80		Mean max—relative humidity (%)	88
	Mean min—relative humidity (%)	70		Mean min—relative humidity (%)	78
September	Wind velocity (m/s) (10 m off the ground)	3	October	Wind velocity (m/s) (10 m off the ground)	3.5
	Wind angle (°)	0		Wind angle (°)	315
	Mean max—air temperature (°C)	26		Mean max—air temperature (°C)	21
	Mean min—air temperature (°C)	20		Mean min—air temperature (°C)	14
	Mean max—relative humidity (%)	90		Mean max—relative humidity (%)	88
	Mean min—relative humidity (%)	80		Mean min—relative humidity (%)	78
November	Wind velocity (m/s) (10 m off the ground)	3.5	December	Wind velocity (m/s) (10 m off the ground)	3
	Wind angle (°)	0		Wind angle (°)	315
	Mean max—air temperature (°C)	18		Mean max—air temperature (°C)	10
	Mean min—air temperature (°C)	11		Mean min—air temperature (°C)	4
	Mean max—relative humidity (%)	87		Mean max—relative humidity (%)	87
	Mean min—relative humidity (%)	77		Mean min—relative humidity (%)	77

In addition, the other basic settings of the simulation are shown in Table 6.

Table 6. The other basic settings of the simulation.

Location	Data
Longitude	113.109° E
Latitude	28.235° N
Beginning time and total simulation time	Data
Beginning time	8:00 am
Total simulation time	24 h

3. Results and Discussion

According to the average meteorological data of the 12 months in 2020, 60 (5 × 12) models were simulated, where 5 indicated the 5 greening scenarios and 12 meant the 12 months. In total, 1440 (24 × 5 × 12) indoor air temperatures were collected. A group of the simulated models of the five scenarios is shown in Figure 4, representing the indoor air temperatures at 6:00 in August 2020.

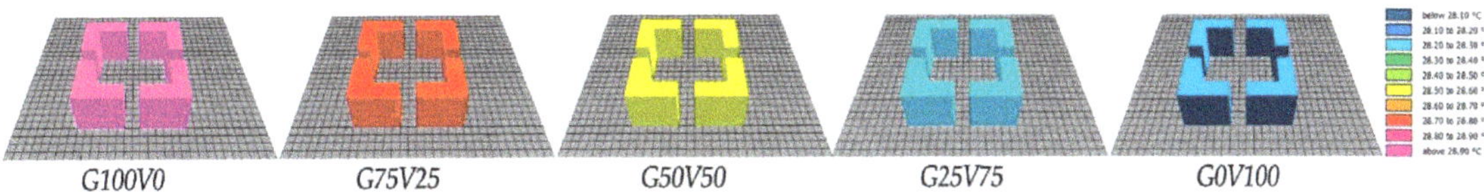

Figure 4. Example of the simulated results. G represents the ratio of ground greening to the total greening, and V represents the ratio of vertical greening to the total greening.

3.1. Cooling Differences of Greening Scenarios in Each Month

Based on the average meteorological data of the 12 months, 60 (5 × 12) scenarios were simulated. The indoor daily temperatures of each month were analyzed one by one and are presented in Figure 5. In Figure 5, the five lines in each subgraph correspond to the five greening scenarios.

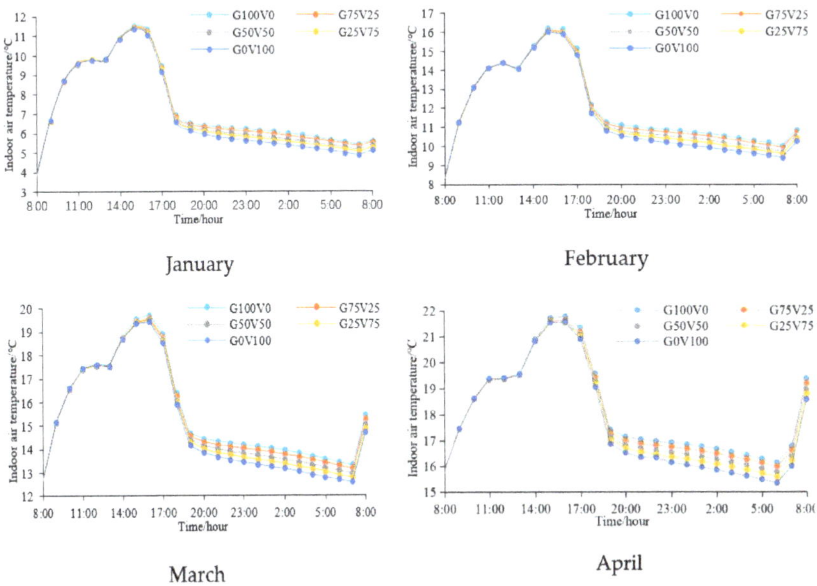

Figure 5. *Cont.*

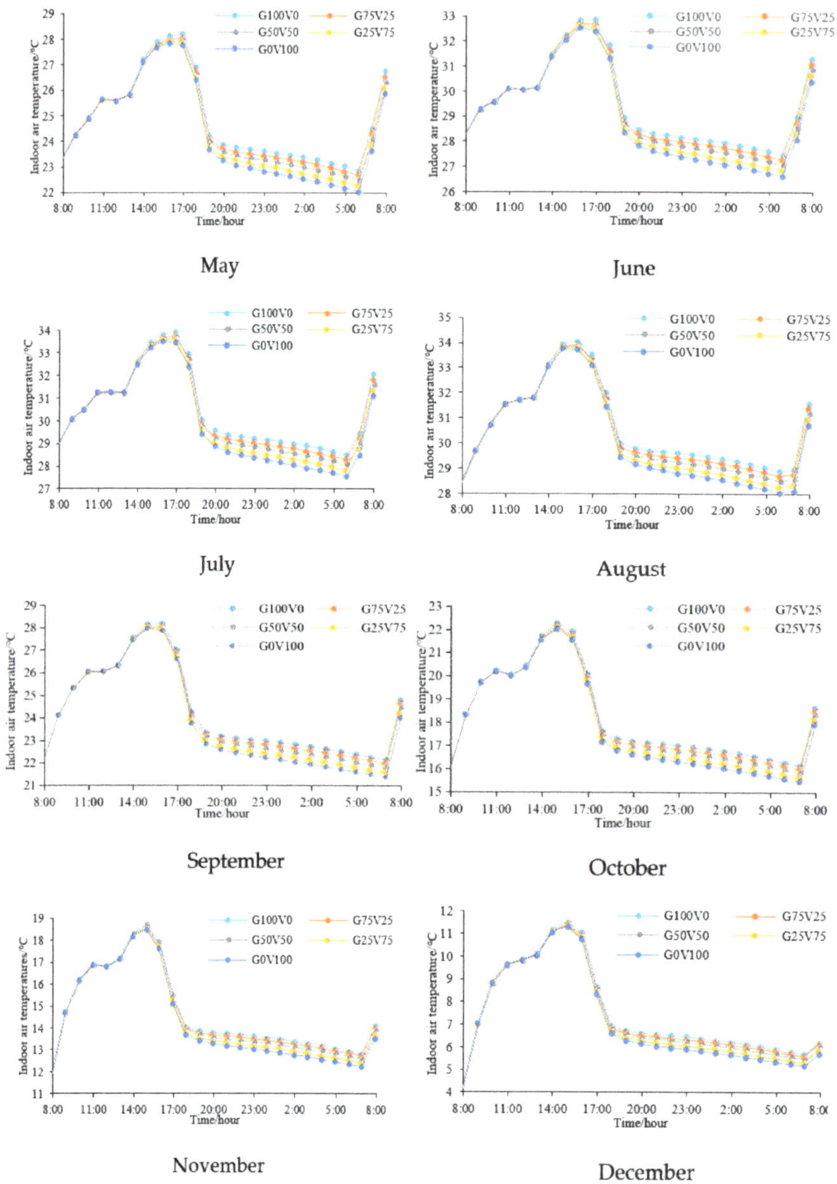

Figure 5. Daily variations of simulated indoor air temperature in the 12 months.

The variations in the indoor air temperatures in the 12 months under the five greening scenarios illustrate that vertical greening cools the indoor air temperature, as the higher the vertical greening rate, the lower the indoor temperature throughout the year. Besides this, the indoor cooling of vertical greening at night was greater than that of the daytime. In addition, no matter which month, the maximum indoor cooling of vertical greening occurred at 6:00. The specific mean indoor air temperature of the average day in each month under the five greening scenarios is shown in Table 7.

Table 7. Mean indoor air temperatures under the five greening scenarios of the 12 months.

Month / Scenario	G100V0 (°C)	G75V25 (°C)	G50V50 (°C)	G25V75 (°C)	G0V100 (°C)
January	7.29	7.21	7.12	7.03	6.93
February	11.99	11.89	11.79	11.69	11.58
March	15.51	15.38	15.27	15.16	15.04
April	18.11	17.98	17.86	17.75	17.63
May	24.86	24.72	24.60	24.48	24.35
June	29.47	29.32	29.20	29.08	28.94
July	30.47	30.31	30.18	30.05	29.91
August	30.61	30.48	30.36	30.24	30.11
September	24.24	24.12	24.01	23.89	23.77
October	18.22	18.11	18.00	17.89	17.77
November	14.69	14.60	14.50	14.41	14.30
December	7.47	7.39	7.31	7.22	7.13

Statistical results revealed that the max-cooling effects of the five greening scenarios in the 12 months were 0.36 °C (January), 0.42 °C (February), 0.47 °C (March), 0.48 °C (April), 0.52 °C (May), 0.52 °C (June), 0.56 °C (July), 0.51 °C (August), 0.47 °C (September), 0.45 °C (October), 0.38 °C (November), and 0.34 °C (December). The analytical results verified that high outdoor temperature improved the indoor cooling performance of vertical greening, which means that, in the previous studies, the cooling performances of vertical greening on a typical summer day represented its best performance rather than the average cooling performance throughout the year. This validates the idea that selecting a typical summer day as the evaluation cycle exaggerates the cooling performance of vertical greening.

Besides this, even in the hottest season, under the windowed model, the reduction of indoor air temperature caused by vertical greening was still much lower than those results of 4 °C in the previous studies with windowless models [17,18].

3.2. Daily Distribution of an Indoor Thermal Environment around the Year

In each greening scenario, the average indoor air temperatures in the 12 months were analyzed, but the monthly mean indoor air temperature hardly indicated the successive impacts of vertical greening on an indoor thermal distribution throughout the year. Here, the Kriging model was employed as the translation tool to quantify indoor thermal improvements of the five vertical greenings. The Kriging model has been confirmed feasible for temperature interpolation [47,48]. The Kriging model used in this research was supported by Surfer software [49]. With the help of the Kriging model, the continuous variations of indoor air temperatures throughout the year are drawn in Figure 6A–E, respectively, indicate the annual indoor temperature distributions of the G100V0, G75V25, G50V50, G25V75, and G0V100 scenarios. For each subgraph in Figure 6, the X coordinate means the months, and the Y coordinate represents the daily time, with the Z value indicating the indoor air temperature at that moment. The Z values were calculated based on the monthly average meteorological data at the corresponding moment. For example, the indoor air temperature at 8:00 in January was simulated with the average meteorological data at 8:00 of the 31 days in January 2020. In this way, a total of 1440 (24 h × 12 months × 5 scenarios) items of indoor air temperature were calculated. Besides this, the improvement of thermal perception was an important index to evaluate the role of vertical greening. Referring to the previous studies [50–52], the research took 8, 12, 19, 26, 30 and 34 °C as the perception boundaries of indoor temperature to quantify the improvements of vertical greening on indoor thermal perception. The thermal perception standard was very cold (below 8 °C), cold (8–12 °C), slightly cold (12–19 °C), comfortable (19–26 °C), slightly hot (26–30 °C), hot (30–34 °C), and very hot (above 34 °C).

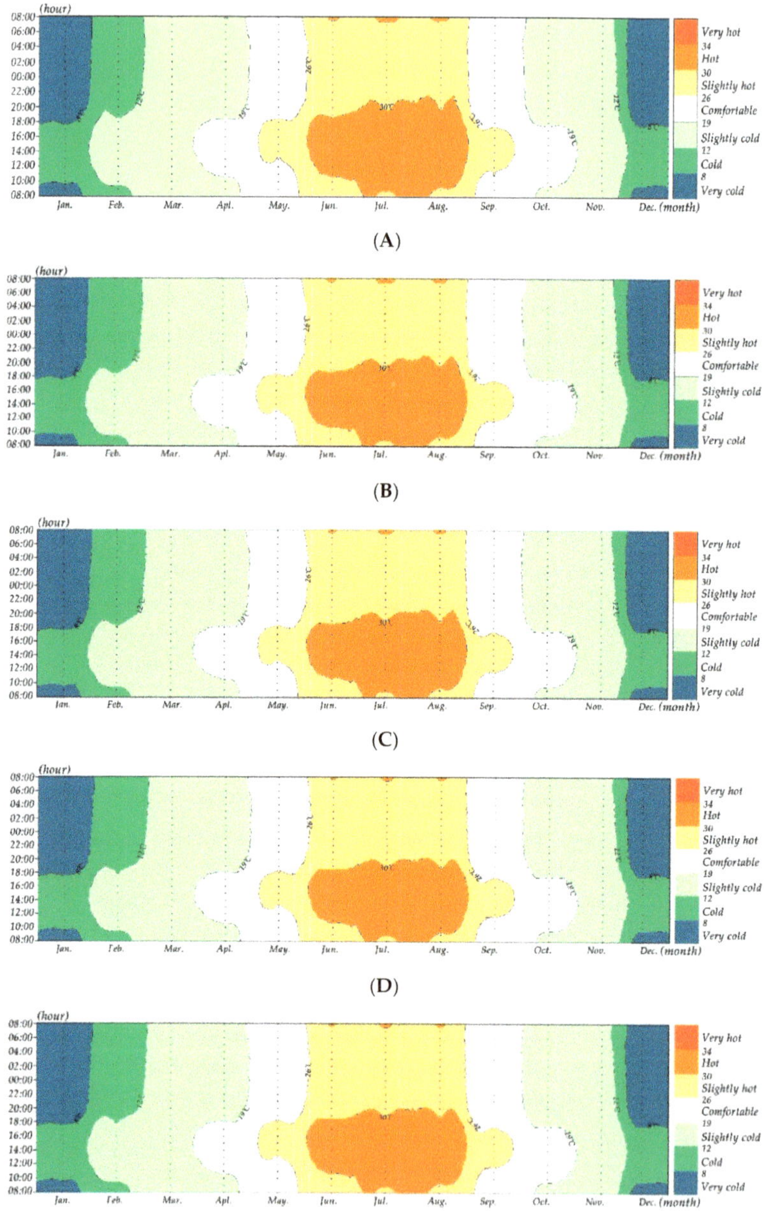

(E)

Figure 6. Improvements caused by vertical greening on annual thermal perceptions. (**A**) Annual indoor thermal perceptions of the G100V0 scenario. (**B**) Annual indoor thermal perceptions of the G75V25 scenario. (**C**) Annual indoor thermal perceptions of the G50V50 scenario. (**D**) Annual indoor thermal perceptions of the G25V75 scenario. (**E**) Annual indoor thermal perceptions of the G0V100 scenario.

From the annual thermal distribution, the hottest time of the year occurred at noon in July, and the coldest time of the year happened at night from December to January. Meanwhile, for the five vertical greening scenarios, the differences in indoor thermal perceptions throughout the year were not as obvious as the previous study suggested, where the greening wall without windows substantially reduced the thermal perceptions of "cold", "hot", and "very hot" by 5.40%, 17.88%, and 2.01%, respectively [50]. To quantify the full-year influences of vertical greening on an indoor thermal environment under the windowed model, by calculating the area proportions, this research calculated the annual proportion and duration of each thermal perception in every vertical greening scenario. The annual proportion and duration of each thermal perception are presented in Table 8.

Table 8. The proportion and duration of indoor temperature state of each greening scenario.

Scenarios		<8 °C	8–12 °C	12–19 °C	19–26 °C	26–30 °C	30–34 °C
G100V0	Proportion (%)	9.6%	11.7%	29.4%	19.5%	18.6%	11.1%
	Duration (h)	841.3	1024.4	2579.2	1710.5	1631.9	972.7
G75V25	Proportion (%)	9.7%	11.8%	29.5%	19.6%	18.8%	10.6%
	duration (h)	850.39	1030.8	2587.4	1715.3	1647.9	928.2
G50V50	proportion (%)	9.9%	11.8%	29.5%	19.6%	18.9%	10.3%
	duration (h)	863.7	1037.0	2585.1	1716.6	1654.3	903.4
G25V75	proportion (%)	10.0%	11.9%	29.5%	19.6%	19.0%	10.1%
	duration (h)	874.2	1043.3	2581.8	1717.4	1659.9	883.5
G0V100	proportion (%)	10.1%	12.0%	29.4%	19.6%	19.0%	9.9%
	duration (h)	885.70	1050.1	2579.2	1720.0	1659.6	865.3

Vertical greening increased the duration of the cold and reduced the duration of heat throughout the year. The "very hot" perception did not happen in the indoor environment of Changsha due to the climate characters and building materials. The "hot" perception was obviously affected by vertical greening. Compared with the G100V0 scenario, the G0V100 scenario reduced the duration of "hot" perception by 107 h, but it was still far from the result achieved in a previous study where the vertical greening reduced the duration of "hot" perception in a windowless building by 1566 h [50]. Apart from that, the cooling performance was far lower than that in the fictional windowless model in summer. From the view of thermal perception, this research also revealed that vertical greening harmed the indoor thermal environment in winter. The G0V100 scenario extended the annual duration of "very cold" perception by 44 h, which was adverse to indoor thermal comfort. For the other thermal perceptions in the annual cycle, the perceptions of "cold", "slightly cold", "comfortable", "slightly hot", and "hot" changed by vertical greening were, respectively, within 26, 8, 10 and 28 h.

Compared with the previous studies of windowless models, in windowed buildings, the effect of vertical greening on the improvement of the indoor thermal environment was limited. Regarding the best performance, the cooling effects of vertical greening in a windowed building were only 50% and 12.5% compared with the results of 1.1 and 4 °C achieved in the windowless research [17,22]. In addition, from the perspective of the annual cycle, although vertical greening benefited the indoor thermal perception by cooling in summer, the cooling also deteriorated the indoor thermal perception in winter. Therefore, in the windowed building, the role of vertical greening on indoor temperature was not always positive, and the positive effect was not as strong as the previous study suggested.

4. Conclusions

Little attention has been paid to indoor cooling compared with the surface cooling of vertical greening. The few studies on the indoor cooling of vertical greening are almost all conducted in a hot climate area with windowless building models, which is suspected of amplifying the positive effect of vertical greening on an indoor thermal environment. Compared with previous studies, this study used annual periodic evaluation rather than typical summer days. Besides this, this study constructed the vertical greening building model according to the actual window-to-wall ratio, rather than a fictional windowless model. The analytical results revealed that vertical greening cooled the indoor air temperature around the year because the growth of vertical greening rate strengthened the indoor cooling performance in every month. However, the indoor cooling performance varied throughout the year, with the highest cooling performance occurring in summer. Even the highest cooling effect in summer in this research was only 0.56 °C, which is far from the results of 4 °C derived from the windowless models in the previous studies [17,18]. Besides this, from the view of thermal perceptions, the windowed buildings fully covered with vertical greening only reduced the "hot" perception by 107 h, which is also far from the 1566 h achieved in a windowless building in a previous report [50]. However, except for the unsatisfactory performance in the summer, vertical greening also prolonged the "very cold" time of windowed buildings by 44 h throughout the year. Generally, the existing research on vertical greening exaggerated its positive performance on an indoor thermal environment in hot seasons, and conversely, ignored its potential negative impact in cold seasons.

Although this study gives us a realistic understanding of the actual performance of vertical greening, there are still some limitations that need to be explored further.

(1) In this study, the window and door are closed where the building was treated as zero ventilation and zero infiltration. If the buildings are ventilated or infiltrated, what will the results be?
(2) The greening adopted here is modular greening. If other modes of greening are adopted, what will the results be?

Author Contributions: Conceptualization, J.L. and B.Z.; methodology, J.L.; software, J.L.; validation, J.L.; formal analysis, B.Z.; investigation, J.L.; resources, J.L.; data curation, B.Z.; writing—original draft preparation, J.L.; writing—review and editing, J.L.; visualization, J.L.; supervision, B.Z.; project administration, J.L.; funding acquisition, B.Z. All authors have read and agreed to the published version of the manuscript.

Funding: This research was funded by the China Scholarship Council, grant number 202006370126; Hunan Provincial Philosophy and Social Science Planning Fund Office, grant number XSP20ZDI020.

Institutional Review Board Statement: Not applicable.

Data Availability Statement: The data is available on request from the corresponding author.

Acknowledgments: Thanks are given to the Hunan Shangjia Green Company for help in the field experiment.

Conflicts of Interest: The authors declare no conflict of interest.

References

1. Eisenman, T.S.; Churkina, G.; Jariwala, S.P.; Kumar, P.; Lovasi, G.S.; Pataki, D.E.; Weinberger, K.R.; Whitlow, T.H. Urban trees, air quality, and asthma: An interdisciplinary review. *Landsc. Urban Plan.* **2019**, *187*, 47–59. [CrossRef]
2. Shams, Z.I. Changes in diversity and composition of flora along a corridor of different land uses in Karachi over 20 years: Causes and implications. *Urban For. Urban Green.* **2016**, *17*, 71–79. [CrossRef]
3. Ren, Z.; He, X.; Zheng, H.; Zhang, D.; Yu, X.; Shen, G.; Guo, R. Estimation of the relationship between urban park characteristics and park cool island intensity by remote sensing data and field measurement. *Forests* **2013**, *4*, 868–886. [CrossRef]
4. Tiwary, A.; Godsmark, K.; Smethurst, J. Field evaluation of precipitation interception potential of green façades. *Ecol. Eng.* **2018**, *122*, 69–75. [CrossRef]
5. Widiastuti, R.; Zaini, J.; Caesarendra, W. Field measurement on the model of green facade systems and its effect to building indoor thermal comfort. *Measurement* **2020**, *166*, 108212. [CrossRef]
6. Stanley, C.H.; Helletsgruber, C.; Hof, A. Mutual Influences of Urban Microclimate and Urban Trees: An Investigation of Phenology and Cooling Capacity. *Forests* **2019**, *10*, 533. [CrossRef]
7. Coccolo, S.; Kämpf, J.; Mauree, D.; Scartezzini, J.-L. Cooling potential of greening in the urban environment, a step further towards practice. *Sustain. Cities Soc.* **2018**, *38*, 543–559. [CrossRef]

8. Li, J.; Zheng, B.; Shen, W.; Xiang, Y.; Chen, X.; Qi, Z. Cooling and energy-saving performance of different green wall design: A simulation study of a block. *Energies* **2019**, *12*, 2912. [CrossRef]
9. Balany, F.; Ng, A.W.; Muttil, N.; Muthukumaran, S.; Wong, M.S. Green infrastructure as an urban heat island mitigation strategy—A review. *Water* **2020**, *12*, 3577. [CrossRef]
10. Vivian, J.; Chiodarelli, U.; Emmi, G.; Zarrella, A. A sensitivity analysis on the heating and cooling energy flexibility of residential buildings. *Sustain. Cities Soc.* **2020**, *52*, 101815. [CrossRef]
11. Zhang, L.; Deng, Z.; Liang, L.; Zhang, Y.; Meng, Q.; Wang, J.; Santamouris, M. Thermal behavior of a vertical green facade and its impact on the indoor and outdoor thermal environment. *Energy Build.* **2019**, *204*, 109502. [CrossRef]
12. Medl, A.; Stangl, R.; Florineth, F. Vertical greening systems—A review on recent technologies and research advancement. *Build. Environ.* **2017**, *125*, 227–239. [CrossRef]
13. Morakinyo, T.E.; Lai, A.; Lau, K.K.-L.; Ng, E. Thermal benefits of vertical greening in a high-density city: Case study of Hong Kong. *Urban For. Urban Green.* **2019**, *37*, 42–55. [CrossRef]
14. Zhou, W.; Cao, F.; Wang, G. Effects of spatial pattern of forest vegetation on urban cooling in a compact megacity. *Forests* **2019**, *10*, 282. [CrossRef]
15. Raji, B.; Tenpierik, M.J.; van den Dobbelsteen, A. The impact of greening systems on building energy performance: A literature review. *Renew. Sustain. Energy Rev.* **2015**, *45*, 610–623. [CrossRef]
16. Moya, T.A.; van den Dobbelsteen, A.; Ottele, M.; Bluyssen, P.M. A review of green systems within the indoor environment. *Indoor Built Environ.* **2019**, *28*, 298–309. [CrossRef]
17. Olivieri, F.; Olivieri, L.; Neila, J. Experimental study of the thermal-energy performance of an insulated vegetal façade under summer conditions in a continental mediterranean climate. *Build. Environ.* **2014**, *77*, 61–76. [CrossRef]
18. Fernández-Cañero, R.; Urrestarazu, L.P.; Franco Salas, A. Assessment of the cooling potential of an indoor living wall using different substrates in a warm climate. *Indoor Built Environ.* **2012**, *21*, 642–650. [CrossRef]
19. Coma, J.; Pérez, G.; Solé, C.; Castell, A.; Cabeza, L.F. New green facades as passive systems for energy savings on buildings. *Energy Procedia* **2014**, *57*, 1851–1859. [CrossRef]
20. Haggag, M.; Hassan, A.; Elmasry, S. Experimental study on reduced heat gain through green façades in a high heat load climate. *Energy Build.* **2014**, *82*, 668–674. [CrossRef]
21. Seyam, S. The impact of greenery systems on building energy: Systematic review. *J. Build. Eng.* **2019**, *26*, 100887. [CrossRef]
22. Chen, Q.; Li, B.; Liu, X. An experimental evaluation of the living wall system in hot and humid climate. *Energy Build.* **2013**, *61*, 298–307. [CrossRef]
23. Bhatia, A.; Sangireddy, S.A.R.; Garg, V. An approach to calculate the equivalent solar heat gain coefficient of glass windows with fixed and dynamic shading in tropical climates. *J. Build. Eng.* **2019**, *22*, 90–100. [CrossRef]
24. Kumar, K.; Saboor, S.; Kumar, V.; Kim, K.-H.; Ashok Babu, T.P. Experimental and theoretical studies of various solar control window glasses for the reduction of cooling and heating loads in buildings across different climatic regions. *Energy Build.* **2018**, *173*, 326–336.
25. Hami, A.; Abdi, B.; Zarehaghi, D.; Maulan, S.B. Assessing the thermal comfort effects of green spaces: A systematic review of methods, parameters, and plants' attributes. *Sustain. Cities Soc.* **2019**, *49*, 101634. [CrossRef]
26. Yang, F.; Yuan, F.; Qian, F.; Zhuang, Z.; Yao, J. Summertime thermal and energy performance of a double-skin green facade: A case study in Shanghai. *Sustain. Cities Soc.* **2018**, *39*, 43–51. [CrossRef]
27. Daemei, A.B.; Azmoodeh, M.; Zamani, Z.; Khotbehsara, E.M. Experimental and simulation studies on the thermal behavior of vertical greenery system for temperature mitigation in urban spaces. *J. Build. Eng.* **2018**, *20*, 277–284. [CrossRef]
28. Liao, J.; Tan, X.; Li, J. Evaluating the vertical cooling performances of urban vegetation scenarios in a residential environment. *J. Build. Eng.* **2021**, *39*, 102313. [CrossRef]
29. Forouzandeh, A. Prediction of surface temperature of building surrounding envelopes using holistic microclimate ENVI-met model. *Sustain. Cities Soc.* **2021**, *70*, 102878. [CrossRef]
30. Tsoka, S.; Tsikaloudaki, A.; Theodosiou, T. Analyzing the ENVI-met microclimate model's performance and assessing cool materials and urban vegetation applications—A review. *Sustain. Cities Soc.* **2018**, *43*, 55–76. [CrossRef]
31. Laaha, G.; Skøien, J.O.; Nobilis, F.; Blöschl, G. Spatial prediction of stream temperatures using Top-kriging with an external drift. *Environ. Model. Assess.* **2013**, *18*, 671–683. [CrossRef]
32. Chaoqun, C. Researches on application of the renewable energy technologies in the development of low-carbon rural tourism. *Energy Procedia* **2011**, *5*, 1722–1726. [CrossRef]
33. Mi, X.; Liu, R.; Cui, H.; Memon, S.A.; Xing, F.; Lo, Y. Energy and economic analysis of building integrated with PCM in different cities of China. *Appl. Energy* **2016**, *175*, 324–336. [CrossRef]
34. Mohajerani, A.; Bakaric, J.; Jeffrey-Bailey, T. The urban heat island effect, its causes, and mitigation, with reference to the thermal properties of asphalt concrete. *J. Environ. Manag.* **2017**, *197*, 522–538. [CrossRef]
35. Ng, E.; Ren, C. China's adaptation to climate & urban climatic changes: A critical review. *Urban Clim.* **2018**, *23*, 352–372. [PubMed]
36. Osman, M.M.; Sevinc, H. Adaptation of climate-responsive building design strategies and resilience to climate change in the hot/arid region of Khartoum, Sudan. *Sustain. Cities Soc.* **2019**, *47*, 101429. [CrossRef]
37. Priya, U.K.; Senthil, R. A review of the impact of the green landscape interventions on the urban microclimate of tropical areas. *Build. Environ.* **2021**, *205*, 108190. [CrossRef]

38. Li, J.; Zheng, B.; Chen, X.; Zhou, Y.; Rao, J.; Bedra, K.B. Research on Annual Thermal Environment of Non-Hvac Building Regulated by Window-to-Wall Ratio in a Chinese City (Chenzhou). *Sustainability* **2020**, *12*, 6637. [CrossRef]
39. Smyth, M.; Eames, P.; Norton, B. Annual performance of heat retaining integrated collector/storage solar water heaters in a northern maritime climate. *Sol. Energy* **2001**, *70*, 391–401. [CrossRef]
40. Apreda, C.; Reder, A.; Mercogliano, P. Urban morphology parameterization for assessing the effects of housing blocks layouts on air temperature in the Euro-Mediterranean context. *Energy Build.* **2020**, *223*, 110171. [CrossRef]
41. Liu, Z.; Cheng, W.; Jim, C.Y.; Morakinyo, T.E.; Shi, Y.; Ng, E. Heat mitigation benefits of urban green and blue infrastructures: A systematic review of modeling techniques, validation and scenario simulation in ENVI-met V4. *Build. Environ.* **2021**, *200*, 107939. [CrossRef]
42. Simon, H.; Fallmann, J.; Kropp, T.; Tost, H.; Bruse, M. Urban trees and their impact on local ozone concentration—A microclimate modeling study. *Atmosphere* **2019**, *10*, 154. [CrossRef]
43. Bruse, M. *ENVI-met 3.0: Updated Model Overview*; University of Bochum: Bochum, Germany, 2004. Available online: www.envi-met.com (accessed on 12 October 2021).
44. Chen, Y.-C.; Lin, T.-P.; Matzarakis, A. Comparison of mean radiant temperature from field experiment and modelling: A case study in Freiburg, Germany. *Theor. Appl. Climatol.* **2014**, *118*, 535–551. [CrossRef]
45. Salata, F.; Golasi, I.; de Lieto Vollaro, R.; de Lieto Vollaro, A. Urban microclimate and outdoor thermal comfort. A proper procedure to fit ENVI-met simulation outputs to experimental data. *Sustain. Cities Soc.* **2016**, *26*, 318–343. [CrossRef]
46. Yang, X.; Zhao, L.; Bruse, M.; Meng, Q. Evaluation of a microclimate model for predicting the thermal behavior of different ground surfaces. *Build. Environ.* **2013**, *60*, 93–104. [CrossRef]
47. Wu, T.; Li, Y. Spatial interpolation of temperature in the United States using residual kriging. *Appl. Geogr.* **2013**, *44*, 112–120. [CrossRef]
48. Yu, Z.; Song, Y.; Song, D.; Liu, Y. Spatial interpolation-based analysis method targeting visualization of the indoor thermal environment. *Build. Environ.* **2021**, *188*, 107484. [CrossRef]
49. Abdeladim, K.; Razagui, A.; Semaoui, S.; Arab, A.H. Updating Algerian solar atlas using MEERA-2 data source. *Energy Rep.* **2020**, *6*, 281–287. [CrossRef]
50. Li, J.; Zheng, B.; Chen, X.; Qi, Z.; Bedra, K.B.; Zheng, J.; Li, Z.; Liu, L. Study on a full-year improvement of indoor thermal comfort by different vertical greening patterns. *J. Build. Eng.* **2021**, *35*, 101969. [CrossRef]
51. Liu, W.; Zhang, Y.; Deng, Q. The effects of urban microclimate on outdoor thermal sensation and neutral temperature in hot-summer and cold-winter climate. *Energy Build.* **2016**, *128*, 190–197. [CrossRef]
52. Yang, W.; Wong, N.H.; Zhang, G. A comparative analysis of human thermal conditions in outdoor urban spaces in the summer season in Singapore and Changsha, China. *Int. J. Biometeorol.* **2013**, *57*, 895–907. [CrossRef] [PubMed]

MDPI AG
Grosspeteranlage 5
4052 Basel
Switzerland
Tel.: +41 61 683 77 34

Forests Editorial Office
E-mail: forests@mdpi.com
www.mdpi.com/journal/forests

www.ingramcontent.com/pod-product-compliance
Lightning Source LLC
Chambersburg PA
CBHW061519210926
51850CB00008B/1976
9783725815111